rings, groups, and algebras

PURE AND APPLIED MATHEMATICS

A Program of Monographs, Textbooks, and Lecture Notes

LECTURE NOTES IN PURE AND APPLIED MATHEMATICS

Additional Volumes in Preparation

rings, groups, and algebras

edited by

X. H. Cao
East China Normal University
Shanghai, People's Republic of China

S. X. Liu
Beijing Normal University
Beijing, People's Republic of China

K. P. Shum
Chinese University of Hong Kong
Shatin, New Territories, Hong Kong

C. C. Yang
The Hong Kong University of Science and Technology
Kowloon, Hong Kong

CRC Press
Taylor & Francis Group
Boca Raton London New York

CRC Press is an imprint of the
Taylor & Francis Group, an **informa** business

CRC Press
Taylor & Francis Group
6000 Broken Sound Parkway NW, Suite 300
Boca Raton, FL 33487-2742

First issued in hardback 2017

ISBN 13: 978-1-138-40199-0 (hbk)
ISBN 13: 978-0-8247-9733-1 (pbk)

Visit the Taylor & Francis Web site at
http://www.taylorandfrancis.com

and the CRC Press Web site at
http://www.crcpress.com

Library of Congress Cataloging-in-Publication Data

Rings, groups, and algebras / edited by X. H. Cao ... [et al.].
    p. cm. — (Lecture notes in pure and applied mathematics ; v.181)
    ISBN 0-8247-9733-7 (pbk. : alk. paper)
    1. Rings (Algebra). 2. Group theory. 3. Algebra. I. Cao, X. H. II. Series.
QA247.R5753 1996
512'.02—dc20
                                                  96-22831
                                                       CIP

Preface

This project started in 1992 with the goal of setting forth developments and accomplishments in the theories of rings, groups, and algebras in China since the 1950s. It was hoped that this collection of articles would promote exchanges and interactions between Eastern and Western scholars in these areas. However, at an early stage in the project we learned that a Chinese monograph entitled *Group Theory in China* had been edited by Z. X. Wan and S. M. Shi, and would be published by Science Press of China. Thus the focus of the present collection has been shifted to topics in rings and algebras.

Early in the 20th century, China sent many talented young men and women abroad to study Western science and technology. As a result, by the 1930s mathematics in China began to thrive, producing several eminent mathematicians such as HUA Lokeng (deceased) and CHEN Jinren in analysis and number theory. Hua's work in particular has greatly influenced the development of algebra and analysis in China. It is well known that academic activities were almost completely suspended during the so-called "Cultural Revolution," which lasted approximately 10 years, from 1966 to 1976. After the turmoil of that period ended, China reopened her doors and resumed the national program of sending many able scientists, engineers, and graduate students abroad. As before, the objective was to shorten the scientific and technological gaps between China and developed nations and to help China keep abreast of the times in science and technology.

Besides some research notes, most of the articles in this collection are surveys of certain topics, most notably in Hopf algebra and representation theory, in which Chinese mathematicians have made significant contributions. They have been written by leading Chinese experts in China and overseas.

We would like to take this opportunity to thank all the contributors for their efforts and cooperation and LI Ping for his typographical assistance in making the book. Also we want to thank Professor E. Taft and Marcel Dekker, Inc., whose endorsement and support have made our goal attainable.

S. X. Liu   C. C. Yang
刘绍学   杨重骏
X. H. Cao   K. P. Shum
曹锡华   岑嘉评

Contents

Contributors

CHEN Guiyun　陳貴云
　　　　Department of Mathematics, Southwest Normal University, Chongqing, China

CHEN Zhongmu　陳重穆
　　　　Department of Mathematics, Southwest Normal University, Chongqing, China

CHIU Sen　邱森
　　　　Department of Mathematics, Shanghai Institute of Education, Shanghai, China

DAI Zhizhong　戴執中
　　　　Department of Mathematics, Nanchang University, Nanchang, China

DUAN Z. Y.　段澤勇
　　　　Department of Mathematics, Southwest China Normal University, Chongqing
　　　　China

FONG Yuen　方源
　　　　Department of Mathematics, Cheng-Kung University, Tainan, Taiwan

FAN Yun　樊惲
　　　　Department of Mathematics, Hubei University, Wuhan, China

GUO Wenbin　郭文彬
　　　　Department of Mathematics, University of Yangzhou, Jiangsu, China

GUO Yuqi　郭聿琦
　　　　Department of Mathematics, Lanzhou University, Lanzhou, China

LI Huishi　李會師
　　　　Department of Mathematics, Shanxi Normal University, Xian, China

LIU Shaoxue　劉紹學
　　　　Department of Mathematics, Beijing Normal University, Bejing, China

SHI Wujie　施武杰
　　　　Department of Mathematics, Southwest Normal University, Chongqing, China

SHUM Karping　岑嘉評
Department of Mathematics, The Chinese University of Hong Kong, Shatin, Hong Kong

TONG Wenting　佟文庭
Department of Mathematics, Nanjing University, Nanjing, China

WANG Yanmin　王燕鳴
Department of Mathematics, Zhongshan University, Guangzhou, China
and University of Minnesota, U.S.A.

XU Yonghua　許永華
Department of Mathematics, Fudan University, Shanghai, China

XUE Weimin　薛衛民
Department of Mathematics, Fujian Normal University, Fujian, China

ZHANG Pu　章璞
Department of Mathematics, University of Science and Technology of China, Hefei, China

ZENG Guangxing　曾廣興
Department of Mathematics, Nanchang University, Nanchang, China

ZHOU Boxun　周伯塤
Department of Mathematics, Nanjing University, Nanjing, China

rings, groups, and algebras

Structure of Finite Groups

G. Y. Chen, Z. M. Chen, W. J. Shi Mathematics Department, Southwest China Normal University,Chongqing, P. R. China

Introduction

In the past 15 years, research on abstract group theory has been extensive in China. Many results have been found. Research on the stucture of finite groups is focused on three aspects: the first, solvable groups (including nilpotent groups and supersolvable groups); the second, the influence of subgroups, quotient groups and the antomorphism group on a finite group; the third, finite simple groups.

Under what conditions is a finite group a solvable group, a supersolvable group or a nilpotent group? Many mathematicians in group theory have worked on this problem and proved various sufficient conditions and criteria. In this area , Chinese mathematicians have done a lot of research.

The subgroups, quotient groups and automorphism group of a finite group influence greatly the structure of the group. Systematic studies were made of the structure of inner, outer $\Sigma-$ groups and minimal non-Σ-group for the group theoretical property Σ being nilpotency, supersolvability or solvability. These groups are used to find and to prove the conditions for a finite groups as $\Sigma-$group. This manner of research may be viewed as a general method. The influence of the automorphism group of a group on structure of the group was also discussed and some careful results were gotten.

One more aspect of research in China is to characterize finite groups, especially finite simple groups, by pure quantitative conditions, using the innate quantity: the order of a group and the set of the orders of elements of a group. A sequence of interesting results have been achieved. Apart from this, the set of lengths of conjugacy classes of a group, the indices of maximal sugroups of a group, the character table of a group and so on have also been used to characterize a finite simple group.

In this survey article, some results proved by Chinese mathematicians on the aspects mentioned above will be introduced.

I. Solvable Groups, Supersolvable Groups, Nilpotenet Groups

Under what condition, is a finite group solvable, supersolvable, or nilpotent? This is an interesting and important question. From the begining of group theory, many mathematicians have been working on this problem from various directions. Nowadays, the content of group theory is quite abundant. But there are many group theorists still discussinng the conditions under which a group is nilpotent, supersolvable and solvable. In China, some quite good theorems about structure of finite groups and criteria to judge a group nilpotent, supersolvable or solvable have been set up with restrictions on "maximal subgroups", "minimal subgroups", "sections", subgroups etc. for a finite group. Some well-known theorems proved by foreign mathematicians have been extended. Some good methods used to do this research have been summed up. In this section, the main aspects of this research will be introduced, and a few results will be exhibited. In next section, we shall introduce more detailed information case by case.

The first aspect, structure of subgroups of a group, is considered. Surely the restrictions are various, and the subgroups considered are also various.

The maximal subgroups of a finite group and their indices all influence the structure of the group. Here are some theorems as examples.

Shi (1989a) considered the classes of the maximal subgroups with the same order and came to the following theorem.

THEOREM 1.1 A finite group G having exactly two classes of maximal subgroups of the same order is solvable or isomorphic to $G/S(G) \cong L_2(7)$, where $S(G)$ is the maximal normal solvable subgroup of G. (Also see S. R. Li (1990).)

X. H. Li (1994) has studied the finite group having exactly three classes of maximal subgroups of the same order.

R. J. Wang (1989) has generalized J. G. Thompson's result (Thompson (1964)).

THEOREM 1.2 (Extension of Thompson's Theorem) Let M be a maximal subgroup of group G and $P \in Syl_2M$. If M is nilpotent and $\Omega_2(P) \leq Z(P)$, then G is solvable.

Deskins has defined the normal index of the maximal subgroup M of a finite group G, denoted by $\eta(G:M)$, as the order of a principal factor H/K of G, where H is the minimal mormal complement of M in G. A maximal subgroup M of G is called c-maximal if $|G:M|$ is a composite integer.

THEOREM 1.3 Let p be the largest prime factor of the order of G, then G is p−solvable if and only if $[\eta(G: M)]_p = 1$ for each non−nilpotent c-maximal subgroup M of G with $[G:M]_p = 1$. (Guo (1991))

Corresponding to maximal subgroups of a group, one can also consider minimal subgroups of it. But usually we consider not only the mumber or the indices of minimal subgroups. Some special "normalities" are considered so that some important results about stucture of the group can be gotten.

DEFINITION 1.1 Suppose G is a finite group, $H \leq G$.

(1) If HK=KH, for any $K \leq G$, then H is called a quasinormal subgroup.

(2) If HK=KH, for any Sylow subgroup K of G, then H is called an S−quasinormal subgroup.

(3) If for any prime p, there exists a Sylow subgroup K, such that HK=KH, then H is called a weakly S−normal subgroup of G.

(4) If HK=KH, for every subgroup K with $(|H|, |K|)=1$, then H is clled seminormal subgroup of G .

(5) If for prime p, $p \nmid |H|$, $p \mid |G|$, any p−Sylow subgroup K satisfies that HK=KH, then H is called an S−seminormal subgroup of G.

(6) If there exists a subgroup such that G=HB, and HK=KH for any $K \leq B$, then H is called a half-normal subgroup of G.

Here is relation of "normalities" defined above:

$$\begin{array}{ccccc}
\text{normal} \Longrightarrow \text{quasinormal} & & \Longrightarrow \text{S−quasinormal} \Longrightarrow \text{weakly S−normal} \\
\Downarrow & & \Downarrow \\
\text{half-normal} \Longrightarrow \text{seminormal} \Longrightarrow \text{S−seminormal}
\end{array}$$

The purpose of studying the above six "normalities" is to try to weaken the conditions guaranteeing a group nilpotent, supersolvable or solvable, (refer to Ito (1955), Huppert(1967)(TH5.7), Buckley(1970)). Here are some examples to illustrate research in this topic in China.

THEOREM 1.4 (Z. M. Chen(1987a)) If every subgroups of prime order and cyclic group of order 4 in a finite group is seminormal, then G is supersolvable.

THEOREM 1.5 (J. P. Zhang(1987)) Suppose that G is finite group. $H \leq G$ and G/N is supersolvable, then

(1) If subgroups of prime orders in N are all in the superquasicenter, then G is supersolvable.

(2) If subgroups of prime orders in N are S−quasinormal in G, then G is supersolvble if and only if G is not related to D_{2q}, where D_{2q} is a Schmidt group without generator of order 2, whose Sylow 2−subgroup is normal.

THEOREM 1.6 (Z. M. Chen(1994)) Suppose that p is an odd prime, if all minimal p−subgroups of G are seminormal (or S−seminormal), then G is p −supersovable.

THEOREM 1.7 (Z. M. Chen(1994)) If all minimal subgroups of odd order of G are S−seminormal, then G is p−supersolvable for every odd prime p.

THEOREM 1.8 (Z. M. Chen(1994)) If all minimal 2−subgroups of G are S− seminormal, then G is 2−nilpotent if and only if G contains no subgroups of type D_{2q}.

For more information about the research of "normality", please refer to Y. N. Zhen (1990), Z. M. Chen(1987a), Q. X. Li(1993), Y. C. Ren(1990), X. Y. Shu (1988).

Surely we can give restrictions on general subgroups of a group. This is also an important topic for the study of the structure of finite groups, (refer to S. R. Li (1987, 1988a,

1988b, 1988c, 1992). Most results related to this topic will be introduced in next section. Here only several examples are exhibited.

THEOREM 1.9 (Y. M. Wang (1991), W. J. Xiao (1989)) If $N_G(P)/C_G(P)$ is a p−group for prime $p||G|$ and every Sylow p−subgroup P of a group G, then G is nilpotent.

The above theorem has been improved. The following two theorems are given by X. F. Wang(1989).

THEOREM 1.10 Suppose that $N_G(P)/C_G(P)$ is a π−group for each p−subgroup P of a group G with prime p in π. Then G is π'−closed (see next section) if one of the following conditions is satisfied:
(1) Each π−subgroup of G is 2−clossed;
(2) Each π−subgroup of G is $2'$−closed.

THEOREM 1.11 If $N_G(Z(J(P)))$ is π'−closed for a Sylow p-subgroup P of a group G with prime p in π, then so is G.

Here are another two theorems on subgroups:

THEOREM 1.12 (Y. Fan (1986a)) The hypercenter of a group G is the intersection of all maximal nilpotent subgroups of G.

THEOREM 1.13 (X. Y. Guo(1989a)) Let H be a π-Hall subgroup of G, there exists a normal π−complement of H in G if and only if the following two conditions hold:
(a) If two π−elements of H are conjugate in G, then they are conjugate in H.
(b) For every $y \notin H$ and $h \in Z(G)$, if $p \nmid h$, then p does not divide the index $|C_G(h) : C_H(h)|$.

For more information about complements, please refer to X. Y. Guo (1988), Z. M. Chen(1982a).
Now we introduce one interesting condition related to supersolvability.

DEFINITION 1.2 Asumme that $H \leq G$. Let $P_G(H) = <x \in G| <x>H=H<x>>$. We say that G satisfies the permutizer condition if $H \leq P_G(H)$ for every subgroup H of G.

THEOREM 1.14 (X. Y. Guo (1989b), J. P. Zhang (1985), Y. N. Zhen (1986)) If a solvable group G satisfies the permutizer condition, then
(1) G is supersolvable if and only if H had no section S_4.
(2) G is p−supersolvable for every odd prime p.
(3) G is supersolvable if and only if G' satisfies the permutizer condition.

At the end of this section we introduce another way to study the structure of a finite group, which is to consider the act of a group on another group.
X. H. Li(1990) has investigated the group G such that AutG acts doubly transitively on elements of the same order.

THEOREM 1.15 If AutG acts doubly transitively on elements of the same order of group G, then $G \cong Z_2$ or S_3 or an elementary abelian 2–group of order >2.

Z. M. Chen (1989a) solved the well–known conjecture for solvability of a group with a fixed–point–free automorphism group in a broader sense.

THEOREM 1.16 Suppose that a group A acts on a group G, $(|A|,|G|)=1$. If the fixed–point subgroup $C_G(A)$ of G has no section isomorphic to S_3, A_4 or $Sz(2)=5:4$, then G is solvable. In particular G is solvable if $C_G(A)$ is of odd order, or nilpotent.

For other results, refer to G. X. Zhang(1986).

It is worth mentioning that Z. M. Chen(1993) has given one way to extend theorems in group theory. Suppose that N is a normal subgroup of a group G, G/N is a Σ–group, where Σ is a group theoretic property, Chen's way is to consider under what conditions on N is G a Σ–group.

II. The Influence on the Structure of a Group of Its Subgroups, Quotient Groups and Automorphism Group

In Section I, we have introduce a few results about the influence on the structure of a group of its subgroups, quotient groups and automorphism group. In this section, our purpose is to enrich the content of Section I. The book written by Z. M. Chen (1988) will be specially cited for this book sums up the research method for finding a minimal counter-example when the conditions on the group is inherited by its subgroups or quotient groups. At the end of this section, we shall introduce the situation of finite groups with automorphism group given.

DEFINITION 2.1 Suppose Σ denotes a group theoretic property, a group G is called a Σ–group if G has property Σ.

A group G is called an inner Σ–group if G is not a Σ–group, but all its proper subgroups are Σ–groups. A group G is called an outer Σ group if G is not a Σ–group, but every proper quotient group of G is a Σ–group.

A group is called a minimal non-Σ–group if all proper subgroups and all proper quotient groups are Σ–groups but G is not a Σ–group.

The purpose of studying inner Σ–groups and outer Σ–groups is to find the conditions under which a group is a Σ–group, especially when Σ is one of the properties "commutation property", "nilpotency", "p–nilpotency", "supersolvability", "solvability", "p–solvability" and so on.

In the preface of Z. M. Chen (1988) (also see Z. M. Chen (1980)), some general properties of inner, outerΣ–groups and minimal non-Σ–groups have been discussed.

DEFINITION 2.2 Assume σ is a group theoretic property. σ is called " closed for subgroups", if σ satisfies " if a group is a σ–group, then every subgroup is a σ–group."

If a group is a σ–group, then so is every quotient group. Then σ is called " closed for quotient groups".

THEOREM 2.1 Suppose σ, Σ are two group theoretic properties. If property σ is closed for subgroups, and each inner Σ−group is not a σ−group, then a σ−group is a Σ−group.

If σ is closed for quotient groups, and each outer Σ−group is not a σ−group, then a σ−group is a Σ−group.

If σ is closed for subgroups and quotient groups, and each minimal non-Σ−group is not a Σ−group, then a σ−group is a Σ−group.

From this theorem we can see the significance of studying the inner, outer Σ−group and minimal non-Σ−group.

EXAMPLE 2.1 Suppose the normalizer of each abelian subgroup of a group G equals the centralizer, then G is an abelian group.

Proof: The condition σ of the example is closed for subgroups, since an inner−abelian group $\overline{G}$ is solvable, and a maximal normal subgroup M of an inner−abelian group is an abelian group with index prime. But $C_{\overline{G}}(M)=M\neq\overline{G}=N_{\overline{G}}(M)$, so $\overline{G}$ is not a σ−group. The example follows by Theorem 2.1

The following theorem expresses the relation between an inner Σ−group and an outer Σ−group.

THEOREM 2.2 Suppose group theoretic property Σ is closed for quotient groups. If an inner Σ−group G satisfies $\Phi(G)=1$, then G is an outer Σ−group.

Conversely, suppose Σ satisfies

(Δ) Assume $G/\Phi(G)$ is a Σ−group, then so is G.

Then a minimal non-Σ−group is an inner Σ−group with $\Phi(G)=1$, where $\Phi(G)$ is the Frattini subgroup of G.

Remark: All nilpotent groups and supersolvable groups satisfy the condition (Δ).

THEOREM 2.3 Every inner Σ−group can be generated by k elements if and only if "G is a Σ−group, when every subgroup of G, which is generated by k elements, is a Σ−group".

Every outer Σ−group can be generated by k elements if and only if "G is a Σ−group, when every quotient group of G, which is generated by k elements, is a Σ−group".

Remark: Inner−abelian groups, inner−nilpotent groups and inner−regular p-groups can be generated by 2 elements.

Because inner−abelian groups, inner p−nilpotent groups and inner p−closed groups, which are solvable, are special inner−nilpotent groups, Z. M. Chen introduced the inner-nilpotent groups in the first chapter of his book.

2.1 Nilpotent Groups

The following structure theorem for inner−nilpotent groups is an improved theorem by Z. M. Chen.

THEOREM 2.4 Let G be an inner$-$nilpotent group, then

1) $|G|= p^{\alpha}q^{\beta}$, where p, q are primes .

2) G has a normal subgroup Q. Suppose $|Q|=q^{\beta}$, then a p$-$Sylow subgroup P is cyclic. If $P=<a>$, then $\Phi(P)=<a^p>$ and $\Phi(P)\leq Z(G)$.

3) Let N be the maximal proper subgroup of Q, such that $N\trianglelefteq G$, then $|Q:N|=q^b$, where b is the order of q modulo p. Moreover $N=\Phi(Q)=Q'$.

4) Let $c\in Q$, then c is a generator of Q if and only if $[c,a]\neq 1$.

5) If c is a generator of Q, then so is $[c, a]$, and $Q=<c, [c, a], ..., [c, a^{p-1}]>$.

6) If Q is abelian, then Q is an elementary abelian group.

7) If Q is non$-$abelian, then N is an elementary abelian group, $N=Z(Q)$. Moreover if $q\neq 2$, then the exponent of Q is q. Otherwise the exponent of Q is 4.

(8) $Z(Q)=\Phi(P)\times\Phi(Q)=\Phi(G)$.

DEFINITION 2.3 An inner$-$nilpotent group with normal q$-$Sylow subgroup is called a q$-$basic group.

J. P. Zhang(1989b) has studied 2$-$basic group.

EXAMPLE 2 (Extension of Ito's lemma, M. Weinstein (1982), Thm 6.6) If the elements with order a prime or 4 in a group G are contained in Z(G), then G is a nilpotent group.

Proof: The condition is closed for subgroups of G. Since an nner$-$nilpotent group can$n't$ satisfies the condition, G is nilpotent by Theorem 2.1.

An inner$-$abelian group which is not a nilpotent group is a special inner$-$nilpotent group. A nilpotent inner$-$abelian group is a p$-$group with order p^n such that $|Z(G)|=p^{n-2}$, $Z(G)=\Phi(G)$. For more information about inner$-$abelian group, refer to R'eddi(1947).

2.2 $\pi_{\sigma}-$nilpotent group

DEFINITION 2.4 Assume π is a set of primes, π_{σ} is a well$-$ordered set of π with order σ. Suppose G has normal π-complement, and the $\pi_{\sigma}-$Hall subgroup of G is a $\sigma-$Sylow tower group, then G is called a $\pi_{\sigma}-$nilpotent group.

N. Ito studied inner p$-$nilpotent groups. Fathahi Abiadollah studied inner π_{σ}-Sylow tower groups.

THEOREM 2.5 An inner $\pi_{\sigma}-$nilpotent group is a p$-$basic group, where p is the first prime in π_{σ} and $p\big||G|$.

The following two theorems may be viewed as applications of Theorem 2.1.

THEOREM 2.6 Suppose p, q are distinct primes, and $p\neq 2$. If for any x, $y\in G$, $|x|=p$, $|y|$ is a power of q and there exists n such that $[x, \underbrace{y, ..., y}_{n}]= 1$, then G is p$-$nilpotent.

If $p=2$ and $[x, \underbrace{y, ..., y}_{n}]=1$ holds for all $x\in G$ with $|x|=2$ or 4 and $y\in G$, $|y|$ is a power of q, then G is a 2$-$nilpotent group.

This theorem extends Example 1.

THEOREM 2.7 (Z. M. Chen(1984)) If p is a prime, $(N_G(P)/C_G(P))^{p-1}$ is a p−group for every p−subgroup P of a group G, then G is a Sylow tower group.

Proof: An inner−Sylow tower group G_1 is a p−basic group of order $p^\alpha q^\beta$, p<q. Let $P_1 \in Syl_pG$, then $(N_G(P_1)/(C_G(P_1))^{p-1}$ is not a p−group. Hence an inner−Sylow tower group doesn't satisfy the condition. Therefore the theorem follows by Theorem 2.1.

2.3 P−closed Groups

The content of this subsection comes from Z. M. Chen (1986b).

DEFINITION 2.5 Let p be a prime. A group G is called p−closed, if the p−Sylow subgroup of G is normal in G. In particular, G is p−closed if p $\nmid$ |G|.

THEOREM 2.8 An inner p−closed group G satisfies one of the following:
1) $G/\Phi(G)$ is a non−abelian simple group, such that the non−abelian proper simple subgroup is a p'−group.
2) G is a p−basic group of order $p^\alpha q^\beta$.

The following theorem is an application of above the theorem.

THEOREM 2.9 Let G be a p−solvable group, Q a Sylow q−subgroup (q≠p). If the p−elements and p′−elements of $N_G(Q)$ are commutative, then G is a p−closed group.

Inner 2−closed groups and inner 3−closed groups have been carefully classified (refer to S. R. Li (1983b, 1986)). But how about the classification of inner 5−closed groups? This is an open question.

DEFINITION 2.6 If a group G has a normal π−Hall group, then G is called a π−closed group.

X. F. Wang(1989) has studied the structure of inner π'−closed groups. Z. M. Chen(1990) has studied the stucture of inner(π, π')−closed groups and minimal non−(π, π')−closed groups. He came to the following useful conclusion.

THEOREM 2.10 Let π be a set of odd primes, P a Sylow p−subgroup of G. If $N(Z(J(P)))$ has a normal π−complement for each p∈π, then G has a normal π−complement.

This theorem is more precise than the result given by Arad, Z. and Chillage, D.(1984).

2.4 Solvable Groups

A minimal simple group is a nonsolvable simple group with all proper subgroups solvable. J. G. Thompson has classified all minimal simple groups.

THEOREM 2.11 If G is a mimimal simple group, then G is one of $L_2(p)$ (p>3, 5 $\not|p^2-1$),
$L_2(2^q)$ (q a prime), $L_2(3^q)$ (q an odd prime), $L_3(3)$, $Sz(2^q)$ (q an odd prime).

THEOREM 2.12 (Z. M. Chen) If G is an inner−solvable group, then $G/\Phi(G)$ is a minimal simple group.

By using Theorem 2.1 and the above two theorems, Z. M. Chen came to the following theorems about "solvability".

THEOREM 2.13 Suppose $n=2^\alpha 3^\beta p_1^{\alpha_1}...p_k^{\alpha_k}$, then all groups of order n are solvable if and only if n satisfies following conditions.
1) For any prime q≤α, $2^{2q} - 1 \not| n$.
2) For any prime q≤α/2, $(2^{2q} + 1)(2^q - 1) \not| n$
3) For any prime q≤β, $2^4.13 \not| n$, if q=3; and $(3^{2q} - 1)/2 \not| n$, if q≠3.
4) For each p_j, if 5 $\not| p_j^2 - 1$, then $(p_j^2 - 1)/2 \not| n$.

THEOREM 2.14 If no subgroup of G is homomorphic to $L_2(7)$ and for every chain of maximal subgroups
$G = M_0 > M_1 > ... > M_n = 1,$
$|M_i/M_{i+1}|$ is a power of a prime, i=0, 1, ..., n-1, then G is solvable.

THEOREM 2.15 Asumme a prime p≥5. If every p′−subgroup of a group G is 2−closed, then G is solvable.

THEOREM 2.16 If no subgroup of G is homomorphic to $L_2(7)$ or $L_3(3)$, and every 3′−subgroup of G is 2−closed, then G is solvable.

2.5 Outer Σ−groups

The following main lemma is the kernal of this section, for several kinds of outer Σ−groups have structure given in the results in the main lemma. For the proof of it , refer to Z. M. Chen(1987b), L. W. Zhang(1986) and J. S. Rose (1987).

MAIN LEMMA Suppose a group G has exactly one minimal normal subgroup N, $|N|=p^\alpha$, $\Phi(G)=1$. Then
1) G=AN, A∩N=1, where A is a maximal subgroup of G.
2) $C_G(N)=N$.
3) A has no subnormal p−subgroup. Hence N is the Sylow subgroup of G if A is nilpotent.
4) Fitting subgroup F(G)=N, N is the subgroup of p−Sylow subgroup of G.
5) Suppose $O_{p'}(A)\neq1$. If H≤G, such that G=HN, then H is conjugate to A in G. If $O_q(A)\neq1$, q≠p, then $p^\alpha \equiv1$ (mod q).
6) Z(A) acts on N fixed−points freely. Z(A) is cyclic.
7) If 1≠H⊴G, then
i) H=BN, B∩N=1, where B=A∩H

ii) If $\Gamma_k(H)=1$, then H=N. Otherwise $\Gamma_k(H)= \Gamma_k(B)N$. If $\Gamma_k(H)$ is p−nilpotent, then p $\nmid$|B|, B is a nilpotent group with nilpotency class$\leq$k, where $\Gamma_k(H)$ is the kth term in the lower central chain of H, $\Gamma_1(H)=H$.

8) The following are equivalent
I) G′ is p−nilpotent, II) G′ is nilpotent,
III) A is abelian, IV) A is cyclic.

9) If A is cyclic, then G is generated by 2 elements. If A is not cyclic, and A is generated by m elements , A is p−solvable, then G is generated by m elements (Y. M. Wang).

THEOREM 2.17 Let G be an outer−supersolvable group, then all results in the main lemma hold for G. Moreover G=AN, A is supersolvable and p−nilpotent. The p−Sylow subgroup of A is abelian, $|N|=p^\alpha$, $\alpha<1$. If G′ is nilpotent, then A cannot be written as A= $A_1 A_2$, such that $A_i N$ is supersolvable, $A_i<A$, i=1, 2.

Presentations of minimal non−supersolvable groups have been given by a Soviet mathematician in 1975.

Suppose G is an inner−supersolvable group, R is a normal Sylow subgroup of G. Doerk (1966) described some properties of $\Phi(R)$. Z. M. Chen has described them more precisely.

THEOREM 2.18 Let R be a Sylow subgroup of an inner−supersolvable group G. Then
1) $R/\Phi(R)$ is a minimal normal subgroup of $G/\Phi(R)$.
2) If R is abelian, then $\Phi(R)=1$.
3) $\Phi(R)=R'$.
4) If R is non-abelian, then$\Phi(R)=R'=Z(R)$.
5) The order of an element of R' is 1 or r.
6) If r>2, then exponent of R is r. If r=2, then exponent of R is at most 4.

About the property of an inner−supersolvable group, there is following theorem.

THEOREM 2.19 If G is an inner−supersolvable group, then G is a Sylow tower group or anti−Sylow tower group. Moreover if r is a prime that is the last prime factor of |G|, then any maximal subgroup containing the r−complement has an index divisible by r^2 in G.

By Theorems 2.1, 2.17, 2.18, 2.19, we can get various sufficient conditions for a group to be supersolvable. There are 46 such theorems in Z. M. Chen(1988). Some of them were founded by Z. M. Chen, J. P. Zhang, L. W. Zhang, Y. Fan etc. Some of them were founded by others, but the proofs are shortened. Here are some choosed as examples.

THEOREM2.20 (Z. M. Chen) Let p be the smallest prime such that p$||$G|. If the number of generating subgroups of G $< p^2 + p + 1$, then G is supersolvable.

THEOREM 2.21 (Z. M. Chen) Suppose A_i, i=1, 2, 3, 4, are subgroups of G, and the indices of A_i are pairwisely coprime. If A_1, A_2, A_3 are supersolvable, and A_4 is metanilpotent, then G is solvable.

If G has three subgroups A, B, C with pairwisely coprime indices without square divisor, such that A, B are supersolvable, and C is metanilpotent, then G is supersolvable.

For more information, refer to Z. M. Chen(1986, 1982b, 1988), Y. Fan (1986c, 1981a, 1981b, J. P. Zhang (1988a,b) and Section I.

THEOREM 2.22 Suppose G is an outer p−supersolvable group which is p−solvable. Then all results in the Main Lemma hold for G. Moreover G=AN, $|N|=p^\alpha$, $\alpha>1$, A is p-supersolvable. If G' is p−nilpotent, then there aren't A_1, A_2, such that A=$A_1 A_2$, $A_1 \leq A$, $A_2 \leq A$, A_i is p−supersolvable, i=1, 2.

The above theorem is similar to Theorem 2.17.

The presentations of minimal p−supersolvable groups which are p−solvable were found in 1975.

Surely we can also obtain some conditions for a group to be p−supersolvable. Please refer to Z. M. Chen(1988), Y. Fan (1986b), J. P. Zhang (1989a).

We have introduced the application of the so-called "inner, outer Σ−method" to discuss the conditions for a group to be abelian, nilpotent, solvable, supersolvable, p−solvable and p−supersolvable. In Z. M. Chen(1988) some other cases have been discussed, for example, group formations, infinite groups, p−nilpotent groups. We could not introduce more for lack of space. Refer to Z. M. Chen (1984, 1987b, 1989b, 1988), R. C. Ren (1989), please.

At the end of this section, we introduce a topic related to action of a group on another group, to sudy the automorphism group to get the structure of a finite group. Henineken, H. and Liebeck. H. (1973) discussed p−groups with odd order automorphism group. Jonah, D. and Konvisser, M. (1975) gave examples of non−abelian p−groups with abelian automorphism groups. D. MacHale (1981, 1983) characterized some finite groups with rare automorphism groups. Afterwards, G. Y. Chen (1990a,b, 1992a,b, 1994a) has done some research on this topic. G. L. Ban and S. X. Yu(1992, 1993) have done some studies on how the automorphism groups of p−groups influence the structure of the groups. Simultanously, they have come to some improved results on LA−groups (S. X. Yu(1990). Some other research on this topic was done by S. R. Li (1994). Also refer to Martin(1986), M. J. Curran(1988), please.

III. Finite Simple Groups

In this section, we are going to introduce the quantitative structure of groups, especially quantitative structure of finite simple groups, and related topics .

The order of a finite group and the orders of its elements are the most fundamental concepts in group theory, and they play an important role in the quantitative structure of groups. For example, Sylow's Theorems and Lagrange's Theorem are well-known in group theory. Cauchy's Theorem says that if a prime p divides the order of a finite group, then the group has an element of order p. This theorem expresses the relation between the order of a group and the orders of its elements. Surely there are other theorems related to the order of a group, such as the theorem on the odd order solvable groups, $p^a q^b$−theorem and so on. The quantitative structure , introduced here, is focused on the set of orders of elements of a group (denoted by $\pi_e(G)$). It has been proved that a lot of finite simple groups and some nonsolvable groups can be characterized by the set $\pi_e(G)$.

The history of investigating the influence the of the set of orders of elements of a finite group on the structure of the group goes back to the work by W. Burnside(1902), B. H.

Neumann(1937) and M. Suzuki(1962). W. J. Shi(1986) began this research by investigating a special class of finite groups G with $\pi_e(G)=\{1$ or a prime$\}$ and came to the following interesting results:

THEOREM 3.1 Let G be a finite group. Then $G \cong A_5$ if and only if $\pi_e(G)=\{1, 2, 3, 5\}$.

Set $\pi_e(G)=\{1\} \cup \pi'_e(G) \cup \pi''_e(G)$, where $\pi'_e(G)$ is the subset of primes in $\pi_e(G)$ and $\pi''_e(G)$ is the subset of composite numbers in $\pi_e(G)$.

THEOREM 3.2 If G is a finite group, $|\pi''_e(G)| \leq 1$, then one of the following holds:
(1) G is one of Z_p, $L_2(q)$, $L_3(4)$, Sz(8), $L_2(3^n)$, $L_2(2^m)$, where p, $(3^n-1)/2$, $(3^n+1)/4$, 2^n-1, $(2^m+1)/3$ are primes, q=5. 7, 8, 9, 11, 13 or 16.
(2) $G/N \cong A_5$, where N is an elementary abelian 2-group, $\pi_e(G)=\{1, 2, 3, 4, 5\}$.

Corollary 3.1 Let G be a finite group and $|\pi''_e(G)|=k$, if k=0, then $|\pi(G)| \leq 3$, and if $|\pi(G)|=3$, then G is simple. If k =1, then $|\pi(G)| \leq 4$, and if $|\pi(G)|=4$, then G is simple, where $\pi(G)$ denotes the set of prime divisors of $|G|$.

Question 3.1 Let G be a finite group. Does there exist a general relation among $|\pi(G)|$, $|\pi''_e(G)|$ and the simplicity of group G?

For every $a \in \pi''_e(G)$, if $p \nmid a$, p=2 or 3, then W. J. Shi and C. Yang obtained the structure of the group G and we have the following theorem (see W. J. Shi(1992a))

THEOREM 3.3 Let G be a finite group. Then $G \cong L_2(p^n)$, p=2 or 3, $n \geq 3$, if and only if $\pi_e(G)=\pi_e(L_2(p^n))$.

It was proved the above conclusion is also true for $L_2(q)$, q odd, $q \neq 9$ (refer to R. Brandl (1994)), while searching for the converse of the well-known Dickson's Theorem.

THEOREM 3.4 Let G be a finite group, then $G \cong L_2(q)$, q odd, $q \neq 9$, if and only if $\pi_e(G)=\pi_e(L_2(q))$.

Question 3.2 Suppose G is a finite group and $p \nmid a$, where p is a prime, $p \neq 2, 3$, for all $a \in \pi_e(G)$, how about the structure of group G?

We have mentioned the influence of the order of a finite group and the orders of its elements on the structure of the group. In the following we shall talk about this topic in more detail.

3.1 On Orders Of Simple Groups

As early as the 1940s R. Brauer and H. F. Tuan(1945) characterized some projective special linear groups by using the condition of the order of a group and the simplicity of a group. After them many mathematicians characterized many simple groups by various quantitative conditions. M. Herzog(1968) came to the following theorem.

THEOREM 3.5 Suppose G is a finite simple group. If $|\pi(G)|=3$, then G is isomorphic to one of the following groups: A_5, $L_2(7)$, $L_2(8)$, A6, $L_2(17)$, $L_3(3)$, $U_3(3)$ or $U_4(2)$.

The simple groups in the above theorem are called simple K_3−groups.

W. J. Shi(1991a) has classified all simple K_4−groups, that is the simple groups with order divisible by 4 distinct primes only.

THEOREM 3.6 Let G be a simple K_4−group. Then G is one of following groups: A_n, n=7, 8, 9, 10; M_{11}, M_{12}, J_2; $L_2(q)$, q=16, 25, 49, 81; L3(q), q=4, 5, 7, 8, 17; $L_4(3)$; $O_5(q)$, q=4, 5, 7, 9; $O_7(2)$, $O_8^+(2)$; $G_2(3)$; $U_3(q)$, q=4, 5, 7, 8, 9; $U_4(3)$; $U_5(2)$, $^3D_4(2)$, $^2F_4(2)'$, Sz(8), Sz(32) and

$L_2(r)$, r being a prime and satistfying
$$r^2 - 1 = 2^a 3^b u^c, \ a{\geq}1, \ b{\geq}1, \ u \ a \ prime, \ u{>}3; \qquad (1)$$

$L_2(2^m)$, m satisfying
$$2^m - 1 = u$$

m≥1, u, t, primes, t>3, b≥1; $\qquad\qquad (2)$
$$2^m + 1 = 3t^b$$

$L_2(3^m)$, m satisfying
$$3^m + 1 = 4t$$
$$3^m - 1 = 2u^c \qquad\qquad (3)$$

or
$$3^m + 1 = 4t^b$$
$$3^m - 1 = 2u \ , \ where \ m{\geq}1, \ u, \ odd \ primes, \ c{\geq}1, \ b{\geq}1. \qquad (4)$$

Because there are four indefinite equations included in above theorem, so we have

QUESTION 3.3 What is the number of solutions of equations (1)-(4) ? Is the number of simple K_4−groups finite or infinite?

E. Artin(1955) studied the orders of the known types of finite simple groups. He pointed out that it is interesting to study a simple group G that contains a Sylow subgroup of order greater than $|G|^{1/3}$. R. Brauer and W. F. Reynolds (1958) solved the simplest case.

THEOREM 3.7 Let G be a finite simple non−cyclic group with order divisible by a prime p>$|G|^{1/3}$. Then G is one of $L_2(P)$, p>3, p is a prime , or $L_2(p-1)$, p>3, p is a Fermat prime.

W. J. Shi(1992b) has solved E. Artin's problem by using a classification theorem for finite simple groups.

THEOREM 3.8 Let G be a non−abelian simple group. If $p^k\big|\big||G|$ (i. e. $p^k\big||G|$, p^{k+1} $\nmid|G|$), where p is a prime, $|G|<p^{3k}$, then G is isomorphic to one of the following grcups:

(1) A simple proup of Lie type of characteristic p;

(2) A_5, A_6, A_8, A_9, A_{10}, M_{12}, M_{22}, J_2, HS , M_{24}, Suz, Ru, Fi_{22}, Co_2, Co_1, B, $L_2(8)$, $U_5(2)$.

(3) $L_2(r)$, where r is a Fermat prime or a Mersenne prime .

(4) $L_2(\text{p-1})$, p is a Fermat prime.

Making use of above theorem, Shi (1993a) proved the followng theorem:

THEOREM 3.9 Let G be a group and $|G|=|M|$, where M is a simple group. Then one of the following holds:
(1) If $|M|=|A_8|=|L_3(4)|$, then $G{\cong}A_8$ or $L_3(4)$.
(2) If $|M|=|B_n(q)|$, where $n{\geq}3$, q odd, $G{\cong}B_n(q)$ or $C_n(q)$.
(3) If $|M|$ is not in the above case (1) or (2), then $G{\cong}M$.

This result was also obtained by W. Kimmerle et al. (1990). The above theorem has been improved by G. Y. Chen (1994b):

THEOREM 3.10 If G is a finite group, M is a non−abelian simple group, then $G{\cong}M$ if and only if G, M have the same character tables.

THEOREM 3.11 If G, M are simple groups, $|G|=|M|$, $N(G)=N(M)$, where $N(G)$ denotes the set of lengths of conjugacy classes of elements in G, then $G{\cong}$ M.

3.2 On Quantitative Characterization

In this section quantitative characterization (abbr. QC) is used to mean characterization of a finite group by quantitative conditions. The most work we have done is on QC of finite simple groups by order of a group and the set of orders of elements of the group. W. J. Shi(1989a,1992c,1990a,1991b) has come to the following theorem:

THEOREM 3.12 Let G be group, M one of following simple groups:
(1) A cyclic group of order a prime;
(2) A alternating group A_n, $n{\geq}5$;
(3) A Lie type group except B_n, C_n, D_n, ^2A_n and ^2D_n;
(4) A simple group with order $<10^8$.
Then $G{\cong}M$ if and only if (a) $\pi_e(G)=\pi_e(M)$, (b) $|G|=|M|$.

By Theorem 3.12, we believe the following conjecture is correct.

CONJECTURE 3.1 Let G be a group, M a finite simple group. Then $G{\cong}M$ if and only if (a) $\pi_e(G)=\pi_e(M)$, (b) $|G|=|M|$.

For proving Theorem 3.1 we may first consider the set of orders of elements and discuss the simple section by using the classification theorem and the results on the prime graph components of finite simple groups. We may also consider the order of group G first and then discuss the group theoretic property by introducing some special number-functions.

Besides the finite simple groups we may characterize some unsolvable group G using only the condition on $\pi_e(G)$ and $|G|$. For example J. X. Bi (1990) proved the following theorem.

THEOREM 3.13 Let G be a group. Then $G{\cong}S_n$, $n{\geq}3$, (S_n denotes the symmetric group of order n) if·and only if (a) $\pi_e(G)=\pi_e(M)$, (b) $|G|=|S_n|$.

Surely we cannot characterize general unsolvable groups, even p–groups by using the order of the group and the set of orders of its elements. From this point of view we may say the simple groups are indeed simple.

3.3 Characterization of Groups by Using $\pi_e(G)$ Only

Because in Theorem 3.12 there are two conditions (a), (b), and (b) seems too strong, we now consider only the condition (a). First we have

DEFINITION 3.1 For a set m of positive integers, let h(m) be the number of finite groups G that are not isomorphic to each other and satisfy $\pi_e(G)$=m.

We have the following Theorem (refer to R. Brandl and W. J. Shi (1991, 1994); H. L. Li (1989, 1993); C. E. Praeger(1994); W. J. Shi(1985, 1987a,b, 1989c, 1990b, 1991b, 1992d, 1994).

THEOREM 3.14 Let G be one of the following groups:
(1) A_n, n=5, 7, 8, 9, 11, 13 or S_7;
(2) A sporadic simple group except J_2, Co_1;
(3) $L_3(4)$, $U_3(4)$, $U_4(3)$, $U_5(2)$, $U_6(2)$, M_{10}, $L_3(4)<\beta>$, where β is a unitary automor·phism of $L_3(4)$;
(4) $L_2(q)$, q>3, q≠9; $Sz(2^{2m+1})$, $R(3^{2m+1})$, m≥1;
Then $h(\pi_e(G))$=1. Moreover $h(\pi_e(J_2))$=h$(\pi_e(S_8))$=∞.

Recently V. D. Mazurov (1994) proved the following result: If m=$\pi_e(L_3(5))$, then h(m)=2. But we can not find such m satisfying h(m)=3. Therefore we put forward the following problem:

PROBLEM 3.1 Does there exist a positive integer k such that h(m)∈{0, 1, 2, ... , k, ∞} for all subsets m of positive integers? If it exists, what is such k ?

Z. M. Chen proved the following lemma (refer to W. J. Shi(1994)).

LEMMA 3.1 Let N be a minimal normal subgroup of G and exp(N)∈π_e(N), where exp(N) is the exponent of N. Then h$(\pi_e(G))$=∞.

A group is called a C_{pp}–group if the centralizer of any p–element is a p–group. Recently we have proved h(m)∈{1, ∞} holds for C_{22}-groups and the groups whose element orders ≤12.

3.4 Thompson's Two Problem

A problem related to conjecture 3. 1 is the following open problem put forward by J. G. Thompson (a letter to W. J. Shi, 1987).
For each finite group G and each integer d≥1, let G(d)= {x∈G$|x^d$=1 }.

DEFINITION 3.2 G_1 and G_2 are called groups of the same order type if and only if $G_1(d)|=|G_2(d)|$, d=1, 2,...

PROBLEM 3.2 Suppose G_1 and G_2 are groups of the same order type, and G_1 is solvable. Is it true that G_2 is necessarily solvable?

In Thompson's letter he pointed out that "The problem arose initially from the study of algebraic fields and is of considerable interest. "

If G_1 and G_2 are finite groups of the same order type, then $\pi_e(G_2)=\pi_e(G_2)$, $|G_1|=|G_2|$. Thus if Conjecture 3.1 is proved, G_1, G_2 are of the same order, G_1 is solvable, then G_2 is not simple.

Another problem of Thompson aims at characterizing all finite simple groups by conditions related to quantity.

CONJECTURE 3.3 Let G be a finite group with $Z(G)=1$, M a non−abelian simple group and $N(G)=N(M)$, where $N(G)=\{M\in N|G$ has a conjugacy class C, such that $|C|=n\}$. If $N(G)=N(M)$, then $G\cong M$.

On this conjecture G. Y. Chen (1994c) has dealt with the situation that M is a finite simple group with nonconnected prime graph. He has come to

THEOREM 3.15 Suppose G is a finite group, $Z(G)=1$, M is a simple group with nonconnected prime graph, $N(G)=N(M)$, then $G\cong M$.

3.5 Some Related Topics

In this subsection some related work is going to be introduced.

(a) Finite Groups Whose Element Orders Are Consecutive Integers

DEFINITION 3.3 Let n be a positive integer. Then a group G is called an OC_m−group if every element of G has order $\leq n$, and for each $m\leq n$ there exists an element of G having order m.

R. Brandl and W. J. Shi(1991) have come to the following theorem.

THEOREM 3.16 Let G be a finite OC_m−group. Then $n\leq 8$.

Moreover they have classified all finite OC_n−groups.

After their paper was published. D. Macttal proposed the problem: To classify all the finite groups in which the orders of proper subgroups are consecutive integers.

DEFINITION 3.4 A group G is called a $OS_m−$ group if the orders of all proper subgroups of G are 1, 2, . . . , n.

A group G is called a $OS_{n,m}$ group, m>n+1, if the orders of all proper subgroups of G are 1, 2, . . . , n, m.

A group G is called a OS_{n,m_1,m_2} group, $m_1 > m_2 > n+1$, if the orders of all proper subgroups of G are 1, 2, . . . , n, m_1, m_2.

J. H. Li(1993) has classified all OS_{n_1,m_1,m_2} groups, $OS_{n,m}$ groups and OS_m-groups.

M. C. Xu(1994) has classified all $OC_{n,p}$ groups where an $OC_{n,p}$ group is a finite group with $\pi_e(G)=\{1, 2, . . . , n; p_1, . . ., p_n\}$, such that $p_i > n+1$, p_i are primes, i=1, 2, . . . , s.

(b) Finite groups in which the number of the maximal order elements is given

This is a topic related to the groups of the same order type. C. Yang (1993) has the following result.

THEOREM 3.17 Let G be a finite group and M(G) be the set of maximal order elements. If one of the following conditions holds, then G is solvable.

(1) $2 \nmid |M(G)|$;
(2) $|M(G)|\leq 4$;
(3) $|M(G)|=2p$, p is a prime;
(4) $|M(G)|=\phi(k)$,where $\phi(k)$ is the Eulerian function of the maximal order k.

Moreover if $|M(G)|=2, 4$, or an odd mumber, then G is supersolvable.

(C) A general discussion of the set of orders of elements of a group

If G_1, G_2 are two groups, $G=G_1\times G_2$, then $\pi_e(G)=\{\frac{n_1n_2}{(n_1,n_2)}\big| n_1\in\pi_e(G_1), n_2\in\pi_e(G_2)\}$. C. E. Praeger has the following question.

QUESTION 3.4 What properties does the set $X=\{m \mid$ there exists a finite group G with $\pi_e(G)=m\}$ have?

On this question, we know just a few trivial properties. At the end of this section, there are a few words we wish to say. We believe that there are many problems related to "orders" worth discussing. And we hope that this work can be continued and the results of this work could be applied to other branches of mathematics.

Bibliography

[1] Arad, Z. and D. Chillag, D. (1984). A criterion for the existence of normal $\pi-$ complement in finite group. J. Algebra , 87: 472–452.

[2] Artin, E. (1955). The order of the classical simple groups, Comm. Pure Appl. , 8: 355–365

[3] Ban, B. L. and Yu,S. X. (1992). On the order of automorphism groups of a class of p–groups, Acta. Math. Sin. , 35(4).

[4] Ban, G. L. and Yu, S. X. (1993). On LA-groups and related theorems , J. Guang Xi Univ.

[5] J. X. Bi, J. X. (1990). A characterization of symmetric groups (Chinese), Acta Math. Sinica , 33: 70-77.

[6] Brandl, R. and Shi, W. J. (1991) The finite groups whose element orders are consecutive integers, J. Algebra, 143: 388-400

[7] Brandl, R. and Shi, W. J. (1994) The characterization of PSL(2, q) by its element orders, J. Algebra, 163: 109-114

[8] Brauer, R. and Reywolds, W. F. (1958). On a problem of E. Artin , Ann. Math. , 68: 713-720

[9] Brauer, R. and Tuan, H. F. (1945). On simple groups of finite order, Bull. Amer. Math. Soc. , 51: 765-766

[10] Buckley, J. (1970). Finite groups whose minimal subgroups are normal, Math. Zeit, 116 : 15-17

[11] Burnside , W. (1902). On an unsettled question in the theory of discontinuous groups Quart. J. Pure Appl. Math. , 33: 230-338

[12] Chen, G. Y. (1990). Finite groups with automorphism group having an order of product of three distinct primes . Proc. R. Ir. Acad. , 90A(1): 57-62

[13] Chen, G. Y. (1990). Finite groups with automphism group having an order $p_1 p_2 ... p_n$ or pq^2 (Chinese), J. Southwest China Teachers Univ. , 15(1): 21-28

[14] Chen, G. Y. (1992). Finite groups with Schmidt group as automorphism group. Chin. Ann. of Math. , 13B(1): 105-109

[15] Chen, G. Y. (1992). Finite groups with automorphism group having all Sylow sub-groups cyclic. Proc. R. Ir. Acad. , 92A(1): 37-40

[16] Chen, G. Y. (1994). On Thompson's Conjecture, Sichuan Univ. Ph. D. Thesis.

[17] Chen, G. Y. (1994). Finite groups with automorphism group an extension of a cyclic group by an abelian group, Acta. Math. Sinica, 37(5): 645-652

[18] Chen, G. Y. (1994). A new characterization of finite simple groups, Chin. Sci. Bull. 39(6):1448-1451

[19] Chen, Z. M. (1981). Inner $\Sigma-$ groups, Acta Math. Sin. , 23(2): 239-243 ; 24(3): 331-335

[20] Chen, Z. M. (1982). A Theorem about a finite group having normal complement (Chinese) , Chin. Adv. Math. , 11(4): 308-320

[21] Chen, Z. M. (1982). Some characterizations of finite supersolvable groups, Chin. Ann. Math. , 3(5): 561-566

[22] Chen, Z. M. (1984). Sufficient conditions for a finite group being an inner-supersolvable or outer-supersolvable group, Acta. Math. Sin. , 27(5): 694-703

[23] Chen, Z. M. (1986). Some theorems about finite supersolvable groups (Chinese) , J. southwest China Teachers Univ. (Nat. Sc. Ed.), No. 2: 1-6

[24] Chen, Z. M. (1986). On inner p-closed groups, Chin. Adv. Math. (Chinese), 15(14): 385-388

[25] Chen, Z. M. (1987). A generalization of Sirinivasan's Theorem (Chinese) , J. Southwest China Teaachers Uni. (Nat. sci. Ed.), No. 1

[26] Chen, Z. M. (1987). Outer $\Sigma-$ groups of finite order. Chin. Ann. Math. , 8B(1): 109-119

[27] Chen, Z. M. (1988). Inner, Outer $\Sigma-$ groups. Publishing House of Southwest China Teachers Univ. ,Chongqing,PRC

[28] Chen, Z. M. (1989). (with Y. M. Wang). Minimal non-solvable groups with an acting group (Chinese), Koxue Tong Bao. , 35(22): 1691-1693

[29] Chen, Z. M. (1989). On Bp-groups (chinese). Acta Math. Sin. , 32(6): 834-840

[30] Chen, Z. M. (1990). Inner-(π, π')-closed group (Chinese), Chin. Ann. Math. , 11A(5) 576-582

[31] Chen, Z. M. (1993). One way to generalize theorems in group theory, Chin. Sci. Bull. 38(6): 491-493

[32] Chen, Z. M. (1994). Finite groups whose minimal p-subgroups are seminormal, Acta. Ann. Sin. , (New Ser.) 10(Spc. Iss.): 69-73

[33] Curran, M. J. (1988). Automorphisms of certain p-groups(p odd), Bull. Austra. Math. Soc. , 38 : 299-305

[34] Doek, D. K. (1966). Minimal nicht 'uber auflösbare endiche Gruppern. Math. Zeit. , 189-205

[35] Fan, Y. (1981). (with Y. D. Zhang) On supersolvability of groups of order n, Chin. J. Math. , 1(1): 86-95

[36] Fan, Y. (1981). A criterion for supersolvability of groups(chinese), J. Wuhan Univ. (Nat. Sci. Ed.), spec. Issue of Math:103-109

[37] Fan, Y. (1986). A note on hypercenters and hypergeneralized centers (Chinese). Chin. J. Math. , 6(2): 215-220

[38] Fan, Y. (1986). On p-supersolvable groups, Chin. Adv. Math. , 15(2): 201-204

[39] Fan, Y. (1986). On groups of odd order and rank≤2, Chin. ann. Math. , 7B(3) : 350 -364

[40] Guo, X. Y. (1988). A note of normal p-complement(Chinese). J. of Northeast China Math. , 4: 446-452

[41] Guo, X. Y. (1989). (with B. L. Zhang) Normal π-complements in finite groups. Comm. Algebra, 17(7): 1601-1606

[42] Guo, X. Y. (1989). On relative normal p-complement in finite groups(Chinese). Chin. Ann . Math. 10A: 1601-1606

[43] Guo, X. Y. (1991). The normal index of maximal subgroups in finite groups(Chinese) , Acta Math. Sin. , 34: 203-212

[44] Heineken, H and Liebeck, H. (1973). On p-groups with odd order automorphism group, Arch. Math. , 24: 8-16

[45] Herzog, M. (1968). On finite simple groups of orders divided by three primes only, J. Algebra: 383-388

[46] Huppert, B. (1967). Endliche Gruppen I. Springer-Verlag, Berlin Heidelberg, New York

[47] Ito, N. (1955). Uber eine zur Frattini Gruppe duale Bildung, Nagoya J. Math. , (9): 123-127

[48] Jonah, D. and Konvisser, M. (1975). Some non-abelian p-groups with abelian auto-morphism groups, Arch. Math. , 26: 131-133

[49] Kimmerle W.,Lyons, R. ,Sandling, R. and Teague, D. N. (1990). Composition factors from the group ring and Artin's Theorem on orders of simple groups. Proc. Lond. Math. Soc. , 60(63): 89-122

[50] Li, H. L. (1989). (with W. J. Shi) A characteristic property of M_{12} and PSU(6, 2) (Chinese), Acta. Math. Sin. , 32:758-764

[51] Li, H. L. (1993) (with W. J. Shi) A characteristic property of some sporadic somple groups (Chinese), Chin. Ann. Math. , 14A(2): 144-151

[52] Li, J. H. (1993). Finite group whose proper subgroups' orders are consecutive integers except two (Chinese). J. Southwest Chin. Teachers Univ., 18(4): 393- 401

[53] Li, Q. X. (1993). Several types of groups and the semi-normality(Chinese), J. Southwest China Teachers Univ. , 18(3): 265-268

[54] Li, S. R. (1983). Finite groups in which every non-maximal proper subgroup of every order is 2-closed (Chinese), Chin. ann. Math. , 4B: 199-206

[55] Li, S. R. (1983). Finite groups in which every non-maximal proper subgroup of even order is 2'-close(Chinese), Chin. J. Math. , 3(1): 1-6

[56] Li, S. R. (1986). Finite groups in which every non- maximal proper subgroup is 3 -closed (Chinese), Acta. Math. Sin. , 29(4): 498-503

[57] Li, S. R. (1987). A class of finite non-solvable groups (Chinese). Chin. Adv. Math. , 16(3): 289-293

[58] Li, S. R. (1988). Finite non-solvable groups with supersolvable second maximal 3d -subgroups(Chinese), Chin. Ann. Math., 9A(1): 32-37

[59] Li, S. R. (1988). Finite groups in which every supersolvable subgroup is either 2 -closed or a Schmidt Group(Chinese), Acta. Math. Sin. , 31(3): 341-347

[60] Li, S. R. (1988). (with Y. Q. Zhao) Some finite non-solvable groups characterized by their solvable subgroups(Chinese), Acta Math. Sin. 31(4): 5-13

[61] Li, S. R. (1990). Finite groups having exactly two same classes of non- invariant maximal subgroups (Chinese), Acta Math. Sin. , 33(3): 389-392

[62] Li, S. R. (1992). A theorem of the solvability of finite factorizable groups (Chinese), Acta. Math. Sin., 30(3): 388-392

[63] Li, S. R. (1994). Automorphism groups of some finite groups. Sci. in Chin., 37(3) : 295-303

[64] Li, X. H. (1990). On a problem of finite groups (Chinese), J. Southwest China Teachers Univ., 15(1): 144-146

[65] Li, X. H. (1994). Finite groups having three chasses of maximal subgroups of the same order, Acta Mcth. Sin. , 37(1): 108-115

[66] MacHale, D. (1981). (with D. Flannery) Some finite groups which are rarely automorphism groups I, Proc, R, Ir, acdd, (Ser. A), 81(1): 189-196

[67] MacHale, D. (1983). Some finite groups which are rarely automorphism groups II, Proc. R. Ir. Acad. (Ser. A), 83(1): 184-196

[68] Martin, U. (1986). Almost all p-group have automorphism group a p-groups, Bull. Amer. Math. Soc. , 15(1)

[69] Mazurov, V. D. (1994). On the set of orders of elements of a finite group. Algebra and Logic, 33(1)

[70] Neumann, B. H. (1937). Groups whose elements have bounded order, J. London Math. Soc. , 12: 195-198

[71] Praeger, C. E. (1994). (with W. J. Shi) A characterization of some alternative and symmetric groups, Comm. in Algebra,22(5): 1507-1530

[72] R'edei, L. (1947). Das schiefe product in der grupentheorie. Conn. Math. Helv, . 20:255-264

[73] Ren, Y. C. (1989). Several remarks for p-nilpotent groups (Chinese), J. Sichuan Univ., 26(1):35-38

[74] Ren, Y. C. (1990). Acta Math. Sin. , 33(6): 798-803

[75] Rose, J. S. (1978). A course on group theory, Camb. Univ. Press

[76] Satry, N. S. and Deskins, W. E. (1978). Influence of normality conditions on almost minimal subgroups of a finite group. J. Algebra, 52: 346-377

[77] Shi, W. J. (1985). A characterization of some special linear groups, Chin. J. Math., 5: 191-200

[78] Shi, W. J. (1986). (with W. Z. Yang) The finite groups all of whose elements are of prime power orders (Chinese), J. Yunan Educational College, (1): 2-10

[79] Shi, W. J. (1987). A characteristic property of J_1 and PSL(2, 2^n) (Chinese) , Adv. in Math. 16: 397-401

[80] Shi, W. J. (1987). A characteristic property of A_8, Acta Math. Sin. , New Ser. , 3:92-96

[81] Shi, W. J. (1989). A new characterization of some simple groups of Lie type, Contemp. Math. 82: 171-180

[82] Shi, W. J. (1989). Finte groups with two same order classes of maximal subgroups (Chinese). Chin. Ann. Math. , 10A(15): 523-537

[83] Shi, W. J. (1989). A characterization of Conway simple group Co_2 (Chinese) , Chin. J. Math. , 9: 171-172

[84] Shi, W. J. (1990). (with J. X. Bi) A characteristic property for each finite projective special linear group. Lecture Notes in Math. , 1456:171-180

[85] Shi, W. J. (1990). A characterization of the Higman-Sims groups. Houston J. Math., 16: 597-602

[86] Shi, W. J. (1991). On simple K_4−groups(Chinese), Chin. Sci. Bull. 36: 1281-1283

[87] Shi, W. J. (1991). (with J. X. Bi) A characterization of Suzuki-Ree groups. Sci. in Chin. (Ser. A), 34: 14-19

[88] Shi, W. J. (1992). (with C. Yang), A class of special finite groups, Chinese Sci. Bull. 37: 252-253

[89] Shi, W. J. (1992). On problem of E. Artin(Chinese), Acta Math. Sin. , 35(2): 262-265

[90] Shi, W. J. (1992). (with J. X. Bi) A new characterization of the alternating groups, SEA Bull. Math. 16(1): 81-90

[91] Shi, W. J. (1992). A characterization of Suzuki's simple groups, Proc. Amer. Math. Soc. , 114(3): 589-591

[92] Shi, W. J. (1993). On the order of the finite simple groups (Chinese), Chin. Sci. Bull., 38(4): 296-298

[93] Shi, W. J. (1994). The characterization of the sporadic simple groups by their elements orders, Algebra Colloq. (PRC), 1(2): 159-166

[94] Shu, X. Y. (1988) Half-normal subgroups of finite groups (Chinese), Chin. J. Math. , (8):5-9

[95] Suzuki, M. (1962). On a class of doubly transitive groups. Ann. Math. 75: 105-145

[96] Suzuki, M. (1981). Finite groups with nilpotent contralizer. Trans. Amer. Math. Soc. 99: 425-470

[97] Thompson, J. G. (1964). Normal p-complement of finite groups, J. Algebra, (1): 43-46

[98] Wang, R. J. (1989). On generalized A−groups(Chinese), J. Math. Res. and Expos., 9(4): 509 -528

[99] Wang, X. F. (1989). On inner-π'-closed groups and normal π-complements(Chinese), Chin. Ann. Math., 10B(3): 323-331

[100] Wang, Y. M. (1991). On Zassenhaus' Conjecture (Chinese), Chin. Sci. Bull. , 36(5)

[101] Weinstein, M. (1982). Between nilpotent and solvable. Polygonal Publishing House.

[102] Xiao, W. J. (1989). On Zassenhaus' Conjecture(Chinese), Koxue Tong Bao, 34(4): 244- 246

[103] Xu, M. C. (1994). Finite groups whose element orders are consecutive integers except some primes(Chinese), J. Southwest China Teachers Univ. , 19(2): 119- 122

[104] Yang. C. (1993). Finite groups with a given number of the maximal order elements (Chinese), Chin. Ann. Math. , 14A(5): 561-567

[105] Yu, S. X. (1990). On LA-groups having some characteristic subgroups. J. Guang Xi Univ. (Nat. Sci. Ed.), 15(2)

[106] Zhang, G. X. (1986). On two theorems of Thompson, Proc. Amer. Math. Soc. , 98(4): 579- 582

[107] Zhang, J. P. (1985). On finite groups which satisfy the permutizer condition(Chinese) Koxue Tong Bao, 14: 1048-1049

[108] Zhang, J. P. (1987). (with L. W. Zhang) The supersolvability of a class of finite groups (Chinese), Acta Math. Sin. , 30(5): 622-625

[109] Zhang, J. P. (1988). On Syskin's problem of finite groups(Chinese), Koxue Tong Bao. 33A(2): 124-128

[110] Zhang, J. P. (1988). On the supersolvability of QCLT-groups, Acta Math. Sci., 32(1): 29-32

[111] Zhang, J. P. (1989). On p-solvability of the finite groups with a T. I. Sylow p-subgroup (Chinese). Koxue Tong Bao. , 34(3): 177-179

[112] Zhang, J. P. (1989). Influence of S-quasinormality condition on almost minimal subgroups of a finite group(Chinese), Acta Math. Sin. , (New series) 3(2): 125- 132

[113] Zhang, L. W. (1986). The outer structure of formation and its applications. Acta Math. Sin. , New Ser., 2:78-61

[114] Zhen, Y. N. (1986). π-properties of π-solvable groups (Chinese), Chin. J. Math. , 6(3): 297-366

[115] Zhen, Y. N. (1990). Weakly S-normal subgroups (Chinese), Chin. J. Math., 10(1): 33-38

Representation and Cohomology Theory of Modular Lie Algebras

CHIU SEN

Department of Mathematics, Shanghai Institute of Education, Shanghai 200050, China

In [2,3], it was shown that the finite-dimensional restricted simple Lie algebras over algebraically closed fields of characteristic $p > 7$ are of classical or Cartan type. In [1, 11, 12, 24, 25, 28–32], one could get more precise information about the representations and the cohomology groups of modular classical Lie algebras. Investigations on the representation theory of modular Lie algebras of Cartan type began early (cf. e.g., [4]). Much progress in this direction has been made recently. It is the purpose of this article to give a survey of our main results in the representation and cohomology theory of modular Lie algebras.

Throughout this paper let F be an algebraically closed field with char $F = p > 0$.

1. GRADED MODULES

In this section, we give some general properties of a graded module $V = \oplus_{i \geq 0} V_i$ of a graded Lie algebra $L = \oplus_{i \in \mathbb{Z}} L_i$.

Let $L = \oplus_{i \in \mathbb{Z}} L_i$ be a graded Lie algebra. The subspaces L_0, $L^+ := \oplus_{i > 0} L_i$ and $L^- := \oplus_{i < 0} L_i$ are subalgebras of L and $L = L^- \oplus L_0 \oplus L^+$. If $V = \oplus_{i \geq 0} V_i$ a nonzero graded module, by reindexing the subscripts if necessary, we shall always assume $V_0 \neq 0$ and V_0 is called base space of V. A graded module V is called transitive if $\text{Ann} L^- := \{v \in V \mid L^- v = 0\} = V_0$. We have

Theorem 1.1. *[34] Let $V = \oplus_{i \geq 0} V_i$ be a graded module of the graded Lie algebra $L = \oplus_{i \in \mathbb{Z}} L_i$. Then V is irreducible if and only if (1) V_0 is an irreducible L_0-module, (2) V is generated by V_0, (3) V is transitive.*

Theorem 1.2. *[34] Let $L = \oplus_{i \in \mathbb{Z}} L_i$ be a graded Lie algebra. For every irreducible L_0-module V_0, there exists, up to isomorphism one and only one irreducible graded module V with V_0 as its base space.*

Theorem 1.3. *[34] Let $L = \oplus_{i \in \mathbb{Z}} L_i$ be a graded Lie algebra and V be a finite-dimensional irreducible L-module. Then V is isomorphic to a graded module if and only if the elements of L^+ and L^- act nilpotently on V.*

By Theorem 1.3, we can easily show

Supported by National Natural Science Foundation of China .

Theorem 1.4. *[34] Let L be a finite-dimensional restricted graded Lie algebra over F. If the center of L is zero, then (1) Every finite-dimensional irreducible restricted L-module is isomorphic to a graded module. (2) If V is an L-module, let V_0 be the subspace $\text{Ann}L^-$ of V, then $V \mapsto V_0$ is, up to isomorphism, a bijective map of the set of irreducible restricted L-modules onto the set of irreducible restricted L_0-modules.*

2. MIXED PRODUCT

Let $\mathfrak{A}$ be a commutative associative algebra over F, $a \in \mathfrak{A}$. In the following, when no confusion arises, a will also denote the operator of multiplication by a. If $\{D_1, \cdots, D_n\}$ is a set of multually commutative derivations of $\mathfrak{A}$, the set of derivations $K = \{\sum a_i D_i \mid a_i \in \mathfrak{A}\}$ is a Lie algebra. In fact,

$$\left[\sum_i a_i D_i, \sum_j b_j D_j\right] = \sum_i \sum_j \left(a_j D_j(b_i) - b_j D_j(a_i)\right) D_i. \tag{2.1}$$

Let $M(n) = \mathfrak{gl}(n)$ be the Lie algebra of all linear transformations of an n-dimensional linear space V_0 with basis $\{e_1, \cdots, e_n\}$ and E_{ij} be the linear transformation such that $E_{ij}e_k = \delta_{jk}e_j$, $i, j, k = 1, \cdots, n$. Then

$$[E_{ij}, E_{lm}] = \delta_{jl}E_{im} - \delta_{im}E_{lj}, \ i, j, l, m = 1, \cdots, n. \tag{2.2}$$

If L_0 is a Lie subalgebra of $M(n)$, then $L = \{\sum a_i D_i \in K \mid \sum_{i,j} D_i(a_j) \otimes E_{ij} \in \mathfrak{A} \otimes L_0\}$ is a Lie subalgebra of K. L is called the extension of L_0 in K. If L is any subalgebra of K, the holomorph $L \oplus \mathfrak{A}$ will be denoted by $\overline{L}$. In $\overline{L}$, we have

$$[D, a] = D(a), \ D \in L, \ a \in \mathfrak{A}.$$

A representation ρ (resp. a module U) of $\overline{L}$ is admissible if the restriction of ρ (resp. U) on $\mathfrak{A}$ is an associative algebra representation (resp. module) of $\mathfrak{A}$. If L is a subalgebra of $\overline{K} = K \oplus \mathfrak{A}$, $A \in L$, ρ a representation of L in the module U and V is a linear space, set

$$\bar{\rho}_V(A) = \rho(A) \otimes 1_V,$$

which acts on $U \otimes V$. For any element $D = \sum a_i D_i \in K$, write

$$\tilde{D} = \sum D_i(a_j) \otimes E_{ij} \in \mathfrak{A} \otimes M(n).$$

By direct verification, we have

Theorem 2.1. *[33] Let L_0 be a subalgebra of $M(n)$, L the extension of L_0 in K and $\overline{L} = L \oplus \mathfrak{A}$. If ρ_0 denotes a representation of L_0 in the module V, ρ an admissible representation of $\overline{L}$ in the module U and $k \in F$, let*

$$\sigma_k(D + f) = \bar{\rho}_V(D + kf) + (\rho|_{\mathfrak{A}} \otimes \rho_0)(\tilde{D}),$$

where $D \in L$, $f \in \mathfrak{A}$. Then σ_k is a representation of L in $U \otimes V$. Moreover, σ_k is admissible if $k = 0$ or 1.

The representation σ_k (resp. module $U \otimes V$) in Theorem 2.1 is called the mixed product with dilation k of ρ and ρ_0 (resp. U and V) and denoted by $\rho_{(k)} \rtimes \rho_0$ (resp. $U_{(k)} \rtimes V$).

We proceed to construct the mixed products of Lie algebras of Cartan type $W(n, \mathbf{m})$, $S(n, \mathbf{m})$ and $H(n, \mathbf{m})$, adopting the notations of [37].

Let $A(n)$ be the set of n-tuples of non-negative integers. For $\alpha = (\alpha_1, \cdots, \alpha_n) \in A(n)$, let $|\alpha| = \sum_{i=1}^n \alpha_i$. Set $\varepsilon_i = (\delta_{1i}, \cdots, \delta_{ni}) \in A(n)$. Let $\mathfrak{A}(n)$ be the divided power algebra with basis $\{x^{(\alpha)} \mid \alpha \in A(n)\}$ and multiplication

$$x^{(\alpha)} x^{(\beta)} = C_\alpha^{\alpha+\beta} x^{(\alpha+\beta)}, \ \alpha, \beta \in A(n),$$

where $C_\beta^\alpha = \prod_{i=1}^n C_{\beta_i}^{\alpha_i}$ for $\alpha = (\alpha_1, \cdots, \alpha_n)$, $\beta = (\beta_1, \cdots, \beta_n) \in A(n)$.

If $\mathbf{m} = (m_1, \cdots, m_n)$ is an n-tuple of positive integers and $A(n, \mathbf{m}) = \{\alpha \in A(n) \mid \alpha_i < p^{m_i}, \ i = 1, \cdots, n\}$, then $\mathfrak{A} = \mathfrak{A}(n, \mathbf{m}) = \langle x^{(\alpha)} \mid \alpha \in A(n, \mathbf{m}) \rangle$ is a subalgebra of $\mathfrak{A}(n)$. Write $\pi = (p^{m_1} - 1, \cdots, p^{m_n} - 1) \in A(n, \mathbf{m})$. Define the derivations D_i, $i = 1, \cdots, n$ of $\mathfrak{A}(n, \mathbf{m})$ by

$$D_i x^{(\alpha)} = x^{(\alpha - \varepsilon_i)}$$

(We set $x^{(\alpha)} = 0$, if $\alpha \notin A(n)$).

If L is any one of $W(n, \mathbf{m})$, $S(n, \mathbf{m})$ and $H(n, \mathbf{m})$, then $L = \oplus_{i \geq -1} L_i$ is a graded Lie algebra and under the linear map $x^{(\varepsilon_i)} D_j \mapsto E_{ij}$, L_0 is isomorphic to $\mathfrak{gl}(n)$, $\mathfrak{sl}(n)$ and $\mathfrak{sp}(n)$ respectively, where E_{ij} is the matrix whose (k, l)-component is $\delta_{ik} \delta_{jl}$. Let ρ_0 be a representation of L_0 in the module V_0 and $\tilde{V}_0 = \mathfrak{A} \otimes V_0$. If $D = \sum a_i D_i \in L$, then $\tilde{D} := \sum D_i(a_j) \otimes E_{ij} \in \mathfrak{A} \otimes L_0$. Let $\tilde{D} = \sum g_i \otimes l_i$, where $g_i \in \mathfrak{A}$, $l_i \in L_0$. Define a linear transformation $\tilde{\rho}_0(D)$ of $\tilde{V}_0$ by

$$\tilde{\rho}_0(D)(f \otimes v) = D(f) \otimes v + \sum g_i f \otimes \rho_0(l_i) v, \quad f \in \mathfrak{A}, \ v \in V.$$

By Theorem 2.1, $\tilde{\rho}_0$ is a representation of L in $\tilde{V}_0$ and $\tilde{V}_0$ is the mixed product $\mathfrak{A} \rtimes V_0$ of L (We drop the notation (k) and simply write $\mathfrak{A} \rtimes V_0$, since it is independent of k as an L-module). We have (cf. [34])

Theorem 2.2. (1) $\tilde{V}_0 = \oplus_{i \geq 0} V_i$ is a graded L-module where $V_i = \langle x^{(\alpha)} \otimes V_0 \mid |\alpha| = i \rangle$. The base space of $\tilde{V}_0$ is $1 \otimes V_0 \cong V_0$.

(2) $\tilde{V}_0$ is an transitive (i.e., $\text{Ann} L_{-1} = 1 \otimes V_0$).

(3) If V_0 is irreducible L_0-module, then the irreducible graded L-module with base space V_0 is isomorphic to the (unique) minimum submodule $(\tilde{V}_0)_{\min}$ of $\tilde{V}_0$.

(4) If $\mathbf{m} = \mathbf{1}$, i.e., L is restricted, and V_0 is an irreducible restricted L_0-module, then $(\tilde{V}_0)_{\min}$ is the unique irreducible restricted L-module whose base space is isomorphic to V_0.

(5) If $V = \oplus_{i \geq 0} V_i$ is an irreducible graded module then $V \mapsto V_0$ is, up to isomorphism, a bijective map of the class of irreducible graded L-modules onto the class of irreducible L_0-modules.

(6) If L is centerless and restricted, then every irreducible restricted L-module V is graded and $V \mapsto V_0$ is, up to isomorphism, a bijective map of the class of irreducible restricted L-modules onto the class of irreducible restricted L_0-modules.

Adopting the notation of [22,§ 1], we introduce certain induced modules. Let $L = \oplus_{i=-r}^{s} L_i$ be a graded Lie algebra and $\mathcal{L}_i = \oplus_{j \geq i} L_j$. We define the subalgebra $\theta(L, \mathcal{L}_0)$ of $U(L)$ as follows. Let $\{w_1, \cdots, w_k\}$ be a basis of L^-. Then there exist $m_1, \cdots, m_k$ such that

$$(\mathrm{ad} w_i)^{p^{m_i}} = 0, \quad 1 \leq i \leq k.$$

Thus the elements $z_i = w_i^{p^{m_i}}$ belong to the center of $U(L)$ and the subalgebra $\theta(L, \mathcal{L}_0)$ of $U(L)$, which is generated by $U(\mathcal{L}_0)$ and $\{z_1, \cdots, z_k\}$, is isomorphic to $F[z_1, \cdots, z_k] \otimes_F U(\mathcal{L}_0)$, the first factor being a polynomial ring in k indeterminates.

Let V be an $\mathcal{L}_0$-module. The action of $U(\mathcal{L}_0)$ on V can be extended to $\theta(L, \mathcal{L}_0)$ by letting the polynomial algebra $F[z_1, \cdots, z_k]$ operate via its canonical suppplementation. Henceforth all $\mathcal{L}_0$-modules will be considered as $\theta(L, L_0)$-modules in this fashion. Let ζ denote the natural representations of $\mathcal{L}_0$ in $L/\mathcal{L}_0$. Then there exists a unique homomophism $\sigma: U(\mathcal{L}_0) \to F$ of F-algebras such that $\sigma(x) = \mathrm{tr}(\zeta(x))$. We introduce a twisted action on V by setting $x \cdot v := xv + \sigma(x)v$. The new $\mathcal{L}_0$-module will be called V_σ. By [22, Theorem 1.4], there is a natural isomorphism of $U(L)$-modules between the induced module and the coinduced module

$$U(L) \otimes_{\theta(L, \mathcal{L}_0)} V_\sigma \cong \mathrm{Hom}_{\theta(L, \mathcal{L}_0)}(U(L), V).$$

If V is an $\mathcal{L}_0$-module, then we can extend the operations on V to L_0 by letting $\mathcal{L}_1$ act trivially and regard it as an $\mathcal{L}_0$-module.

If L is a graded Lie algebra of Cartan type $W(n, \mathbf{m})$ $(= W)$, $S(n, \mathbf{m})$ $(= S)$, $H(n, \mathbf{m})$ $(= H)$ and V is a finite-dimensional L_0-module, then by [19, Corollary 2.6], we have

$$U(L) \otimes_{\theta(L, \mathcal{L}_0)} V_\sigma \cong \mathfrak{A} \rtimes V.$$

3. IRREDUCIBLE MODULES

In this section, all irreducible graded modules of $L(n, \mathbf{m}) = \oplus_{i \geq -1} L_i$ of Cartan type W, S and H are determined, where char $F > 3$.

Let $\mathbf{h}(L_0)$ be the standard Cartan subalgebra of $L_0 = \mathfrak{gl}(n)$, $\mathfrak{sl}(n)$, or $\mathfrak{sp}(n)$ and Λ_i, $i = 1, \cdots, n$, the linear functions on $\mathbf{h}(\mathfrak{gl}(n)) = \langle E_{11}, \cdots, E_{nn} \rangle$ such that

$$\Lambda_i(E_{jj}) = \delta_{ij}.$$

The restriction of Λ_i on every $\mathbf{h}(L_0)$ will also be denoted by Λ_i. Let

$$\lambda_0 = 0, \quad \lambda_i = \sum_{j=1}^{i} \Lambda_j, \quad i = 1, \cdots, n.$$

Then λ_i $(i = 1, \cdots, n,$ for $L_0 = \mathfrak{gl}(n)$; $i = 1, \cdots, n-1$, for $L_0 = \mathfrak{sl}(n)$; $i = 1, \cdots, r$, for $L_0 = \mathfrak{sp}(n))$ are the fundamental weights of L_0. Every weight of $\mathfrak{gl}(n)$, $\mathfrak{sl}(n)$, or $\mathfrak{sp}(n)$ is a linear combination of the fundamental weights. We have

Proposition 3.1. *[35] If V_0 is L_0-irreducible, then $\tilde{V}_0$ is $L(n, \mathbf{m})$-irreducible unless V_0 is trivial or a highest weight module with a fundamental weight as its highest weight.*

If $\lambda \in (\mathbf{h}(L_0))^*$, denote by $V_0(\lambda)$ the irreducible L_0-module with highest weight λ. We denote $(\tilde{V}_0(\lambda_i))_{\min}$ by $\tilde{M}(\lambda_i)$. Let $d_i \colon \tilde{V}_0(\lambda_i) \to \tilde{V}_0(\lambda_{i+1})$ be the exterior differential operator which is a $W(n, \mathbf{m})$-module homomorphism. We have

Theorem 3.1. *[35] (1) Let char $F = p > 3$, $L = W(n, \mathbf{m})$. Then (i) $\tilde{V}_0$ is reducible if and only if $V_0 = V_0(\lambda_i)$, $i = 0, 1, \cdots, n$; (ii) $\tilde{M}(\lambda_0) = F$, $\tilde{M}(\lambda_i) = d_{i-1}\tilde{V}_0(\lambda_{i-1})$, $i = 1, \cdots, n$; (iii) $\dim \tilde{M}(\lambda_0) = 1$, $\dim \tilde{M}(\lambda_i) = a_i$, $i = 1, \cdots, n$, where $a_i = (p^{|\mathbf{m}|} - 1)C_i^{n-1}$; (iv) the composition factors of $V_0(\lambda_i)$ are $\tilde{M}(\lambda_i)$, $\tilde{M}(\lambda_{i+1})$ and F ($C_i^n - \delta_{i0}$ times), $i = 0, 1, \cdots, n$.*

(2) Let char $F = p > 3$, $L = S(n, \mathbf{m})$ and $U(\lambda_i) = d_{i-1}\tilde{V}_0(\lambda_{i-1})$. Then (i) $\tilde{V}_0$ is reducible if and only if $V_0 = V_0(\lambda_i)$, $i = 0, 1, \cdots, n-1$; (ii) $\tilde{M}(\lambda_0) = F$, $\tilde{M}(\lambda_1) = \sum_{j=0}^{N(\mathbf{m})-2} U(\lambda_1)_j$ (where $N(\mathbf{m}) = \sum_{i=1}^{n}(p^{m_i} - 1)$), $\tilde{M}(\lambda_i) = U(\lambda_i)$, $i = 2, \cdots, n-1$; (iii) $\dim \tilde{M}(\lambda_0) = 1$, $\dim \tilde{M}(\lambda_i) = a_i - \delta_{i1}$, $i = 1, \cdots, n-1$; (iv) the composition factors of $\tilde{V}_0(\lambda_i)$ are $\tilde{M}(\lambda_i)$, $\tilde{M}(\lambda_{i+1})$ and F ($C_i^n - \delta_{i0} + \delta_{i1}$ times), $i = 0, 1, \cdots, n-1$ (if $i = n-1$, then $\tilde{M}(\lambda_{i+1})$ decomposes into two factors, F and $\tilde{M}(\lambda_i)$).

(3) Let char $F = p > 3$, $L = H(n, \mathbf{m})$, $n = 2r$. Then (i) $\tilde{V}_0$ is reducible if and only if $V_0 = V_0(\lambda_i)$, $i = 0, 1, \cdots, r$; (ii) $\tilde{M}(\lambda_0) = F$, $\tilde{M}(\lambda_i) = d_{i-1}\bar{d}_i\tilde{V}_0(\lambda_i)$, $i = 1, \cdots, r$, where $\bar{d}_i \colon \tilde{V}_0(\lambda_i) \to \tilde{V}_0(\lambda_{i-1})$ is an L-module isomorphism (see [35]); (iii) $\dim \tilde{M}(\lambda_0) = 1$, $\dim \tilde{M}(\lambda_i) = (p^{|\mathbf{m}|} - 1)C_{i-1}^{n-2} - C_{i-1}^{n-1}$, $i = 1, \cdots, r$; (iv) the composition factors of $\tilde{V}_0(\lambda_i)$ are $\tilde{M}(\lambda_{i-1})$ ($1 - \delta_{i1}$ times), $\tilde{M}(\lambda_i)$ (2 times), $\tilde{M}(\lambda_{i+1})$ and F ($C_i^n + C_i^{n+1} - 2\delta_{i0} + \delta_{i1}$ times), $i = 0, 1, \cdots, r$ ($\tilde{M}(\lambda_{-1}) = 0$).

By Theorem 2.2 (5), Proposition 3.1 and Theorem 3.1, we have determined all irreducible graded modules of $L(n, \mathbf{m})$ of Cartan type W, S and H. Moreover, by Theorem 2.2 (6), we obtain all irreducible restricted modules for $L(n, (1, \cdots, 1))$ of Cartan type W, S and H.

Note 3.1. (1) In [35], all irreducible graded modules of infinite-dimensional graded Lie algebras $L(n)$ of Cartan type W, S and H are also determined.

(2) In [26], Holmes showes that, with a few exceptions, the irreducible restricted modules for a restricted contact Lie algebra $K(2r + 1, \mathbf{1})$ are induced from those for the homogeneous component of degree zero.

(3) In [36], by imbedding $L = \mathfrak{sl}(2)$ in the Witt algebra W_1 and "twisting" the W_1-module $A = F[x]/(x^p)$, all irreducible L-modules are realized. This construction can be generalized. The case of $\mathfrak{sl}(3)$ is illustrated as an example.

4. Principal Indecomposable Modules

Let L be any one of $W(n, \mathbf{1})$, $S(n, \mathbf{1})$, $H(n, \mathbf{1})$ and $K(n, \mathbf{1})$ over an algebraically closed field F of characteristic $p > 3$. In this section, we shall extend the results [28, 30] on modular representations of classical Lie algebras and semisimple groups to the case of L and obtain some properties of principal indecomposable modules of $u(L)$ which parallel closely those of classical Lie algebras.

Let G be a semisimple, simply connected algebraic group over an algebraically closed field F of characteristic $p > 0$, $\mathbf{g}$ its Lie algebra and $u(\mathbf{g})$ the restricted universal enveloping algebra of **g**. Let $\mathbf{h} = \langle h_1, \cdots, h_l \rangle$ be a Cartan subalgebra of **g**, **b** is a Borel subalgebra such that $\mathbf{h} \subset \mathbf{b}$, Λ denote the collection of p^l restricted weights λ characterized by the conditions $0 \leq \lambda(h_i) < p$, $1 \leq i \leq l$. For each $\lambda \in \Lambda$, we can canonically obtain the one-dimensional **b**-module which is denoted by F_λ. The induced module

$$Z(\lambda) = u(\mathbf{g}) \otimes_{u(\mathbf{b})} F_\lambda$$

is an indecomposable universal highest weight module. Let $V(\lambda)$ denote the restricted irreducible **g**-module of highest weight λ and $Q_{\mathbf{g}}(\lambda)$ the projected cover (= injective hull) of $V(\lambda)$. In [28], Humphreys proved that the principal indecomposable module (PIM) $Q_{\mathbf{g}}(\lambda)$ of $u(\mathbf{g})$ has a filtration with quoients isomorphic to various $Z(\mu)$ and $Z(\mu)$ occurs as many times as $V(\lambda)$ occurs as a composition factor of $Z(\lambda)$ (cf. [28], [30]).

Let **h** (resp. $\mathbf{h}(L_0)$ be the standard Cartan subalgebra of **g** ($:= L_0$), **n** (or $\mathbf{n}^-$) the sum of positive (or negative) root spaces of **g**, $\mathbf{b} = \mathbf{h} \oplus \mathbf{n}$ the Borel subalgebra of **g**, $\mathbf{b}^- = \mathbf{h} \oplus \mathbf{n}^-$, $\mathcal{N} = \mathbf{n} \oplus \sum_{i \geq 1} L_i$, $\mathcal{B} = \mathbf{h} \oplus \mathcal{N}$, $\mathcal{N}^- = \mathbf{n}^- \oplus L_{-1}$ and $\mathcal{B}^- = \mathbf{h} \oplus \mathcal{N}^-$. We denote the lattice of all weights of $\mathbf{h}(L_0)$ by Λ. For each $\lambda \in \Lambda$, it is a linear combination of the fundamental weights. We denote the canonical one-dimensional **b**-module by F_λ and extend the operatations on F_λ to $\mathcal{B}$ by letting L_1 act trivially which is also denoted by F_λ. Denote

$$Z(\lambda) = u(\mathbf{g}) \otimes_{u(\mathbf{b})} F_\lambda.$$

Lemma 4.1. *[9] If $\lambda \in \Lambda$, then*

$$u(L) \otimes_{u(\mathcal{L}_0)} Z(\lambda) \cong u(L) \otimes_{u(\mathcal{B})} F_\lambda,$$

which is indecomposable.

The induced $u(L)$-modules $u(L) \otimes_{u(\mathcal{L}_0)} Z(\lambda)$ play an important role in the description of the PIM's $Q(\lambda)$.

Now we introduce an artificial category of $u(L^e)$-T^e modules, inspired by Jantzen's method in [30].

Let $L = W$, $\mathfrak{A} = \mathfrak{A}(n, \mathbf{1})$, and Aut W and Aut $\mathfrak{A}$ be the automorphism groups of W and $\mathfrak{A}$ respectively. Then

$$\text{Aut } W \cong \text{Aut } \mathfrak{A}.$$

Obviously, Aut $\mathfrak{A}$ is a closed subgroup of $GL(\mathfrak{A})$. Note that any $\phi \in$ Aut $\mathfrak{A}$ is uniquely determined by the action on $\{x^{(\varepsilon_1)}, \cdots, x^{(\varepsilon_n)}\}$. Clearly,

$$\{t \in \text{Aut } \mathfrak{A} \mid t(x^{(\varepsilon_i)}) = t_i x^{(\varepsilon_i)}, \ t_i \in F^*, \ i = 1, \cdots, n\}$$

denoted by $T(W)$, is both a Cartan subgroup and a maximal torus of the algebraic group Aut $\mathfrak{A}$, which is isomorphic to

$$\{\text{diag}(t_1, \cdots, t_n) \mid t_i \in F^*, \ i = 1, \cdots, n\}$$

and whose Lie algebra is $\mathbf{h}(W_0)$. For $a \in F^*$, we define $E_a \in$ Aut $\mathfrak{A}$ by

$$\begin{cases} E_a(x^{(\varepsilon_i)}) = a x^{(\varepsilon_i)}, \ 1 \leq i \leq n, & \text{if } L = W, S \text{ or } H, \\ E_a(x^{(\varepsilon_i)}) = a x^{(\varepsilon_i)}, \ E_a(x^{(\varepsilon_n)}) = a^2 x^{(\varepsilon_n)}, \ 1 \leq i \leq n-1, & \text{if } L = K. \end{cases}$$

Write $T_1 = \{E_a \mid a \in F^*\}$. We set

$$
T(L) = \begin{cases}
\{t \in T(W) \mid t = \mathrm{diag}(t_1, \cdots, t_n), \prod t_i = 1\}, & \text{if } L = S, \\
\{t \in T(W) \mid t = \mathrm{diag}(t_1, \cdots, t_{2r}), \ t_j t_{j+r} = 1, \ 1 \le j \le r\}, & \text{if } L = H, \\
\{t \in T(W) \mid t = \mathrm{diag}(t_1, \cdots, t_{2r}, 1), \ t_j t_{j+r} = 1, \ 1 \le j \le r\} \times T_1, & \text{if } L = K,
\end{cases}
$$

whose Lie algebra is $\mathbf{h}(L_0)$.

Let $L = W$, S, H or K and $T = T(W)$, $T(S)$, $T(H)$ or $T(K)$. Let Δ be the set of simple roots of L_0, $X(T)$ the character group of T (i.e. the group of all homomorphisms $T \to F^*$) which may be identified with the lattice of all weights of T, $X(T)^+$ the set of dominant weights in $X(T)$, and $X_1(T) = \{\lambda \in X(T)^+ \mid 0 \le \langle \lambda, \alpha \rangle < p, \text{ for all } \alpha \in \Delta\}$. More precisely we ought to replace $X_1(T)$ by $X(T)/pX(T)$. Then $X_1(T) = \Lambda$. We denote the Lie algebra of T_1 by $\mathbf{h}_1$. To define a certain partial ordering of weights, we shall extend T. Let $T^e = TT_1$ and $\mathbf{h}^e = \mathbf{h} + \mathbf{h}_1$ (if $L = W$ or K, then $T^e = T$ and $\mathbf{h} = \mathbf{h}^e$). If $L = S$ or H, then let $\chi \in X(T^e)$ such that

$$
\chi(t) = \begin{cases}
1, & \text{if } t \in T, \\
\Lambda_n(t), & \text{if } t \in T_1.
\end{cases}
$$

Then

$$
X(T_1) = \left\{ \sum_{i=1}^n a_i \Lambda_i \ \middle| \ \Lambda_1 = \cdots = \Lambda_n, \ a_i \in \mathbb{Z}, \ i = 1, \cdots, n \right\} \cong \mathbb{Z}\chi.
$$

For convenience, let $\chi = 0$ for $L = W$ or K. Then the character group of T^e is

$$
X(T^e) \cong X(T) \oplus \mathbb{Z}\chi.
$$

Let $L^e = L + \mathbf{h}_1$, $\mathcal{B}^e = \mathcal{B} + \mathbf{h}_1$ and $\mathcal{L}_0^e = \mathcal{L}_0 + \mathbf{h}_1$. Then the adjoint L^e-module is also a T^e-module and z $(= x^{(\alpha)}D_j, D_{i,j}(x^{(\alpha)}), D_H(x^{(\alpha)})$ or $D_K(x^{(\alpha)})$ is not only a weight vector relative to $\mathbf{h}^e$ but also a weight vector relative to T^e. Let $u = z_1 \cdots z_k \in u(L^e)$, we define

$$
\mathrm{Ad}\,(t)(u) = \mathrm{Ad}\,(t)(z_1) \cdots \mathrm{Ad}\,(t)(z_k) = tut^{-1}, \text{ for } t \in T^e.
$$

Then $u(L^e)$ is also a T^e-module.

A finite dimensional vector space V is called a $u(L^e)$-T^e-module (for convenience, we shall just call it a $\hat{u}(L)$-module), if V is both a $u(L^e)$-module and a T^e-module and satisfies:

(a) The actions of $\mathbf{h}^e$ coming from L^e and from T^e coincide;

(b) $t \cdot (u \cdot v) = (\mathrm{Ad}\,(t)u) \cdot (t \cdot v)$, for $v \in V$, $t \in T^e$, $u \in u(L^e)$.

We can canonically define the category of $\hat{u}(L)$-modules.

Let $V = \oplus_\lambda V^\lambda$ be the weight space decomposition (relative to T^e). Then (a) means that for $h \in \mathbf{h}^e$, $v \in V^\lambda$, $h \cdot v = \lambda(h)v$, where $\lambda \in X(T^e)$ induces the weight $\lambda \in X(T^e)/pX(T^e)$ (relative to $\mathbf{h}^e$), while (b) means that $u \cdot V^\lambda \subseteq V^{\lambda+\mu}$ for $\lambda \in X(T^e)$ and $u \in u(L^e)^\mu$. Obviously, L^e and $u(L^e)$ are $\hat{u}(L)$-modules.

Now we shall define a partial ordering on $X(T^e)$. Let $\mathbb{Z}^n$ be the set of n-tuples of integers, which is ordered lexicographically. For $\mathbf{a} = (a_1, \cdots, a_n) \in \mathbb{Z}^n$, write $|\mathbf{a}| = \sum_{i=1}^n a_i$.

We define a partial ordering on $X(T^e)$: (a) For $L = W$ or S, $\sum a_i \Lambda_i < \sum b_i \Lambda_i$ if and only if $|(a_1, \cdots, a_n)| < |(b_1, \cdots, b_n)|$ or $|(a_1, \cdots, a_n)| = |(b_1, \cdots, b_n)|$ and $(a_1, \cdots, a_n) < (b_1, \cdots, b_n)$. (b) For $L = H$ or K, let $\lambda, \mu \in X(T^e)$ and $\lambda|_T = \sum_{i=1}^r a_i \Lambda_i$, $\mu|_T = \sum_{i=1}^r b_i \Lambda_i$, then $\lambda < \mu$ if and only if $\lambda(E_2) < \mu(E_2)$ or $\lambda(E_2) = \mu(E_2)$ and $(a_1, \cdots, a_r) < (b_1, \cdots, b_r)$.

An arbitray $\lambda \in X(T^e)$ (resp. $X(T)$), viewd as a homomorphism $\lambda: u(\mathbf{h}^e)$ (resp. $u(\mathbf{h}))$ $\rightarrow F$, can be extened to a homomorphism $\lambda: u(\mathcal{B}^e)$ (resp. $u(\mathcal{B})) \rightarrow F$ by setting $\lambda(x_i) = 0$ for $i = 1, \cdots, s$. So via λ we can give F the structure of 1-dimensional $u(\mathcal{B}^e)$ (resp. $u(\mathcal{B})$) -module denoted by F_λ. Define $\hat{Z}(\lambda) = u(L^e) \otimes_{u(\mathcal{B}^e)} F_\lambda$ (In the case $\lambda \in X_1(T)$, its restriction to $u(L)$ is essentially the same as the previous $u(L) \otimes_{u(\mathcal{L}_0)} Z(\lambda) \cong u(L) \otimes_{u(\mathcal{B})} F_\lambda$. But here λ can be arbitray in $X(T^e)$). Then $\hat{Z}(\lambda)$ is a $\hat{u}(L)$-module of highest weight λ. Then there is a unique maximal $\hat{u}(L)$-module and a unique irreducible quotient which is denoted by $\hat{M}(\lambda)$. For any arbitray $\hat{u}(L)$-module V, let $[V: \hat{M}(\lambda)]$ be the number of times $\hat{M}(\lambda)$ occurs as a composition factor of V.

In the category of $\hat{u}(L)$-modules, standard arguments show that each irreducible $\hat{u}(L)$-module $\hat{M}(\lambda)$ has an indecomposable projective cover $\hat{Q}(\lambda)$ with $\hat{M}(\lambda)$ as its unique irreducible quotient and each indecomposable projective $\hat{u}(L)$-module is isomorphic to some $\hat{Q}(\lambda)$.

A $\hat{u}(L)$-module V is a $\hat{Z}$-filtered module, if there is a filtration

$$0 = V_0 \subset V_1 \subset \cdots \subset V_r = V$$

such that the filtration quotients $V_i/V_{i-1} \cong \hat{Z}(\mu_i)$ for some $\mu_i \in X(T^e)$, $i = 1, \cdots, r$. The above filtration is called a $\hat{Z}$-filtration and the number of indies with $\mu_i = \mu$ is denoted by $(V: \hat{Z}(\mu))$.

We can show that every projective $\hat{u}(L)$-module has a $\hat{Z}$-filtration. Then we have the reciprocity theorem for the category of $\hat{u}(L)$-modules.

Theorem 4.1. *[9] Let $\lambda, \mu \in X(T^e)$. Then*

$$m_{u(\mathcal{N}^-)}(\lambda - \mu)(\hat{Q}(\lambda): \hat{Z}(\mu)) = m_{u(\mathcal{N})}(\mu - \lambda)[\hat{Z}(\mu): \hat{M}(\lambda)],$$

where $m_V(\lambda)$ is the multiplicity of λ as a T^e-weight of V.

For $\lambda \in X_1(T)$, $\hat{Z}(\lambda)$ is essentially $u(L) \otimes_{u(\mathcal{L}_0)} Z(\lambda)$. Its restriction to $u(L)$ is just $u(L) \otimes_{u(\mathcal{L}_0)} Z(\lambda)$. For the irreducible quotient $\hat{M}(\lambda)$ of $\hat{Z}(\lambda)$, we denote its restriction to $u(L)$ by $\hat{M}(\lambda)|_{u(L)}$. We can show that every irreducible $\hat{u}(L)$-module is $u(L)$-irreducible. Then $\hat{M}(\lambda)|_{u(L)}$ is $u(L)$-irreducible, for $\lambda \in X_1(T)$ (cf. [9]).

For convenience, we write $M(\lambda) = \hat{M}(\lambda)|_{u(L)}$, for arbitrary $\lambda \in X(T^e)$. For $\lambda \in X_1(T)$ and $\mu \in pX(T) + \mathbb{Z}\chi$, we have (cf. [9])

$$\begin{cases} M(\lambda + \mu) \cong M(\lambda), \\ \hat{Z}(\lambda + \mu)|_{u(L)} \cong u(L) \otimes_{u(\mathcal{L}_0)} Z(\lambda) \text{ (as } u(L)\text{-modules).} \end{cases}$$

It follows that each composition series of a $\hat{u}(L)$-module V is also a composition series of the $u(L)$-module V, with multiplicity of $M(\lambda)$ as $u(L)$-composition factors given by

$$[V: M(\lambda)] = \sum_{\mu \in pX(T) + \mathbb{Z}\chi} [V: \hat{M}(\lambda + \mu)], \text{ for } \lambda \in X_1(T).$$

As for $\hat{u}(L)$, the category of $u(L)$-modules has enough projective modules. For $\lambda \in X_1(T)$, let $Q(\lambda)$ be the PIM corresponding to $M(\lambda)$. Then $Q(\lambda)$ is an indecomposable projective $u(L)$-module with quotient $M(\lambda)$ (i.e., $Q(\lambda)$ is a projective cover of $M(\lambda)$). Let $\lambda \in X_1(T)$. Then we have

$$\hat{Q}(\lambda) \cong Q(\lambda) \text{ (as } u(L)\text{-modules)}.$$

Finally, we have

Theorem 4.2. [9] Let $\lambda \in X_1(T)$. Then

(a) There is a filtration $0 = Q_0 \subset Q_1 \subset \cdots \subset Q_r = Q(\lambda)$ with $Q_i/Q_{i-1} \cong u(L) \otimes_{u(\mathcal{L}_0)} Z(\mu_i)$, for some $\mu_i \in X_1(T)$.

(b) The set $\{u(L) \otimes_{u(\mathcal{L}_0)} Z(\mu_i) \mid i = 1, \cdots, r\}$ of the filtration quotients (counted with multiplicity) in (a) is uniquely determined by $Q(\lambda)$.

(c) Let $(Q(\lambda): u(L) \otimes_{u(\mathcal{L}_0)} Z(\mu))$ denote the multiplicity of $u(L) \otimes_{u(\mathcal{L}_0)} Z(\mu)$ ($\mu \in X_1(T)$) as filtration quotient of the filtration of $Q(\lambda)$. Then

$$(Q(\lambda): u(L) \otimes_{u(\mathcal{L}_0)} Z(\mu)) = \sum_{\tau \in pX(T) + \mathbb{Z}\chi} (\hat{Q}(\lambda): \hat{Z}(\mu + \tau)) \geq [u(L) \otimes_{u(\mathcal{L}_0)} Z(\mu): M(\lambda)].$$

5. COHOMOLOGY GROUPS

In this section, we first give a general discussion of the cohomology of $L(n, \mathbf{m}) = \oplus_{i \geq -1} L_i$ of Cartan type W, S and H. Then we reduce the computation of $H^1(W(n, \mathbf{m}), \tilde{V}_0)$ to the computation of $H^1(\mathfrak{sl}(n), V_0)$. Thus we determine the structures of $H^1(W(n, \mathbf{m}), V)$ (resp. $H^1_*(W(n, \mathbf{m}), V)$, $\mathbf{m} = \mathbf{1}$), for $n = 2, 3$, where V is an irreducible (resp. irreducible restricted) $W(n, \mathbf{m})$-module. We also determine the structures for $S(3, \mathbf{m})$ and $H(2, \mathbf{m})$ (see [6, 10]).

Let $L = W(n, \mathbf{m})$, $S(n, \mathbf{m})$ or $H(n, \mathbf{m})$. Then we have

Proposition 5.1. [10] Let V_0 be an irreducible L_0-module. If V_0 is not a highest weight module or V_0 is an irreducible highest weight module which is not integral, then

$$H^*(L, \tilde{V}_0) = 0.$$

Let char $F = p > 2$ and $\mathcal{W}_i = \oplus_{j \geq i} W_j$. By [10, Lemma 2.1 and Lemma 3.1], we have

Lemma 5.1. If V_0 is a nontrivial irreducible $\mathfrak{gl}(n)$-module with a highest weight, then

$$H^1(W, \tilde{V}_0) = H^1(\mathcal{W}_0, V_0).$$

If $V_0 = F$ with trivial action, then

$$H^1(W, \tilde{V}_0) \cong \bigoplus_{i=1}^{n} \langle [\beta_i] \rangle \oplus H^1(\mathcal{W}_0, F),$$

where

$$\beta_i(D_j) = \begin{cases} x^{(0, \cdots, p^{m_i} - 1, \cdots, 0)}, & \text{if } j = i, \\ 0, & \text{if } j \neq i \end{cases}$$

such that $[\beta_i] \in H^1(W_{-1}, \mathfrak{A})$, $i, j = 1, 2, \cdots, n$.

Using the cohomology five-term sequence, we can reduce the computation of $H^1(\mathcal{W}_0, V_0)$ to the computation of $H^1(\mathfrak{sl}(n), V_0)$. Then by [1, Corollary 5.5] and Theorem 3.1, we have

Theorem 5.1. *Suppose char $F = p \geq n + 2$ for $n = 2, 3$. Let V be an irreducible $W(n, \mathbf{m})$-module. Then*

$$H^1(W(n, \mathbf{m}), V) \cong \begin{cases} F^{n+1}, & \text{if } V \cong \tilde{M}(\lambda_1), \\ F^{C_i^n}, & \text{if } \tilde{V} \cong \tilde{M}(\lambda_i) \text{ for } i \geq 2, \\ F, & \text{if } V \cong \tilde{V}(2\lambda_1 + \lambda_{n-1}) \text{ and the action of } I \text{ on} \\ & \quad V(2\lambda_1 + \lambda_{n-1}) \text{ is the scalar multilplication by } 1, \\ F \oplus F, & \text{if } n = 2 \text{ and } V \cong \tilde{V}(p - 2) \text{ and the action of } I \text{ is} \\ & \quad \text{trivial}, \\ F, & \text{if } n = 3 \text{ and } V \cong \tilde{V}((p - 2)\lambda_1 + \lambda_2), \text{ and the action} \\ & \quad \text{of } I \text{ is trivial}, \\ H^0(\lambda_1)^{(1)}, & \text{if } n = 3 \text{ and } V \cong \tilde{V}((p - 2)\lambda_1 + \lambda_2) \text{ and the} \\ & \quad \text{action of } I \text{ is trivial}, \\ H^0(\lambda_2)^{(1)}, & \text{if } n = 3 \text{ and } V \cong \tilde{V}(\lambda_1 + (p - 2)\lambda_2) \text{ and the} \\ & \quad \text{action of } I \text{ is trivial}, \\ 0, & \text{otherwise.} \end{cases}$$

Note 5.1. (1) In the above theorem, if $W(n, \mathbf{m}) = W(n, \mathbf{1})$ and V is an irreducible restricted $W(n, \mathbf{1})$-module, then the first restricted cohomology group $H^1_*(W(n, \mathbf{1}), V) \cong H^1(W(n, \mathbf{1}), V)$ (see [7]).

(2) In [23], Feldvoss extends the above theorem to the case of char $F = p > 7$.

6. CENTRAL EXTENSIONS AND $H^1(L, L^*)$

In [14, 16, 17], Dzhumadil'daev and Farnsteiner determined the structures of $H^2(L, F)$, where $L = W(n, \mathbf{m})$ for $p \geq 3$, $S(n, \mathbf{m})$ for $p > 3$ and $n = 3$, $H(n, \mathbf{m})$ for $p > 3$, or $K(n, \mathbf{m})$ for $p > 3$. In this section, we propose a new unifying approach, which is based mainly on the computation of $H^1(L, L^*)$. We first give a description of the dual adjoint module L^* by means of the mixed product and the coinduced module. Thus the computation of $H^1(L, L^*)$ is reduced to the computation of the cohomology of L_0 which is the Lie algebra of a certain reductive algebraic group and we exploit certain techniques in the representation theory of reductive algebraic groups. We determine the structures of $H^1(L, L^*)$ and $H^2(L, F)$, where $L = W(n, \mathbf{m})$ for $p > 0$, $S(n, \mathbf{m})$ for $n \geq 3$ and $p > 2$, $H(n, \mathbf{m})$ for $p > 2$, or $K(n, \mathbf{m})$ for $p > 2$ and $n + 3 \not\equiv 0 \pmod{p}$, with which we are able to determine the central extensions of these graded Lie algebras of Cartan type for all cases, except $L = S(n, \mathbf{m})$ for $p = 2$, or $K(n, \mathbf{m})$ for $p > 2$ and $n + 3 \equiv 0 \pmod{p}$. Then we obtain new central extensions of L. For the remaining case of $K(n, \mathbf{m})$, we discuss some cohomology properties and obtain an estimate of the dimension of $H^2(K(n, \mathbf{m}), F)$ (see [7]).

Let N_0 be an n-dimensional linear space with basis $\{e_1, \cdots, e_n\}$. On N_0, we have the natural representation ν_0 of $\mathfrak{gl}(n)$:

$$\nu_0(E_{ij})e_k = \delta_{jk}e_i.$$

Let T be the trace representation on N_0 and $\sigma_0 = \nu_0 + T$. Thus σ_0 is a representation on N_0 by

$$A \cdot v = Av + (\text{Tr}A)v, \quad \forall A \in \mathfrak{gl}(n), \ v \in N_n.$$

Then we have

Lemma 6.1. *Let char $F = p > 0$. Then the dual module $W(n, \mathbf{m})^*$ is isomorphic to the coinduced module $\text{Hom}_{\theta(W, W_0)}(U(W), N_0)$, where N_0 is the module of the representation σ_0 of W_0.*

Since $S(n, \mathbf{m})$ is isomorphic to the minimum submodule of $W(n, \mathbf{m})$ (regarded as an $S(n, \mathbf{m})$-module), $S(n, \mathbf{m})^*$ is isomorphic to the irreducible quotient module of $W(n, \mathbf{m})^*$. By Lemma 6.1, we have

Lemma 6.2. *Let char $F = p > 2$. Then*
$$S(n, \mathbf{m})^* \cong \tilde{M}(\lambda_2).$$

Lemma 6.3. *Let char $F = p > 2$. Then $H(n, \mathbf{m})^* \cong H(n, \mathbf{m}) \cong \mathfrak{A}'/F \cdot 1$ (regarded as $H(n, \mathbf{m})$-modules), where $\mathfrak{A}' := \oplus_{\alpha \neq \pi}\langle x^{(\alpha)}\rangle$, and there are three exact sequences:*
$$0 \to \mathfrak{A}' \to \mathfrak{A} \ (\cong \text{Hom}_{\theta(H, \mathcal{H}_0)}(U(H), V_0(\lambda_0))) \to F \to 0,$$
$$0 \to F \to \mathfrak{A}' \to \mathfrak{A}'/F \cdot 1 \to 0,$$
$$0 \to \mathfrak{A}'/F \cdot 1 \to \text{Hom}_{\theta(H, \mathcal{H}_0)}(U(H), V_0(\lambda_1)).$$

Let V be the $\langle x^{(\varepsilon_i + \varepsilon_j)} \mid 1 \leq i, j \leq 2r\rangle$-module (i.e., $\mathfrak{sp}(2r)$-module) with the highest weight λ, such that the action of $x^{(\varepsilon_n)}$ on V is scalar multiplication by c. We denote the $\mathcal{K}_0$-module V by $V(\lambda, c)$ and extend the operations on $V(\lambda, c)$ to $\mathcal{K}_0$ by letting $\mathcal{K}_1$ act trivially.

Lemma 6.4. *(1) Suppose that $n + 3 \not\equiv 0 \pmod{p}$. Then*
$$K(n, \mathbf{m})^* \cong \text{Hom}_{\theta(K, \mathcal{K}_0)}(U(K), V(0, n + 3)).$$

(2) Suppose that $n + 3 \equiv 0 \pmod{p}$. Then
$$K(n, \mathbf{m})^* \cong \text{Hom}_{\theta(K, \mathcal{K}_0)}(U(K), V(0, 0))/F,$$

where F is the one-dimensional trivial K-module.

Using the techniques in §5 and the Hochschild-Serre spectral sequence, we can obtain the following theorems.

Theorem 6.1. *Let char $F = p > 0$. Then*
$$\dim_F H^1(W(n, \mathbf{m}), W(n, \mathbf{m})^*) = \begin{cases} m_1, & \text{if } n = 1, \ p = 2, \\ m_1 - 1, & \text{if } n = 1, \ p = 3, \\ 1, & \text{if } n = 1, \ p > 3, \\ 2, & \text{if } n = 2, \ p = 3, \\ 3, & \text{if } n = 3, \ p = 2, \\ 0, & \text{otherwise,} \end{cases}$$

and
$$H^2(W(n, \mathbf{m}), F) \cong H^1(W(n, \mathbf{m}), W(n, \mathbf{m})^*).$$

Theorem 6.2. *Suppose that $p > 2$. Then*

$$\dim_F H^1(S(n, \mathbf{m}), S(n, \mathbf{m})^*) = \begin{cases} |\mathbf{m}| + 1, & \text{if } n = 3, \\ n(n-1)/2, & \text{if } n > 3, \end{cases}$$

and

$$\dim_F H^2(S(n, \mathbf{m}), F) = \begin{cases} |\mathbf{m}|, & \text{if } n = 3, \\ n(n-1)/2, & \text{if } n > 3. \end{cases}$$

Theorem 6.3. *Suppose that $p > 2$. Then*

$$\dim_F H^1(H, (n, \mathbf{m}), H(n, \mathbf{m})^*) = |\mathbf{m}| + 2,$$

and

$$\dim_F H^2(H(n, \mathbf{m}), F) = \begin{cases} |\mathbf{m}| + 2, & \text{if } n + 4 \equiv 0 \pmod{p}, \\ |\mathbf{m}| + 1, & \text{otherwise.} \end{cases}$$

Theorem 6.4. *Suppose that $p > 2$. Then*
(1) $H^1(K(n, \mathbf{m}), K(n, \mathbf{m})^*)$

$$\cong \begin{cases} \langle x^{(p^i \varepsilon_j)} \mid i = 1, \cdots, m_j - 1, j = 1, \cdots, n \rangle, \\ \quad \text{if } n + 5 \not\equiv 0 \pmod{p}, \\ 0, \quad \text{if } n + 3 \not\equiv 0 \pmod{p} \text{ and } n + 5 \not\equiv 0 \pmod{p}, \end{cases}$$

and if $n + 3 \equiv 0 \pmod{p}$, then

$$\dim_F H^1(K, K^*) = \dim_F((U(K) \otimes_{\theta(K, \mathcal{K}_0)} V(\lambda_1, -1)_\sigma / K^*)^K \geq n + 1.$$

(2) $\dim_F(H^2(K(n, \mathbf{m}), F)$

$$= \begin{cases} |\mathbf{m}| - n, & \text{if } n + 5 \equiv 0 \pmod{p}, \\ 0, & \text{if } n + 3 \not\equiv 0 \pmod{p} \text{ and } n + 5 \not\equiv 0 \pmod{p}, \end{cases}$$

and

$$\dim_F H^2(K(n, \mathbf{m}), F) = \dim_F((U(K) \otimes_{\theta(K, \mathcal{K}_0)} V(\lambda_1, -1)_\sigma / K^*)^K$$
$$\geq n + 1, \quad \text{if } n + 3 \equiv 0 \pmod{p}.$$

REFERENCES

[1]. H. Andersen and J. Jantzen, *Cohomology of induced representations for algebraic groups*, Math. Ann. **269** (1984), 487–525.

[2]. R. Block and R. Wilson, *The restricted simple Lie algebras are of classical or Cartan type*, Proc. Natl. Acad. Sci. USA (1984), 5271–5274.

[3]. R. Block and R. Wilson, *Classification of the restricted simple Lie algebras*, J. Algebra **114** (1988), 115–259.

[4]. Chang Ho-Jui, *Über Wittsche Lie Ringe*, Hamgurg Abhandl **14** (1941), 151–184.

[5]. S. Chiu, *Second cohomology of the Witt algebra*, (Chinese), Chin. Ann. of Math. **9A(5)** (1988), 524–529.

[6]. S. Chiu, *Cohomology of graded Lie algebras of Cartan type $S(n, \mathbf{m})$*, Chin. Ann. of Math. **10B(1)** (1989), 105–114.

[7]. S. Chiu, *Central extensions and $H^1(L, L^*)$ of the graded Lie algebras of Cartan type*, J. Algebra **149** (1992), 46–67.

[8]. S. Chiu, *Derivations of the graded Lie algebras of Cartan type*, Chin. Ann. of Math. **13B: 2** (1992), 196–204.

[9]. S. Chiu, *Principal indecomposable representations for restricted Lie algebras of Cartan type*, J. Algebra **155** (1993), 142–160.

[10]. S. Chiu and G. Yu. Shen, *Cohomology of graded Lie algebras of Cartan type of characteristic p*, Abh. Math. Sem. Univ. Hamburg **57** (1987), 139–156.

[11]. E. Cline, B. Parshall and L. Scott, *Cohomology, hyperalgebra and representations*, J. of Algebra **63** (1980), 98–123.

[12]. C. Curtis, *Representations of Lie algebras of classical type with applications to linear groups*, J. Math. Mech. **9** (1960), 307–326.

[13]. A. Dzhumadil'daev, *On the cohomology of modular Lie algebras*, Math. USSR Sbornik (1984), 127–143.

[14]. A. Dzhumadil'daev, *Central extensions and invariant forms of Cartan type Lie algebras of positive characteristic*, Functional Anal. Appl. **18** (1984), 331–332.

[15]. A. Dzhumadil'daev, *Central extensions of the Zassenhaus algebra and their irreducible representation*, (Russian), Mat. Sb. **126** no. 168 (1985), 473–489.

[16]. R. Farnsteiner, *Central extensions and invariant forms of graded Lie algebras*, Algebras, Groups, Geom. **3** (1986), 431–455.

[17]. R. Farnsteiner, *Dual space derivations and $H^2(L, F)$ of modular Lie algebras*, Canad. J. Math. **39** (1987), 1078–1106.

[18]. R. Farnsteiner, *Lie theoretic methods in cohomology theory*, Lect. Notes in Math. *1373*, Springer-Verlag, 1989, pp. 93–110.

[19]. R. Farnsteiner, *Extension functors of modular Lie algebras*, Math. Ann. **288** (1990), 713–730.

[20]. R. Farnsteiner, *Recent developments in the cohomology theory of modular Lie algebras*, Nonassociative algebras and related topics (Hiroshima, 1991) (1991), 19–46, World Sci. publishing, River Edge, NJ.

[21]. R. Farnsteiner, *On indecomposable representations of modular Lie algebras with triangular decomposition*, preprint.

[22]. R. Farnsteiner and H. Strade, *Shapiro's lemma and its consequences in the cohomology theory of modular Lie algebras*, Math. Z. **206** (1991), 153–168.

[23]. J. Feldvoss, *Homologische aspekte der darstellungstheorie modularer Lie-Alegebren*, Dissertation, Universität Hamburg (1989).

[24]. E. Friedlander and B. Parshall, *Cohomology of Lie algebras and algebraic groups*, Amer. J. Math. **108** (1986), 235–253.

[25]. E. Friedlander and B. Parshall, *Modular representation theory of Lie algebras*, Amer. J. Math. **110** (1988), 1055–1094.

[26]. R. Holmes, *Simple restricted modules for the restricted contact Lie algebras*, preprint.

[27]. R. Holmes and D. Nakano, *Brauer-type reciprocity for a class of graded assoiciative algebras*, to appear, J. Algebra.

[28]. J. Humphreys, *Modular representations of classical Lie algebras and semisimple groups*, J. Algebra **19** (1971), 51–79.

[29]. J. Humphreys, *Ordinary and modular representations of Chevalley groups*, Lect. Notes in Math. *528*, Springer–Verlag, 1976.

[30]. J. Jantzen, *Über Darstellungen höherer Frobenius-Kerne halbeinfacher algebraischer Gruppen*, Math. Z. **164** (1979), 271–292.

[31]. J. Jantzen, *Darsteillungen halbeinfacher Gruppen und ihrer Frobenius-Kerne*, J. Reine Angew. Math. **317** (1980), 157–199.

[32]. J. Jantzen, *Representations of algebraic groups*, Academic Press, Orlando, FL, 1987.

[33]. Shen Guangyu, *Graded modules of graded Lie algebras of Cartan type (I)-mixed product of modules*, Scientia Sinica (Ser. A) **29** no. 6 (1986), 570–581.

[34]. Shen Guangyu, *Graded modules of graded Lie algebras of Cartan type (II)-positive and negative graded modules*, Scientia Sinica (Ser. A) **29** no. 10 (1986), 1009–1019.

[35]. Shen Guangyu, *Graded modules of graded Lie algebras of Cartan type (III)-irreducible modules*, Chin. Ann. of Math. **9B(4)** (1988), 404–417.

[36]. Shen Guangyu, *Realization of irreducible modules of $\mathfrak{sl}(2)$ of characteristic p*, Northeastern Math. J. **6(2)** (1990), 151–156.

[37]. H. Strade and R. Farnsteiner, *Modular Lie algebras and their representations*, Dekker 116, New York, 1988.

Valuation Theory and Generalizations of Hilbert's 17th Problem

DAI ZHIZHONG and ZENG GUANGXING, Department of Mathematics, Nanchang University, Nanchang, Jiangxi, P. R. China

I . VALUATION THEORY

In the middle 1950s, Dai began his studies on complete valued fields in the sense of Krull. In [D1], he gave a characterization of the complete fields by means of pseudo-convergent sequences. As a main theorem, Dai gives the following result:

Theorem 1. Let (F,v) be a valued field of arbitrary rank, and A the valuation ring. Then (F,v) is complete if and only if every pseudo-convergent sequence in A with prime breadth has at least one pseudo-limit in F.

The same result as above was rediscovered by P. Ribenboim some years later (1958).

In a subsequent paper [D2], algebraic extensions of a complete valued field are considered, and the following result is established:

Theorem 2. Assume that F is a complete field with respect to a non-trivial valuation v of arbitrary rank, and L is an algebraic extension of F with v^* as a prolongation of v to L. Then the following statements are true:

(1) Let L be separable over F. Then (L,v^*) is complete if and only if L is finite over F.

(2) If L is purely inseparable over F with $[L:F] < \infty$, then (L, v^*) is complete.

(3) Let L be purely inseparable over F. If there are elements in L with arbitrary high exponents, then (L, v^*) is not complete.

In [D3], the notion of discrete valuation of finite rank is generalized to valuations of arbitrary rank. A valuation v and its valuation ring A of arbitrary rank will be called discrete, if for each pair of immediately consecutive prime ideals $P' \supset P$ of A, the corresponding isolated subgroups $\Delta \supset \Delta'$ of the value group of v satisfy the condition that $\Delta / \Delta' \cong$ the additive group of integers. According to this generalized notion of discrete valuation, the following theorem has been established:

Theorem 3. A discrete valued field (F,v) is maximally complete if and only if it is complete.

When a field is not separably closed, it is well-known that F can have only one henselian valuation of rank one up to equivalence. This fact was also generalized by Dai to the situation of arbitrary rank, see [D4]. In [D4], the following theorems have been proved:

Theorem 4. If a field F possesses two independent henselian valuations of arbitrary rank, then F is separably closed.

Theorem 5. Let (F,v) be a complete valued field in the sense of Krull. If F has also a henselian valuation v' independent of v, then F is algebraically closed.

Theorem 6. Let A be a finite rank henselian valuation ring of F, and E a subfield of F. If for every prime ideal P of A different from the valuation ideal M, A_P / PA_P is not algebraically closed, then the restriction $A \cap E$ is a henselian valuation ring of E.

In a paper [CQ] of Chen and Qi, an investigation of the restriction of a henselian valuation ring to subfields has been made. The authors prove the following theorems which improve Theorem 6 listed above:

Theorem 7. Suppose that L/F is an algebraic extension satisfying $[L:F]_s < \infty$, and L is not separably closed. If B is a henselian valuation ring of L of arbitrary rank with residue class field not algebraically closed, the restriction $A = B \cap F$ is a henselian valuation ring of F.

Theorem 8. Suppose that L/F is a normal algebraic extension and L is not separably closed. If B is a henselian valuation ring of the first kind of L, then $A = B \cap F$ is also henselian of the first kind.

In [D5], the author considers a conjecture of Krull in a generalized form given by A. H. Clifford: Every archimedean partially orderd commutative group is isomorphic to a subgroup of an archimedean vector group. This general conjecture of Krull has already been settled in the negative by M. Nagata. One can still try to find some sufficient conditions on·an archimedean ordered group to make the conjecture valid. The author gives a proof of the conjecture in the case Γ is an archimedean latticeordered group (necessarily commutative) with finitely many maximal closed t-ideals. The proof is based on an elementary valuation-theoretic lemma, namely: "Let s be a semi-valuation of some field with value group Γ, and $v_1,...,v_r$ valuations associated with s. Then $v_1, ..., v_r$ are pair-wise independent, if Γ is archimedean".

In a real field henselian valuations are compatible with every ordering of the field. This is a known fact, but the converse need not be true. In note [D6] there is a simple counter-example.

The book [D] is a comprehensive exposition of several fundamental topics in the theory of valuations, intended mainly for graduate students and those interested in this field. The contents of the book are well described by the following headings: Chap. 1. Absolute values; Chap. 2. Valuations, valuation rings and places; Chap. 3. Algebraic extensions of valued fields; Chap. 4. Henselian valued fields; Chap. 5. Complete discretely valued fields. Both Chapters 4 and 5 are concerned with rank one valuations only.

II. GENERALIZATIONS OF HILBERT'S 17TH PROBLEM

One way of generalizing Hilbert's 17th problem is to alter coefficient fields of polynomials and rational functions or other objects. In [Z1], Zeng considers polynomials over real valuation rings with core, and obtains the following result:

Theorem 1. Let (R,C) be a real valuation ring with core C, and let M be the maximal ideal of R. Then, for $f \in R[X_1, ...,X_n]$, the following statements are equivalent:

(i)   f is positive semidefinite on every real extension of (R,C).

(ii) f is positive semidefinite on every real closure of (R,C).

(iii) f can be represented in the form

$$[a^2 f^{2s} + \sum_i c_i(1-\eta_i \Phi_i)g_i^2]f = \sum_j c_j'(1-\eta_j' \Phi_j')g_j'^2$$

where $s \in \mathbb{N}$, $a \in R$, $c_i, c_j' \in C$, $\eta_i, \eta_j' \in M$, $\Phi_i, \Phi_j', g_i, g_j' \in R[X_1,...,X_n]$.

In [Z2], Hilbert's 17th problem for (not necessarily symmetric) matrices, whose entries are polynomials over a field, is considered. Actually, as a more refined solution, the Stellensätze for matrices are established as follows:

Theorem 2. Let (K,S) be a preordered field and let A be a matrix over K[$X_1,...,X_n$]. Then we have:

(i) The eigenvalues of A in every real closure of (K,S) vanish if and only if there exists an expression:

$$A^{2k} + \sum_i s_i G_i^2(A) = 0$$

where $k \in \mathbb{N}$, $s_i \in S$, $G_i(Y) \in K[X_1,...,X_n][Y]$.

(ii) The eigenvalues of A in every real closure of (K,S) are positive if and only if there exists an expression:

$$[E + \sum_i s_i G_i^2(A)] A = E + \sum_j s_j' G_j'^2(A)$$

where s_i, $s_j' \in S$, $G_i(Y)$, $G_j'(Y) \in K[X_1,...,X_n][Y]$.

(iii) The eigenvalues of A in every real closure of (K,S) are nonnegative if and only if there exists an expression:

$$[A^{2k} + \sum_i s_i G_i^2(A)] A = A^{2k} + \sum_j s_j' G_j'^2(A)$$

where $k \in \mathbb{N}$, s_i, $s_j' \in S$, $G_i(Y)$, $G_j'(Y) \in K[X_1,...,X_n][Y]$.

By Theorem 2, Artin's theorem characterizing totally positive elements may be generalized to matrices over a field, and the representations of positive definite and semidefinite (symmetric) matrices with polynomial entries as a sum of squares may be obtained.

Another way of generalizing Hilbert's 17th problem is to characterize those fields for which Hilbert's 17th problem is affirmative; this may be considered as the inverse of Hilbert's 17th problem. For ease of representation, in imitation of K. McKenna, an ordered field (K,P) (respectively a preordered field (K,S)) is said to have the weak Hilbert property if every polynomial $f(X_1, ...,X_m)$ over K in m variables, which is positive semidefinite on (K,P) (respectively on (K,S)) in the sense that $f(\alpha) \in P$ (respectively $f(\alpha) \in S$) for any $\alpha \in K^m$, can be written in the form $f = \sum_{i=1}^n r_i h_i^2$, where $h_i \in K(X_1, ...,X_m)$, and $r_i \in P$ (respectively $r_i \in S$), i=1, ...,n. Thereby the essence of the inverse of Hilbert's 17th problem is to characterize the class of fields with the weak Hilbert property. In 1975, K. McKenna considered the inverse of Hilbert's 17th problem for ordered fields, and showed the important result: Let (K,P) be an ordered field with real closure K. Then (K,P) has the weak Hilbert property if and only if K is dense in K. Afterwards, in 1979, A. Prestel discussed formally real fields with the weak Hilbert property, and obtained the following result: Let K be a formally real field which has only a finite number of orderings. Then K (i.e.$(K, \sum K^2)$) has the weak Hilbert property if and only if K is dense in every real closure of K. In [Z3], Zeng establishes the following result in a simpler but new way, which unifies the result of McKenna and that of Prestel.

Theorem 3. Let (K,C) be a formally real field with core C which has only a finite number of orderings. Then (K,C) has the weak Hilbert property if and only if K is dense in every real closure of (K,C).

When C is just an ordering of K or C={1}, we can obtain McKenna's theorem or Prestel's theorem from Theorem 3 immediately.

In view of the results of McKenna and Prestel, it seems that "(K,S) has the weak Hilbert property" implies " K is dense in every real closure of (K,S) for an arbitrary preordered field (K,S)". Unfortunately, this implication is not true. In [Z4], Zeng obtains a sufficient condition for a preordered field to have the weak Hilbert property as follows:

Theorem 4. Let (K,S) be a preordered field. If there exists a subfield E of K such that (i) E is dense in each real closure of $(E, S \cap E)$, and (ii) K,S are finitely generated over E, $S \cap E$ respectively, then (K,S) has the weak Hilbert property.

By Theorem 4, it is clear that the preordered field $(\mathbb{R}(t), \sum \mathbb{R}(t)^2)$, where t is an indeterminate over the field $\mathbb{R}$ of real numbers, has the weak Hilbert property. However, it is easy to see that $\mathbb{R}(t)$ is not dense in each of its real closures. Thereby, in general, "density" is a stronger concept than "the weak Hilbert property". The question naturally arises: What is a new notion that is similar to density and is exactly equivalent to the weak Hilbert property? In [Z5], Zeng introduces a new notion, the so-called local density, as follows: A preordered field (K,S) is said to be locally dense, if, for every finite real extension K^* of (K,S), α, $\beta \in K^*$ and $\alpha \neq \beta$, there exists some $a \in K$ such that $\alpha <_{P^*} a <_{P^*} \beta$ or $\beta <_{P^*} a <_{P^*} \alpha$ for some ordering P^* of K^* with $S \subset P^*$. With the help of the notion of local density, the following important result may be obtained:

Theorem 5. For a preordered field (K,S), the following statements are equivalent:

(i) (K,S) has the weak Hilbert property;

(ii) (K,S) is locally dense;

(iii) Every positive semidefinite polynomial in one variable can be expressed as a sum of squares with coefficients in S.

For a preordered field (K,S), denote the set of orderings P of K such that $S \subset P$ by $X_K(S)$. As an application of Theorem 5, the following result may be established:

Theorem 6. If (K,S) is a preordered field with the weak Hilbert property, and $P \in X_K(S)$ is isolated in the Harrison topology, then (K,P) is dense in its real closure.

When a preordered field (K,S) has only a finite number of orderings, every point in $X_K(S)$ is obviously isolated. Therefore, Theorem 3 may again be derived as a direct consequence of Theorem 6.

In order to further investigate preordered fields with the weak Hilbert property, another new notion, the so-called strongly local density, is introduced in [Z6]. A preordered field (K,S) is said to be strongly locally dense, if, for every finitely generated real extension K^* of (K,S), $\alpha, \beta \in K^*$ and $\alpha \neq \beta$, there exists some $a \in K$ such that $\alpha <_{P^*} a <_{P^*} \beta$ or $\beta <_{P^*} a <_{P^*} \alpha$ for some ordering P^* of K^* with $S \subset P^*$. By strong local density, Zeng gives again a characterization of preordered fields with the weak Hilbert property as follows:

Theorem 7. A preordered field (K,S) has the weak Hilbert property if and only if (K,S) is strongly locally dense.

Theorem 7 may be apllied to prove the following result, which affirms the validity of the ascent of the weak Hilbert property from a preordered field to its finitely generated extensions.

Theorem 8. Let both (K,S) and (K^*,S^*) be preordered fields, and let (K^*,S^*) be a finitely generated extension of (K,S). If (K,S) has the weak Hilbert property, then (K^*,S^*) has also this property.

Nevertheless, as is pointed out in [Z7], the descent of the weak Hilbert property for a preordered field and its finitely generated extensions is invalid. Actually, the following result can be established:

Theorem 9. Let both (K,S) and (K^*,S^*) be preordered fields, and let (K^*,S^*) be a finitely generated extension of (K,S). If K^* is transcendental over K, then (K^*,S^*) has the weak Hilbert property.

Furthermore, in [Z8], the class of fields with the weak Hilbert property is studied via real valuations. Let (K,S) be a preordered field. A Krull valuation v of K is called a canonical valuation of (K,S), if, for some $P \in X_K(S)$, the associated valuation ring is the "convex hull" A(P) of $\mathbb{Q}$ in K with respect to P, i.e. A(P)=$\{a \in K \mid$ for some $r \in \mathbb{Q}$, $-r <_P a <_P r\}$. Denote the set of canonical valuations of (K,S) by $V_K(S)$. Then, based on valuation theory, Zeng proves the following results:

Theorem 10. Let (K,P) be an ordered field with real closure K. Then the following statements are equivalent:

(i) (K,P) has the weak Hilbert property;

(ii) For every real valuation v of K, we have the assertion: for any α $\in$ K, there exists some a$\in$ K such that v(α -a)=0.

(iii) For some nontrivial or canonical valuation v of K, we have the assertion: for any $\alpha \in$ K, there exists some a$\in$ K such that v(α -a)=0.

Theorem 11. Let (K,S) be a preordered field. Then the following statements are equivalent:

(i) (K,S) has the weak Hilbert property;

(ii) For every finite real extension K^* of (K,S) and any $\alpha \in K^*$, there is some a$\in$ K such that v(α -a)=0 for some v$\in$ $V_{K^*}(\sum SK^{*2})$.

(iii) For every finite real extension K^* of (K,S) and any α_1, ..., $\alpha_n \in$ K^*, there are some a_1, ..., $a_n \in$ K such that v($\alpha_i - a_i$)=0, i=1, ...,n, for some v$\in$ $V_{K^*}(\sum SK^{*2})$.

As a consequence of the theorems above, the following fact is shown:

Corollary. Let (K,S) be a preordered field which possesses only a finite number of archimedean orderings, and let $V_K(S)$ be a finite set. Then (K, S) has the weak Hilbert property if and only if K is dense in every real closure of (K,S).

[DZ] is a monograph on Hilbert's 17th problem, written jointly by Dai and Zeng. It gives a systematic presentation of this problem stated in both its qualitative and quantitative aspects. The chapter headings are: 1. Real fields, ordered fields and preordered fields; 2. Orderings and reality under field extensions; 3. Real closed fields; 4. The uniqueness of the real closure; 5. Real valuations and real places; 6. The Artin-Lang theory; 7. Hilbert's 17th problem; 8. Semialgebraic Nullstellensatz, Nichtnegativstellensatz and Positivstellensatz; 9. Fields with Hilbert property; 10. Preordered fields with weak Hilbert property; 11. The local density of preordered fields and its relation to weak Hilbert property; 12. The theory of quadratic forms related to the quantitative aspect of Hilbert's 17th problem; 13. A lower bound of p(R(X₁,...,Xₙ)); 14. An upper bound of p(R(X₁,...,Xₙ)); Appendix 1. An alternative proof of a counter-example due to D. W. Dubois; Appendix 2. A historical survey. § 11 contains a complete solution of the inverse problem, which is a work of the second author. The booklet may also be used as an introduction to the theory of real fields.

Ⅲ. REAL ALGEBRA.

In the later 80's, Dai extended the notion of real valuation from fields to commutative rings with identity, and obtained some basic relations between real places (or real valuations) and orderings of the ring considered.

Let R be a commutative ring with identity. A valuation pair (A,M) of R is called a real place if the residue ring A/M has real quotient field. Accordingly, the valuation determined by a real place is then called a real valuation of R. Many properties of real valuations could be carried over from real fields to real (or semireal) commutative rings. Dai's first paper [D7] in this direction was not published until 1994, but a summary had been published years before in the Journal of Jiangxi University vol. 13, 1989. The main results of [D7] and its successor [D8] are as follows:

Theorem 1. Let R be a commutative ring with identity. Then we have
    (1) R is semireal if and only if R possesses a real place (or real valuation) which may be trivial.
    (2) A valuation pair (A,M) is a real place of R if and only if (A,M) is compatible with some ordering of R.
    (3) If a valuation pair (A,M) is compatible with a preordering S of R. Then (A,M) is also compatible with some ordering $P \supset S$.
    (4) A real place (A,M) is compatible with the preordering S of R if and only if the following identity holds:

$$(((A \backslash M) \cap S) + M) \cap (-S) = \emptyset.$$

    (5) A real place (A,M) is fully compatible with the preordering S if and only if the following identity holds:

$$((A \backslash M) \cap S) + M \subset \cap_P (P \backslash \text{supp}(P)).$$

where P runs over all orderings over S.

Making use of the real place defined above, it is possible to study real Prufer rings just as we have done for real Prufer domains. Let R be a Prufer ring in the sense of M. Griffin, and K its total quotient ring. If R is real, and for each regular maximal ideal M, $(R_{[M]}, [M]R_{[M]})$ is a real place for K, then R is defined as a real Prufer ring in [D9]. A characterization for real Prufer ring is as follows:

Theorem 2. Let R be a real ring with K as its total quotient ring. Then the following statements are equivalent:
    (i) R is convex with respect to some preordering T of K;
    (ii) R is semi–integrally closed in the preordered ring (K,T);
    (iii) R is a real Prufer ring.

Other results of [D9] here are:

Theorem 3. Let R be a real Prufer ring. If S is an overring of R in K, then S is a real Prufer ring too.

Theorem 4. Let K be a real total quotient ring with a given preordering T, and let S be the subset $\{1/1+x^2 \mid x \in K\}$. Then the T-convex hull R of the subring generated by S is a real Prufer ring in K; further, R is contained in every T-convex subring of K.

As for real Prufer valuation rings, the following result was first obtained by Wen Fengping in his MSc thesis [W]:

Theorem 5. Let R be a real ring with total quotient ring K, and P be an ordering of K. Then the following statements are equivalent:
   (i) R is P-convex;
   (ii) Every overring of R in K is P-convex;
   (iii) R is a P-convex Prufer ring in K;
   (iv) R is a valuation ring of K compatible with P.

On the other hand, in [Z9], Zeng has succeeded in constructing real places on a commutative ring with 1 through its orderings by introducing the notion of Δ-subrings. Let P be an ordering of a commutative ring R with 1. A subring S of R is called a Δ-subring of (R,P), if the following conditions are satisfied: (i) $S \cap \mathrm{supp}(P) = \{0\}$; and (ii) For every $x \in R$ which is infinitely large over S (i.e., $|ux|_P >_P |v|_P$ for any u, $v \in S$), there are some $x' \in R$ and s_1, $s_2 \in S$ such that $|s_1|_P >_P |x'x|_P >_P |s_2|_P$. With the help of Δ-subrings, the following result on construction of real places may be obtained:

Theorem 6. For a (commutative) ring R (with 1), every real place (A, M) on R is exactly of the form

$$A = A(S,P) = \{r \in R| \quad |s_1 r|_P \le_P |s_2|_P \text{ for some } s_1, s_2 \in S\}.$$

$$M = M(S,P) = \{r \in R| \quad |s_1 r|_P <_P |s_2|_P \text{ for any } s_1, s_2 \in S\}.$$

where P is an ordering of R, and S is an Δ-subring of (R,P).

Moreover, in [Z11], prolongation of real places is also considered. As a sufficient and necessary condition for extensibility of real places on a ring, the following theorem is established:

Theorem 7. Let (A,M) be a real place on a ring R with core ρ. and R^* a ring extension of R. Then (A,M) can be extended to a real place of R^*, if and only if R^* possesses an ordering P^* such that (i) $P^* \cap R$ is an ordering of R compatible with (A,M); (ii) over some Δ-subring of (K^*,P^*), every element in $R^* \rho + M$ is infinitely small; and (iii) for any $\eta \in R^* \rho$ and $a \in A \backslash M$, $|\eta|_{P*} <_{P*} |a|_{P*}$.

When R^* is integral over R, Theorem 7 can be improved as follows:

Theorem 8. Let (A,M) be a real place on a ring R, and R* an integral ring extension of R. Then (A,M) can be extended to a real place (A*,M*) of R*, if and only if R* possesses an ordering P* such that P*∩R is an ordering of R compatible with (A,M). Moreover, in this case, (A*,M*) is the only real extension of R* compatible with P*.

Ⅳ. SEMIALGEBRAIC GEOMETRY

In semialgebraic geometry, the Nullstellensatz, Positivstellensatz, and Nichtnegativstellensatz are important results. These Stellensatze characterize polynomial functions which are zero, positive or nonnegative on certain kinds of semialgebraic sets. Various versions of these Stellensatze can be founded in the literature. In [Z10], the so-called homogeneous Stellensatze are established so that the usual Stellensatze may be obtained as direct consequences in some special cases.

Now let R be a real closed field, let K be an ordered subfield (with the inherited ordering), and denote $K^+=\{a\in K \mid a \geqslant 0\}$. Let I be an X-homogeneous ideal of K[X,Y], where $X:=(X_0,X_1,...,X_n)$, $Y:=(Y_1,...,Y_m)$ are indeterminates, let f, u_1, ..., u_s, w_1, ..., w_t be X-homogeneous forms in K[X,Y], let U be the multiplicative subsemigroup of K[X,Y] generated by the u_i, and let W be the multiplicative subsemigroup of K[X,Y] generated by the u_i and the w_j. Then the homogeneous Stellensatze obtained in [Z10] here is:

Theorem 1. Let the notations be as above. Then we have
(i) f is vanishing in R with respect to {U>0; W>0; I=0} if and only if there is an X-homogeneous inclusion

$$uf^{2s}+\sum a_v w_v g_v^2 \in I.$$

(ii) f is positive in R with respect to {U>0; W>0; I=0} if and only if there is an X-homogeneous inclusion

$$(\sum_v a_v w_v g_v^2)f \equiv u+\sum_z a_z' w_z' g_z'^2 \pmod{I}$$

(iii) f is nonnegative in R with respect to {U>0; W>0; I=0} if and only if there is an X-homogeneous inclusion

$$(uf^{2s}+\sum_v a_v w_v g_v^2)f \equiv \sum_z a_z' w_z' g_z'^2 \pmod{I}$$

In the inclusions above, $s\in \mathbb{N}$, a_v, $a_z'\in K^+$, g_v, $g_z'\in K[X,Y]$, u is a product of the u_i, and the w_v, w_z' are (not necessarily distinct) products of the u_i and the w_j.

Moreover, in [Z10]. some quantitative aspects related to the homogeneous Stellensatze are investigated. Thereby the existence of upper bounds for these Stellensatze is shown by the following result:

Theorem 2. Given n, m, s, t and $d \in \mathbb{N}$, there exist $\alpha, \beta \in \mathbb{N}$ depending only on (n,m,s,t,d) such that in the X-homogeneous inclusions of Theorem 1, the number of summands and the (total) degrees of all appearing forms may be taken to be bounded by α and β respectively, whenever none of the (total) degrees of the u_i, w_j and generators of I exceeds d.

As a special example, an open problem posed by Stengle is answered affirmatively.

In [Z11], the following result concerning dimensions of semialgebraic sets is established.

Theorem 3. Let S be a semialgebraic set in $\mathbb{R}^n$ of dimension d. Then there is a linear surjection φ of $\mathbb{R}^n$ into $\mathbb{R}^d$ such that φ is uniformly finite over S and $\dim(\varphi(S))=d$.

From the theorem above, a description of the dimension of a semialgebraic set can be deduced as follows:

Theorem 4. Let S be a semialgebraic set in $\mathbb{R}^n$. Then $\dim(S) = \min\{k \mid$ there is a linear mapping φ of $\mathbb{R}^n$ into $\mathbb{R}^k$ such that φ is uniformly finite over $S\} = \min\{k \mid$ there is a linear mapping φ of $\mathbb{R}^n$ into $\mathbb{R}^k$ such that φ is finite over $S\}$.

Accordingly, we say that φ is uniformly finite or finite over S if the number of inverse images of each of $\varphi(S)$ in S is bounded by some fixed $m \in \mathbb{N}$ or is finite.

BIBLIOGRAPHY

Books.
[D] Dai Zhizhong: Elements of Valuation Theory, Higher Eduction Press, Beijing, 1981.
[DZ] Dai Zhizhong and Zeng Guangxing: Hilbert's 17th Problem, Jiangxi Education Press, Nanchang, 1991.

Papers.
[CQ] Chen Binhui and Qi Zhinan: On Henselian valuation rings, *Kexue Tongbao* (in Chinese), Vol. 26, 1981, pp. 1161-1163.
[D1] Dai Zhizhong: Pseudokonvergenz und die Perfektheit bewerteter Korper, (in Chinese, German summary), *Acta Math. Sinica*, vol. 5, 1955, pp. 489-496.

[D2] -----: Algebraic extensions of a complete field, (in Chinese, English summary), *Acta Scientiarum Naturalum Scholarum Superiorum Sinensium*, vol. 2, 1966, pp. 43–51.

[D3] -----: On discrete valuations, (in Chinese), *Acta Math. Sinica*, vol. 13, 1963, pp. 17 – 22.

[D4] -----: On multiply complete valued fields, (in Chinese), *Shuxue Jinzhan*, vol. 9, 1966, pp. 401–406.

[D5] -----: On a conjecture of W. Krull, (in Chinese), *Jour. of Jiangxi Univ.* (Natural Sc.), vol. 9, 1985, pp. 1-4.

[D6] -----: Notes on ordered fields, (in Chinese), *Jour. of Math. Res. & Exp.*, vol. 2, No. 2, 1982, pp. 7-10.

[D7] -----: Real places and real valuations on a commutative ring, (in English), *Acta Math. Sinica*, vol. 10, Special Issue, 1994, pp. 24–30.

[D8] -----: Preorderings and real places on a commutative ring, (in Chinese), *Acta Math. Sinica*, vol. 36, 1993, pp. 441–450.

[D9] -----: Real Prufer rings with zero divisors, (in Chinese, English summary), *Jour. of Nanchang Univ.* (Natural Sc.), vol. 17, No. 2, 1993, pp. 1-6.

[W] Wen Fengping: On real Prufer valuation rings, Master's thesis, Jiangxi University, 1988.

[Z1] Zeng Guangxing: On polynomials over real valuation rings with core, (in Chinese), *Acta Math. Sinica*, vol. 31, No. 5, 1988, pp. 634–644.

[Z2] -----: Semialgebraic Stellensatze for matrices, *Comm. Algebra* vol. 18, No. 2, 1990, pp. 413–439.

[Z3] -----: A new proof of McKenna's theorem, *Proc. Amer. Math. Soc.* vol. 102, No. 4, 1988, pp. 827–830.

[Z4] -----: Positive definite functions ove formally real fields with core, (in Chinese), *Shuxue Jinzhan*, vol. 17, No. 3, 1988, pp. 285–289.

[Z5] -----: A chrraracterization of preordered fields with the weak Hilbert property, *Proc. Amer. Math. Soc.* vol. 104, No. 2, 1988, pp. 335–342.

[Z6] -----: On preordered fields related to Hilbert's 17th problem, *Math. Z.* vol. 206, 1991, pp. 145–151.

[Z7] -----: A problem about the weak Hilbert property, to appear.

[Z8] -----: Some valuation-theoretic characterizations of fields with the weak Hilbert property, (in Chinese), *Acta Math. Sinica*, vol. 32, No. 5, 1989, pp. 690–701.

[Z9] -----: Construction and prolongation of real places on a commutative ring, (in Chinese), *Acta Math. Sinica*, vol. 34, No. 3, 1991, pp. 343–351.

[Z10] -----: Homogeneous Stellensatze in semialgebraic geometry, *Pacific J. Math.* vol. 136, No. 1, 1989, pp. 103–122.

[Z11] -----: Dimensions of semialgebraic sets and real linear mapping theorem, (in Chinese), *Acta Math. Sinica*, vol. 35, No. 5, 1992, pp. 774–779.

Modules Over Hyperfinite Groups

Z. Y. Duan Department of Mathematics, Southwest China Normal University, Chongqing, 630715

Definition 1   A group G is called a hyperfinite group if G has an ascending normal series $1 = G_0 \leqslant G_1 \leqslant \cdots \leqslant G_\alpha = G$ in which each factor $G_{\beta+1}/G_\beta (\beta < \alpha)$ is finite.

Definition 2   A $\mathbb{Z}G$-module A is said to be artinian if A satisfies the minimal condition on its $\mathbb{Z}G$-submodules.

Definition 3   A $\mathbb{Z}G$-module A is said to be noetherian if A satisfies the maximal condition on its $\mathbb{Z}G$-submodules.

In 1986, Prof. D. I. Zaitsev proved that ([14]): If G is a hyperfinite locally solvable group, then any artinian $\mathbb{Z}G$-module A has an f-decomposition, that is, $A = A^f \oplus A^{\bar{f}}$, where A^f is a $\mathbb{Z}G$-submodule of A whose irreducible $\mathbb{Z}G$-factors as abelian groups are all finite and the $\mathbb{Z}G$-submodule $A^{\bar{f}}$ has no such $\mathbb{Z}G$-factors. Using this result, he proved that ([14]): Any extension E of an abelian group A by a hyperfinite locally solvable group G splits conjugately over A if A is an artinian $\mathbb{Z}G$-module satisfying $A = A^{\bar{f}}$, where E is said

. This project is finacially supported by the Chinese State Education Committee, the Science Committee of Congqing, and the National Natural Science Foundation of China.

to split conjugately over A if E = AB and all such B are conjugate in E.

Later, in 1988, he considered the dual case and, without the f-decomposition result for A, proved that ([15]): Any extension E of an abelian group A by a hyperfinite locally solvable group G splits conjugately over A if A is a noetherian $\mathbb{Z}G$-module satisfying A = A^I.

Is the locally solvable condition necessary for the above results? Is there an f-decomposition for a noetherian $\mathbb{Z}G$-module A over a hyperfinite locally solvable group G? These questions arose naturally and, since then, a series of researches have been carried out (see [1] to [11] and [16]).

1. f-decompositions

Over a hyperfinite group G, D. I. Zaitsev proved in [15] that: if a noetherian $\mathbb{Z}G$-module A has a maximal $\mathbb{Z}G$-submodule B such that B = B^f and A/B = $(A/B)^I$ (or, respectively, B = B^I and A/B = $(A/B)^f$), then A has an f-decomposition. For a complete result, dual to the artinian case, the following steps give an outlien of the proof ([1] or [6]).

Step 1: The important lemmas

Lemma 1.1 ([15])   Let H be a hyperfinitely embedded subgroup of a group G and A a noetherian $\mathbb{Z}G$-module. If $C_A(H) = 0$, then H contains a subgroup K and A contains a nonzero $\mathbb{Z}G$-submodule B such that K is normal in G, $C_B(K) = 0$, and $|K/C_K(B)| < \infty$.

Lemma 1.2 ([1])   Let G be a locally finite group and A a torsion-free noetherian $\mathbb{Z}G$-module. Then pA < A and $\bigcap_{i=1}^{\infty} p^i A = 0$ for any prime p.

Lemma 1.3 ([1])   Let G be a hyperfinite locally solvable group and A a noetherian $\mathbb{Z}G$-module with pA = 0 for some prime p. If all irreducible $\mathbb{Z}G$-factors of A are finite, then A is finite.

Lemma 1.4 ([1])   Let G be a group, A a $\mathbb{Z}G$-module, and M a $\mathbb{Z}G$-submodule of A such that the factor module A/M is a p-group for some prime p. If $H = C_G(A/M)$ contains a nontrivial finite subgroup K that is a q-group for some prime $q \neq p$, then A = $C_A(x)$ + M for any x ∈ K. Further, A = $C_A(K)$ + M.

Step 2: Reduce A to be either torsion-free or an elementary abelian p-group for some prime p.

Proposition 1: ([1])   Let G be a hyperfinite locally solvable group, A a noetherian $\mathbb{Z}$G-module, and B a $\mathbb{Z}$G-submodule of A such that each irreducible $\mathbb{Z}$G-factor of B is finite (recp. infinite) and A/B contains no finite (resp. infinite) irreducible $\mathbb{Z}$G-factors. Then B has a complement in A, i.e., $A = B \oplus C$ for some $\mathbb{Z}$G-submodule C of A.

Step 3: Reduce A to be torsion-free with all finite irreducible $\mathbb{Z}$G-factors being p-groups for some fixed prime p.

Proposition 2: ([1])   Let G be a hyperfinite locally solvable group and A a noetherian $\mathbb{Z}$G-module. If A is torsion or if A contains at least two finite irreducible $\mathbb{Z}$G-factors such that one is a p-group and another a q-group, where p and q are distinct primes, then A has an f-decomposition.

Step 4: Discuss the properties of torsion-free noetherian $\mathbb{Z}$G-modules.

Proposition 3: ([1])   Let G be a hyperfinite locally solvable group, A a torsion-free noetherian $\mathbb{Z}$G-module, and $A_{ii} = p^i A$, where p is a prime and $i = 0, 1, 2, \cdots\cdots$. Then

(1) for any $0 \leqslant j < i$, A_{jj}/A_{ii} has an f-decomposition

$$A_{jj}/A_{ii} = A_{ij}/A_{ii} \oplus A_{ji}/A_{ii},$$

where A_{ij} is the $\mathbb{Z}$G-submodule of A_{jj} such that $A_{ij} \geqslant A_{ii}$ and $A_{ij}/A_{ii} = (A_{jj}/A_{ii})^f$ and the $\mathbb{Z}$G-submodule $A_{ji}(\geqslant A_{ii})$ is such that $A_{ji}/A_{ii} = (A_{jj}/A_{ii})^I$;

(2) $A_{ij} \leqslant A_{ik}$ and $A_{ij} \leqslant A_{sj}$, where $k \leqslant j$ and $s \leqslant i$;

(3) $A_{ii} = A_{ik} \cap A_{si}$, where $k \leqslant i$, $s \leqslant i$, and $i = 0, 1, 2, \cdots\cdots$;

(4) $A_{ij} = A_{ik} \cap A_{sj}$, where $k \leqslant j$, $s \leqslant i$, and $i = 0, 1, 2, \cdots\cdots$;

(5) $A_{ij}/A_{kk} = A_{kj}/A_{kk} \oplus A_{ik}/A_{kk}$, $A_{kj}/A_{kk} = (A_{ij}/A_{kk})^f$, and $A_{ik}/A_{kk} = (A_{ij}/A_{kk})^I$, where $k \geqslant i, j$;

(6) $A_{ij}/A_{sk} = A_{sj}/A_{sk} \oplus A_{ik}/A_{sk}$, $A_{sj}/A_{sk} = (A_{ij}/A_{sk})^f$, and

$$A_{ik}/A_{sk} = (A_{ij}/A_{sk})^{f}, \text{ where } k \geq j, s \geq i, \text{ and } i, j = 0, 1, 2, \cdots\cdots;$$

(7) $A_{ij}/A_{i,j+t} \cong_{\mathbb{Z}G} A_{ks}/A_{k,s+t}$ and $A_{ij}/A_{i+t,j} \cong_{\mathbb{Z}G} A_{ks}/A_{k+t,s}$ where $i, j, k, s, t = 0, 1, 2, \cdots\cdots;$

(8) $p^{k}A_{ij} = A_{i+k,j+k}, i, j, k = 0, 1, 2, \cdots\cdots;$

Furthermore, let $A_{\infty i} = \bigcap_{j} A_{ji}$ and $A_{i\infty} = \bigcap_{j} A_{ij}$ for $i = 0, 1, 2, \cdots\cdots,$ then

(9) $p^{k}A_{\infty i} = A_{\infty,i+k}$ and $p^{k}A_{i\infty} = A_{i+k,\infty}, i, k = 0, 1, 2, \cdots\cdots;$

(10) $A_{\infty k} = A_{\infty j} \bigcap A_{ik}$ and $A_{k\infty} = A_{j\infty} \bigcap A_{ki}, k \geq j, \text{ and } i = 0, 1, 2, \cdots\cdots;$

(11) $A_{\infty j}/A_{\infty k} \cong_{\mathbb{Z}G} (A_{\infty j} + A_{kk})/A_{kk} \leq A_{kj}/A_{kk}$ and
$A_{j\infty}/A_{k\infty} \cong_{\mathbb{Z}G} (A_{j\infty} + A_{kk})/A_{kk} \leq A_{jk}/A_{kk}, k \geq j = 0, 1, 2, \cdots\cdots;$ and

(12) $A_{i\infty}$ (resp. $A_{\infty i}$) has no finite (resp. infinite) irreducible $\mathbb{Z}G$-factors that are p-groups, $i = 0, 1, 2, \cdots\cdots.$

Step 5: Complete the proof.

<u>Proposition 4</u>: ([1])   Let G be a hyperfinite locally solvable group and A a noetherian $\mathbb{Z}G$-module. If all finite irreducible $\mathbb{Z}G$-factors of A are p-groups for some fixed prime p, then A has an f-decomposition.

Combining all the above, we have

<u>Theorem A</u> ([1])   If G is a hyperfinite locally solvable group, then any noetherian $\mathbb{Z}G$-module A has an f-decomposition.

Further, in [5], Z. Y. Duan and M. J. Tomkinson jointly proved

<u>Theorem B</u> ([5])   If G is a hyperfinite locally solvable group, then any minimax $\mathbb{Z}G$-module A has an f-decomposition, where A is called a minimax module if A has a finite series of $\mathbb{Z}G$-submodules $A_0 = A_1 \leq A_2 \leq \cdots \leq A_n = A$ in which each $\mathbb{Z}G$-factor A_i/A_{i-1} is either an artinian $\mathbb{Z}G$-module or a noetherian $\mathbb{Z}G$-module, $i = 1, 2, \cdots, n.$

By considering a subset S consisting of elements of order p for some fixed prime p instead of the finite minimal normal subgroup K of G, we may remove the locally solvability condition from all the above results (see [7], [9] and [10]). For this purpose, we have

<u>Lemma 1.5</u> ([7])   Let G be a group, x an element of G, and H a subgroup of G contained in the centralizer $C_G(x)$ of x in G. If A is a $\mathbb{Z}G$-module, then, for any $\mathbb{Z}H$-submodule B of A, $B(x-1)$ and $C_B(x)$ are $\mathbb{Z}H$-submodules of A and $B/C_A(x) \cong_{\mathbb{Z}G} B(x-1)$.

<u>Lemma 1.6</u> ([7])   Let G be a hyperfinite group and A an artinian $\mathbb{Z}G$-module such that $pA = 0$ for some prime p. If the irreducible $\mathbb{Z}G$-factors of A as abelian groups are all finite, then A is finite.

<u>Theorem C</u> ([7])   If G is a hyperfinite group, then any artinian $\mathbb{Z}G$-module A has an f-decomposition.

For the noetherian case, the same result holds, but the proof is much hard than that in the artinian case.

<u>Lemma 1.7</u> ([9])   Let G be a group, H a subgroup of finite index in G, and A a $\mathbb{Z}G$-module. Then

(a) A is a noetherian (resp. artinian) $\mathbb{Z}G$-module if and only if A is a noetherian (resp. artinian) $\mathbb{Z}H$-module;

(b) A has an irreducible $\mathbb{Z}G$-factor B/C if and only if A as a $\mathbb{Z}H$-module has an irreducible $\mathbb{Z}H$-factor U/V; moreover, some B/C is finite, infinite, or a p-group for some prime p if and only if some U/V is so;

(c) if D is a $\mathbb{Z}H$-submodule of A then the $\mathbb{Z}G$-submodule $D^G(=\Sigma_{g \in G} Dg)$ of A has a finite (resp. infinite) irreducible $\mathbb{Z}G$-factor if and only if D has a finite (resp. infinite) irreducible $\mathbb{Z}H$-factor;

(d) A contains no nonzero $\mathbb{Z}G$-submodule with an f-$(\mathbb{Z}G)$-decomposition if and only if A as a $\mathbb{Z}H$-module contains no nonzero $\mathbb{Z}H$-submodule with an f-$(\mathbb{Z}H)$-decomposition;

(e) if D is a $\mathbb{Z}H$-submodule of A then D^G has an f-$(\mathbb{Z}G)$-decomposition if and only if D has an f-$(\mathbb{Z}H)$-decomposition;

(f) A has an f-($\mathbb{Z}G$)-decomposition if and only if A has an f-($\mathbb{Z}H$)-decomposition.

<u>Lemma 1.8</u> ([9])   Let G be a group, F a nontrivial finite normal subgroup of G, A a $\mathbb{Z}G$-module, and B a $\mathbb{Z}G$-submodule of A. If A/B has an f-decomposition and $F \leqslant C_G(B)$ but F is not contained in $C_G(A)$, then A has a nonzero $\mathbb{Z}G$- submodule D with an f-decomposition, too. Furthermore, D can be chosen such that:

(1) if $(A/B)^f = 0$ then $D^f = 0$, and
(2) if $(A/B)^f = 0$ then $D^f = 0$.

<u>Lemma 1.9</u> ([9])   Let H be a normal subgroup of a hyperfinite group G, A a noetherian $\mathbb{Z}G$-module, and B a nonzero $\mathbb{Z}G$-submodule of A satisfying $B = B^f$ and $A/B = (A/B)^f$ (resp. $B = B^f$ and $(A/B)^f = A/B$). If $C_B(H) = 0$, A/C has an f-decomposition for any nonzero $\mathbb{Z}G$-submodule C of B, and A has no nonzero $\mathbb{Z}G$-submodule D with $D = D^f$ (resp. $D = D^f$), then there is a $K \leqslant H$ and a nonzero $\mathbb{Z}G$-submodule $B^* \leqslant B$ such that K is normal in G, $A/B^* = B/B^* \oplus A^*/B^*$ for some $\mathbb{Z}G$-submodule A^* of A, $C_{B^*}(KC_G(A^*)/C_G(A^*)) = 0$, and $KC_G(A^*)/C_G(A^*)$ is a finite normal characteristically simple subgroup of $G/C_G(A^*)$. Furthermore, if q is a prime factor of the order of $KC_G(A^*)/C_G(A^*)$, then $C_{B^*}(SC_G(A^*)/C_G(A^*)) = 0$ for $S = \{g \in K; gC_G(A^*)$ is of order q in $KC_G(A^*)/C_G(A^*)$ for the prime q$\}$.

<u>Lemma 1.10</u> ([9])   Let G be a group, A a $\mathbb{Z}G$-module, and M a $\mathbb{Z}G$-submodule of A such that A/M is a p-group for some prime p. If $H = C_G(A/M)$ contains a non-empty subset S consisting of finitely many elements of order q^n for some integer n and some prime q other than p, then $A = C_A(x) + M$ for any $x \in S$. Further, $A = C_A(S) + M$.

One more crucial lemma is the following

<u>Lemma 1.11</u> ([9])   Let G be a hyperfinite group and A a noetherian $\mathbb{Z}G$-module with $pA = 0$ for some prime p. If all irreducible $\mathbb{Z}G$-factors of A are finite, then A is finite.

Using the above lemmas, we just follow the proof used in the case that G is hyperfinite and locally solvable and we get

<u>Theorem D</u> ([9])   If G is a hyperfinite group, then any noetherian $\mathbb{Z}G$-module A has an f-decomposition.

Joining Theorem C and Theorem D, it follows that

<u>Theorem E</u> ([10])   If G is a hyperfinite group, then any minimax $\mathbb{Z}$G-module A has an f-decomposition.

Some other results which belong to this section are the following:

<u>Theorem F</u> ([16])   Let G be a hypercyclic group, then any artinian $\mathbb{Z}$G-module A has an f-decomposition.

<u>Theorem G</u> ([3])   If G is a hyper-(cyclic or finite) locally solvable group, then any periodic artinian $\mathbb{Z}$G-module A has an f-decomposition.

<u>Theorem H</u> ([11])   If G is a hyper-(cyclic or finite) locally solvable group, then any artinian $\mathbb{Z}$G-module A has an f-decomposition.

<u>Theorem I</u> ([4])   If G is a hyper-(cyclic or finite) locally solvable group, A is a noetherian $\mathbb{Z}$G-module, and B is a $\mathbb{Z}$G-submodule of A such that B is finite and A/B has no nonzero finite $\mathbb{Z}$G-factors (or A/B is finite and B has no nonzero finite $\mathbb{Z}$G-factors). Then A has an f-decomposition.

<u>Question 1.1</u>   Does any noetherian $\mathbb{Z}$G-module A over a hyper-(cyclic or finite) locally solvable group G have an f-decomposition?

<u>Question 1.2</u>   Does the locally solvable condition in the above question can be removed again?

2. The structure of A^f and $A^{\bar{f}}$

It is evident that the structure of a $\mathbb{Z}$G-module A over a hyperfinite group G is dicided by the structure of A^f and $A^{\bar{f}}$. For A^f, we have a clear description, but the structure of $A^{\bar{f}}$ is known only when A is noetherian, G is a periodic abelian group and $\pi(G) = \{$prime p; G has an element of order p $\}$is a finite set (see [2], [5], [7], [9] and [10]).

We first consider A^f. In fact, we have:

<u>Lemma 2.1</u> ([5])   Let G be a hyperfinite locally solvable group and A an artinian $\mathbb{Z}$G-module with pA = 0 for some prime p. If $A = A^f$, then A is finite.

<u>Lemma 2.2</u> ([5]) Let G be a hyperfinite p'-group and A an artinian $\mathbb{Z}$G-module with pA = 0 for the prime p. If A = A^f, then A is finite.

<u>Lemma 2.3</u> ([7]) Let G be a hyperfinite group and A an artinian $\mathbb{Z}$G-module such that pA = 0 for some prime p. If A = A^f, then A is finite.

<u>Theorem 2.4</u> ([5]) Let G be a locally solvable hyperfinite group and let Λ be an artinian $\mathbb{Z}$G-module such that every irreducible $\mathbb{Z}$G-factor of A is finite. Then A as an abelian group is Chernikov and $G/C_G(A)$ is finite.

<u>Theorem 2.5</u> ([7]) If G is a hyperfinite group and A is an artinian $\mathbb{Z}$G-module all of whose irreducible $\mathbb{Z}$G-factors are finite, then A as an abelian group is Chernikov and $G/C_G(A)$ is finite.

<u>Lemma 2.6</u> ([1]) Let G be a hyperfinite locally solvable group and A a noetherian $\mathbb{Z}$G-module with pA = 0 for some prime p. If all irreducible $\mathbb{Z}$G-factors of A are finite, then A is finite.

<u>Lemma 2.7</u> ([1]) Let G be a hyperfinite group and A a noetherian $\mathbb{Z}$G-module with pA = 0 for some prime p. If G is a p'-group and all irreducible $\mathbb{Z}$G-factors of A are finite, then A is finite.

<u>Lemma 2.8</u> ([9]) Let G be a hyperfinite group and A a noetherian $\mathbb{Z}$G-module with pA = 0 for some prime p. If all irreducible $\mathbb{Z}$G-factors of A are finite, then A is finite.

<u>Theorem 2.9</u> ([2]) If G is a hyperfinite locally solvable group and A is a noetherian $\mathbb{Z}$G-module, then A^f is finitely generated as an abelian group and $G/C_G(A^f)$ is finite.

<u>Theorem 2.10</u> ([10]) Let G be a hyperfinite group and A a noetherian $\mathbb{Z}$G-module, then A^f is finitely generated as an abelian group and $G/C_G(A^f)$ is finite.

Now, we point out two results which encourage us to discuss the structure of A^f.

<u>Lemma 2.11</u> ([2]) Let G be a periodic abelian group and let A be a noetherian $\mathbb{Z}$G-module with pA = 0 for some prime p. Then A has a finite $\mathbb{Z}$G-composition series.

<u>Lemma 2.12</u> ([2]) Let G be a hyperfinite p-group and A a noetherian $\mathbb{Z}$G-module with pA = 0, where p is a prime. Then A is finite.

So far, we have the following main results for A^f.

Proposition 2.13 ([2])   Let G be a periodic abelian group with $|\pi(G)| < \infty$, where $\pi(G)$ is the set of primes p such that G has an element of order p, and let A be a noetherian $\mathbb{Z}G$-module. Then A^f has a finite $\mathbb{Z}G$-composition series and has finite exponent.

Proposition 2.14 ([2])   Let G be a hyperfinite p-group for some prime p and let A be a noetherian $\mathbb{Z}G$-module. Then A^f is a p'-torsion-group of finite exponent.

In general, we have the following conjectures.

Conjecture A ([2])   If G is a hyperfinite locally solvable group and A a noetherian $\mathbb{Z}G$-module with $pA = 0$ for some prime p, then A has a finite $\mathbb{Z}G$-composition series.

Conjecture B ([2])   If G is a hyperfinite locally solvable group and A a noetherian $\mathbb{Z}G$-module, then the $\mathbb{Z}G$-submodule A^f of A has finite exponent.

Question 2.1   If the above conjectures are ture, Can the locally solvability condition be removed?

The following is a notable result for A^f.

Proposition 2.15 ([1])   For any finite integer $n > 0$, there exists a noetherian $\mathbb{Z}G$-module A over a periodic abelian group G such that A^f is of exponent n.

Remark   Under some conditions, we can obtain almost all the possible $\mathbb{Z}G$-submodules A^f of a noetherian $\mathbb{Z}G$-module A if G is a Chernikov group (for details, refer to § 4.2 in [1])

3. The splitting extension of A by G

If A is a normal subgroup of a group E such that $E/A \cong G$, then E is called an extension of A by G. For a subgroup B of A, if E has a subgroup E_1 such that $E = AE_1$ and $A \cap E_1 = B$, then we say E_1 is a supplement to A in E and also say that E splits over A modulo B. Further, if all such supplements are conjugate in E modulo B, E is said to split conjugately over A modulo B. In particular, if $B = 1$, then E splits conjugately over A. There are several conditions for A and G such that the extension E of A by G splits conjugately over A. The following results are just the main theorems that closely relate to the topic of this paper.

<u>Theorem 3. 1</u> ([12]) Let E be a group, A an abelian normal subgroup, E/A hypercyclic, and A an artinian $\mathbb{Z}E$-module. Then E splits conjugately over A modulo A^c, where A^c is the $\mathbb{Z}E$-submodule of A with all irreducible $\mathbb{Z}E$-factors as groups being cyclic and A/A^c has no irreducible $\mathbb{Z}E$-factors of this type.

<u>Theorem 3. 2</u> ([13]) Let E be a group, A an abelian normal subgroup, E/A hypercyclic, and A a noetherian $\mathbb{Z}E$-module. If A has no nontrivial cyclic $\mathbb{Z}E$-images then E splits conjugately over A and A has no nontrivial cyclic $\mathbb{Z}E$-factors.

<u>Theorem 3. 3</u> ([14]) Let G be a hyperfinite locally solvable group and A an artinian $\mathbb{Z}G$-module without nonzero finite $\mathbb{Z}G$-submodules, then any extension E of A by G splits conjugately over A and A has no nonzero finite $\mathbb{Z}G$-factors.

<u>Theorem 3. 4</u> ([15]) Let G be a hyperfinite locally solvable group and A a noetherian $\mathbb{Z}G$-module without nonzero finite $\mathbb{Z}G$-images, then any extension E of A by G splits conjugately over A and A has no nonzero finite $\mathbb{Z}G$-factors.

<u>Theorem 3. 5</u> ([8]) Let G be a hyperfinite group and A an artinian $\mathbb{Z}G$-module. If A has no nonzero finite $\mathbb{Z}G$-submodules and, as a group, is divisible, then any extension E of A by G splits conjugately over A and A has no nonzero finite $\mathbb{Z}G$-factors.

<u>Theorem 3. 6</u> ([8]) Let G be a hyperfinite group and A a noetherian $\mathbb{Z}G$-module. If A has no nonzero finite $\mathbb{Z}G$-images and has no nontrivial elements with order being the order of some element of G, then any extension E of A by G splits conjugately over A and A has no nonzero finite $\mathbb{Z}G$-factors.

Remark The last two results above are partially generalized the corresponding ones in [14] and [15]. For a complete generalization, we need to consider the situation in which A as a group is elementary abelian. But in this case the following questions still remain open.

<u>Question 3. 1</u> ([8]) Let G be a hyperfinite group, A an artinian $\mathbb{Z}G$-module with $pA = 0$ for some prime p. Does the extension E of A by G split conjugately over A modulo A^f?

<u>Question 3. 2</u> ([8]) Let G be a hyperfinite group, A a noetherian $\mathbb{Z}G$-module with $pA = 0$ for some prime p. Does there exist an extension E of A by G such that E fails to split conjugately over A modulo A^f?

It is evident that there is still a lot of work to do on the above topic and we are now doing it.

References

[1] Z. Y. Duan, NOETHERIAN MODULES OVER HYPERFINITE GROUPS, Ph D. thesis, University of Glasgow, (1991).

[2] Z. Y. Duan, The structure of noetherian modules over hyperfinite groups, Math. Proc. Camb. Phil. Soc., 112(1992), 21-28.

[3] Z. Y. Duan, Extensions of abelian by hyper-(cyclic or finite) groups I, Comm. Alg., 20:8(1992), 2305-2321.

[4] Z. Y. Duan, Extensions of abelian by hyper-(cyclic or finite) groups II, Rend. Sem. Mat. Univ. Padova, 89(1993), 113-126.

[5] Z. Y. Duan and M. J. Tomkinson, The decomposition of minimax modules over hyperfinite groups, Arch. Math., 81(1993), 340-343.

[6] Z. Y. Duan, The decomposition of noetherian modules over hyperfinite groups, Riceche di Mat., (1995).

[7] Z. Y. Duan, The f-decomposition of artinian modules over hyperfinite groups, Proc. Edinb. Math. Soc., 38(1995), 117-120.

[8] Z. Y. Duan and H. P. Cao, Splitting extensions of abelian by hyperfinite groups, J. Southwest Teachers Univ., 4(1994).

[9] Z. Y. Duan and H. P. Cao, The f-decomposition of noetherian modules over hyperfinite groups, Alg. Colloquium, 1:3(1994), 281-287.

[10] Y. P. Qin, The decomposition of minimax modules over hyperfinite groups, J. Southwest Teachers Univ., 4(1994).

[11] Y. P. Qin, The decomposition of artinian modules over hyper-(cyclic or finite) groups, Proc. Edinb. Math. Soc., (1995).

[12] D. I. Zaitsev, Hypercyclic extensions of abelian groups (Russian), Akad. Nauk Ukrain SSR, Inst. Mat., Kiev, (1979), 16-37.

[13] D. I. Zaitsev, On extensions of abelian groups, (Russian), Akad. Nauk Ukrain SSR, Inst. Mat., Kiev, (1980), 16-40.

[14] D. I. Zaitsev, Splitting of extensions of abelian groups (Russian), Akad. Nauk Ukrain SSR, Inst. Mat., Kiev, (1986), 21-31.

[15] D. I. Zaitsev, Hyperfinite extensions of abelian groups (Russian), Akad. Nauk Ukrain SSR, Inst. Mat., Kiev, (1988), 17-26.

[16] D. I. Zaitsev and V. A. Maznitsenko, On direct decomposition of artinian modules over hypercyclic groups, Ukrain Mat. Z., 43(1991), 930-934.

Pointed Groups and Nilpotent Blocks

Yun FAN

Department of Mathematics, Wuhan University
WUHAN, 430072, P. R. CHINA

ABSTRACT. This is a survey of pointed groups and nilpotent blocks. After some historical background, the theory of pointed groups is introduced and the theory of nilpotent blocks including some recent developments is sketched.

It is fascinating that the study of nilpotent blocks reached a high level of achievement since the late 1970's. One of the important tools for the study is the theory of pointed groups. This paper gives a description of the theory and sketches the theorey of nilpotent blocks including some recent developments.

Let us fix some necessary notations.

All the groups are finite, G always denotes a finite group and p a prime number.

Let $\mathcal{O}$ denote a complete discrete valuation ring with residue field $k = \mathcal{O}/J(\mathcal{O})$ of characteristic p, where by $J()$ we always denote the Jacobson radical. Note that $\mathcal{O}$ and k are of arbitrary size except for some explicit assumptions we may state, and that it is allowed that $\mathcal{O} = k$.

All the $\mathcal{O}$-algebras are unitary and $\mathcal{O}$-free of finite rank; but subalgebras of an algebra are not necessarily unitary and homomorphisms between algebras are not necessarily unitary. By A^* we denote the multiplicative group consisting of all invertible elements of an algebra A. If A is a G-algebra and $H \leq G$, by A^G we denote the subalgebra consisting of all G-fixed elements in A, and by $A_H^G = \mathrm{Tr}_H^G(A^H)$ denote the image of the trace map $\mathrm{Tr}_H^G\colon A^H \to A^G$.

All the modules are unitary left modules and $\mathcal{O}$-free of finite rank unless otherwise stated.

By bar "–" we always denote the residue map modulo $J(\mathcal{O})$ for algebras and modules, e.g. $\overline{\mathcal{O}} = k$, $\overline{A} = A/(J(\mathcal{O})A)$ etc.

The paper consists of nine sections: §§1—2, historical backgrounds; §§3—5, pointed groups; §§6—9, nilpotent blocks.

---

* Supported partially by National Natural Science Foundation .

1. Background from Module Theory

One of the intuitive notions of pointed groups is from the theory about the direct decompositions of modules.

1.1 Components and primitive idempotents.

Let M be an $\mathcal{O}G$-module. An indecomposable direct summand of M is said to be a component of M.

Let $E = \operatorname{End}_{\mathcal{O}}(M)$ be the $\mathcal{O}$-algebra consisting of the all $\mathcal{O}$-endomorphisms of M. Then there is a group homomorphism $\psi : G \to E^*$, where E^* is the multiplicative group of E; such an algebra E is called an interior G-algebra with structural map ψ, and the $\mathcal{O}G$-endomorphism algebra of M is $\operatorname{End}_{\mathcal{O}G}(M) = E^G$.

Let H be a subgroup of G. The $\mathcal{O}H$-endomorphism algebra of the restricted module $\operatorname{Res}_H^G(M)$ is $\operatorname{End}_{\mathcal{O}H}(M) = E^H$.

Associated with every indecomposable direct decomposition

$$\operatorname{Res}_H^G(M) = N_1 \oplus \cdots \oplus N_r$$

there is a unique decompositoin in E^H of 1_E into orthogonal primitive idempotents:

$$1_E = i_1 + \cdots + i_r$$

In this way, every component N of $\operatorname{Res}_H^G(M)$ corresponds to a primitive idempotent i in E^H such that

$$N = iM \qquad \text{and} \qquad \operatorname{End}_{\mathcal{O}}(N) = iEi$$

and the $\mathcal{O}H$-endomorphism algebra of N is

$$\operatorname{End}_{\mathcal{O}H}(N) = iE^H i = (iEi)^H$$

However, such an idempotent i corresponding to N is not unique. In fact, let $j = i + (1 - i)ai$ for any $a \in E^H$, then j is a primitive idempotent in E^H and $N = jM$; and *vice versa*.

Remark. The above holds for any direct summand N of $\operatorname{Res}_H^G(M)$; e.g. there is an idempotent e in E^H such that $N = eM$ and $\operatorname{End}_{\mathcal{O}}(N) = eEe$; etc.

1.2 Isomorphism classes of components and points.

Either from the point of view of representation theory or from the point of view of general module theory, we are concerned with the isomorphism classes of the components. Sometimes a component of a module just means an isomorphism class of indecomposable direct summands of the module. We recall two well-known facts. Let M and E be as above.

(1.2.1). *Two direct summands $N = iM$ and $L = jM$ of $\operatorname{Res}_H^G(M)$ with idempotents i and j in E^H resp. are isomorphic to each other if and only if i and j are $(E^H)^*$-conjugate to each other.*

The following is clearly equivalent to the Krull-Schmidt Theorem.

(1.2.2). *If* $1_E = i_1 + \cdots + i_r = j_1 + \cdots + j_s$ *are two orthogonal primitive idempotent decompositions in* E^H, *then* $r = s$ *and there is an element* $u \in (E^H)^*$ *that transforms* $\{i_1, \cdots, i_r\}$ *to* $\{j_1, \cdots, j_r\}$ *by conjugation.*

It is reasonable to consider the conjuacy classes of primitive idempotents instead of individual idempotents. Hence we define

(1.2.3) Definition. An A^*-conjugacy class of primitive idempotents of an algebra A is said to be a point on A. By $\mathcal{P}(A)$ we denote the set of the points on A.

Thus the isomorphism classes of the comoponents of $\mathrm{Res}_H^G(M)$ correspond one-to-one to the points on E^H. However, we have seen that the points here are associated with subgroups.

For reference we state two facts about points.

(1.2.4). *Let* α *be a point on an algebra* A *and* I *be an ideal of* A. *Then either* $\alpha \subseteq I$ *or* $\alpha \cap I = \emptyset$; *and* $I + J(A) = \sum_\alpha A\alpha A + J(A)$ *with* $\alpha \in \mathcal{P}(A)$ *and* $\alpha \subseteq I$.

(1.2.5). *Assume* $f : A \to B$ *is an algebra homomorphism such that* $B = \mathrm{Im}(f) + J(B)$. *Then for any point* α *on* A *either* $f(\alpha) = \{0\}$ *or* $f(\alpha)$ *is a point on* B, *and the map* $\alpha \mapsto f(\alpha)$ *is a bijection from the points on* A *such that* $f(\alpha) \neq \{0\}$ *onto the points on* B.

In other words, the above f induces an injection from the points on B to the points on A. This follows from the result on lifting of idempotents.

1.3 Change of subgroups.

Let M and E be as in 1.1. Let $K \leq H \leq G$ and L be a component of $\mathrm{Res}_K^G(M)$ and N be a component of $\mathrm{Res}_H^G(M)$. Assume that i is an idempotent of E^H such that $N = iM$, and assume that $i = j_1 + \cdots + j_r$ is an orthogonal primitive idempotent decomposiotion in E^H. In this way it is easy to see that L is isomorphic to a component of $\mathrm{Res}_K^H(N)$ if and only if there is an idempotent j belonging to the point corresponding to L such that $ij = j = ji$.

In fact we have a general result:

(1.3.1) Proposition. *Let* L *be an* $\mathcal{O}$*-free* $\mathcal{O}K$*-module of finite rank. Then* L *is isomorphic to a component of* $\mathrm{Res}_K^H(N)$ *if and only if there is an injective interior* K*-algebra homomorphism* $f : \mathrm{End}_{\mathcal{O}}(L) \to \mathrm{End}_{\mathcal{O}}(N)$ *such that* $\mathrm{Im}(f) = f(1)\mathrm{End}_{\mathcal{O}}(N)f(1)$.

Remark. Such f in the proposition is called an embedding of algebras. For details see 3.4. The essence of the proposition is the following fact. Two $\mathcal{O}$-free $\mathcal{O}G$-modules of finite rank are isomorphic to each other if and only if their $\mathcal{O}$-endmorphism algebras are interior G-algebra isomorphic to each other. In fact we have a more elementary statement: Let N be an $\mathcal{O}$-free $\mathcal{O}G$-module of finite rank, and let $E = \mathrm{End}_{\mathcal{O}}(N)$ and i be a primitive idempotent of E, then Ei is an $\mathcal{O}G$-module through the interior G-structural map $G \to E$ and $Ei \cong N$ as $\mathcal{O}G$-modules.

Another important fact is as follows.

A component N of $\operatorname{Res}_H^G(M)$ is P-relatively projective if and only if its corresponding primitive idempotent $i \in \operatorname{Tr}_P^H(E^P)$, and such a minimal P is said to be a vertex of N. The sources of N at the vertex P are those components of $\operatorname{Res}_P^H(N)$ which are $N_G(P)$-conjugate to each other and correspond to such points γ on E^P (i.e. $(E^P)^*$-conjugacy classes γ of primitive idempotents in E^P) that there is $j \in \gamma$ such that $ij = j = ji$; of course these γ's are conjugate to each other by $N_G(P)$.

How is the information related to the corresponding endomorphism algebras similar to 1.3.1 ? It is natural to look for something in interior algebras similar to the induced modules and to the sources of modules. The solution was due to Puig, see 3.7 and 4.8.

1.4 All the observations convince us that the all components and all direct indecomposable decompositions of all restricted modules of an $\mathcal{O}G$-module M are determined by the conjugacy classes β's of the primitive idempotents of E^H with H running over the subgroups of G, where the $\mathcal{O}$-algebra $E = \operatorname{End}_\mathcal{O}(M)$ has an "interior" G-structure, i.e. a group homomorphism $\psi : G \to E^*$; i.e. E is an interior G-algebra in Puig's notation. Thus the more important thing for us is not the individual group H, it is the pair (H, β) consisting of a group H and a point β on E^H, we denote it by H_β for brevity, and call it a pointed group on the interior G-algebra E. Hence our contention can be stated as:

The information of the pointed groups on the interior G-algebra $\operatorname{End}_\mathcal{O}(M)$ covers the information of the components of all restricted modules of M, yet some concepts, e.g. inductions and sources, should be developed for pointed groups.

2   Historical Remarks from Block Theory

The more streightforward inspiration of the theory of pointed groups is the research on blocks, especially the nilpotent blocks.

Green [14] initiated new research on defect groups of primitive idempotents in certain fixed element subalgebras of a G-algebra with trace maps. His work covers the results on defect groups of blocks and the vertices of modules.

Brauer [3] and [4] studied the blocks with abelian defect groups of inertial index 1 and gave a construction of characters with p-sections.

The study of nilpotent blocks was the key step for the theory of pointed groups.

Recall that the Sylow Theorem is a very fundamental result on finite groups; it provides a framework in p-subgroups. With it a celebrated theorem is:

2.1 Frobenius Theorem. *A finite group G is a p-nilpotent group if and only if $N_G(Q)/C_G(Q)$ is a p-group for every p-subgroup Q of G.*

Here $N_G(Q)/C_G(Q)$ may be regarded as an automorphism group induced by G; and $N_G(Q)/(Q \cdot C_G(Q))$ is just the corresponding exterior automorphism group. Let us say a little more about them. In fact for any two p-subgroups Q and R of G we can consider the exterior homomorphisms from Q to R induced by suitable elements of G.

2.2 *Remark.* Generally speaking, an exterior homomorphism between two algebraic objects (groups, algebras etc.) is an equivalence class of homomorphisms with respect to the equivalence relation that one can be obtained from another by composing with an interior automorphism of the domain and with an interior automorphism of the codomain. An exterior homomorphism is called an "exomorphism" for brevity in Puig's notaion. In fact, from 1.2 we have seen why the so-called "exomorphisms" should be considered. In many cases we are concerned with conjugacy classes of idempotents instead of individual idempotents.

If a composition of two exomorphisms is still an exomorphism, we say the exomorphisms can be composed, and then we have a category of the objects and the exomorphisms. In many cases an exomorphism can be onto-into decomposed, i.e. it is a composition of an epic exomorphism and a monic exomorphism (including the inclusion map usually). For example, group exomorphisms are such ones. These ideas are developed very well in the theory of pointed groups.

Returning to considering the Frobenius Theorem, we show what is going on with the idea.

For two p-subgroups Q and R of G, by $E(Q, R)$ we denote the set of all exomorphisms from Q to R, while by $E_G(Q, R)$ we denote the set of all exomorphisms from Q to R induced by G (which are called G-exomorphisms for short). Now let $N_G(Q, R) = \{x \in G \mid Q^x \subseteq R\}$, it is easy to see that $N_G(Q, R)$ is a union of some double cosets of $(Q \cdot C_G(Q), R \cdot C_G(R))$, and

$$E_G(Q, R) = (Q \cdot C_G(Q))\backslash N_G(Q, R)/(R \cdot C_G(R)) = C_G(Q)\backslash N_G(Q, R)/R$$

And it can be shown that the condition of the Frobenius Theorem is equivalent to the following one:

"$E_G(Q, R) = E_P(Q, R)$ for any two subgroup Q and R contained in P where P is a Sylow p-subgroup of G."

2.3 The p-local structures.

From now on, b denotes an $\mathcal{O}$-block of G. We also say that b is an $\mathcal{O}G$-block, and denote the block idempotent also by b for convenience; hence $\mathcal{O}Gb$ is the block algebras.

Alperin and Broué [1] developed a p-structure theory. The objects in the theory are no longer individual p-subgroups, they are pairs (Q, b_Q) with Q being a p-subgroup and b_Q being an $\mathcal{O}C_G(Q)$-block such that $\mathrm{Br}_Q(b) \cdot \overline{b_Q} = \overline{b_Q}$, where Br_Q is the Brauer homomorphism from $(\mathcal{O}G)^Q$ to $kC_G(Q)$. Such a pair (Q, b_Q) is called a b-Brauer pair. A suitable partial order relation, named inclusion between b-Brauer pairs, can be defined: we write $(Q, b_Q) \subset (R, b_R)$ if $Q \subseteq R$ and there is a primitive idempotent $i \in (\mathcal{O}G)^R$ such that $\mathrm{Br}_Q(i)\overline{b_Q} \neq 0 \neq \mathrm{Br}_R(i)\overline{b_R}$; see [6, 1.9]. And it was proved that the all maximal b-Brauer pairs are G-conjugate to each other; for further information see 4.7. Hence a maximal b-Brauer pair is said to be a Sylow b-Brauer pair. It is clear that this structure is much finer than what is provided by the Sylow Theorem.

Note that [6] extended the p-structure theory to G-algebras.

Starting from the p-structure, Broué and Puig [7] initiated the research on nilpotent blocks.

From the point of view of the Frobenius Theorem it is natural to define:

2.4 Definition. An $\mathcal{O}$-block b of G is said to be nilpotent if $N_G(Q, b_Q)/C_G(Q)$ is a p-group for every b-Brauer pair (Q, b_Q), where $N_G(Q, b_Q) = \{x \in G \mid Q^x = Q \text{ and } (b_Q)^x = b_Q\}$.

Remark. It will be redefined from Puig's new theory; see 2.9 and 6.10.

Using the techniques on characters of height zero in [5], Broué and Puig [7] proved a series of nice properties of nilpotent blocks, one of which is as follows (Note that $\mathcal{O}$ contains the primitive $|G|$th-root of unity in the theorem):

2.5 Thoerem. *If b is a nilpotent block and χ is a character of height zero of G in b, then the map $\eta \mapsto \chi * \eta$ gives an isometry from the group of the virtual characters of P onto the group of the virtual characters of G in b.*

This is of course a satisfactory result; an isometry is not only an isomorphism, but it also preserves scalar products on its domain and codomain.

(2.5.1) Remark on "$*$-structure". A key construction is "$\chi * \eta$"; sometimes it is called the "star–structure" in Broué's notation. It is a construction of characters based on the p-structure of the block b inspired by Brauer [4]. Later on, Puig [19] related it to a construction of characters based on his local pointed groups; see 5.7.

A straightforward consequence of the theorem is as follows: If a nilpotent block b has an abelian defect group of inertial index 1, then the block algebra $\mathcal{O}Gb$ is isomorphic to a matrix algebra over $\mathcal{O}P$. Broué guessed that it is true for any nilpotent blocks; later the guess became an easy consequence of Puig's work [23]; see 2.14 and 9.2.1.

2.6   Soon after, Puig came to a new idea.

The group algebra $\mathcal{O}G$ has an "interior G-structure", i.e. a group homomorphism: $G \to (\mathcal{O}G)^*$ $x \mapsto x$; a block idempotent b is a primitive idempotent of $(\mathcal{O}G)^G$; the Brauer homomorphism can be extended to general G-algebras; the b-Brauer pair (Q, b_Q) is related to certain primitive idempotents in $(\mathcal{O}G)^Q$; and so on.

Puig introduced interior G-algebras, pointed groups, local pointed groups, exomorphisms between pointed groups, in particular, induced algebras and source algebras, etc. Instead of considering p-local structures, Puig worked more often in the p-local categories(see section 6) and founded the basis of the theory. Even from the point of view of module theory, the new theory provided a fresh and concise unification of information about vertices and defect groups (e.g. see 4.5, 4.6). Many additional new results were gotten.

2.7   In the new way Puig extended the Brauer Second Main Theorem, and modified the $*$-structure to create a construction of characters based on local pointed groups; see 5.7.

2.8   The Puig's construction of characters is in fact dependent only on the so-called local pointed elements, see 5.6; in other words, just on the cyclic p-groups if one looks only at the groups in the pointed groups. Broué and Puig considered the possibility of construction something with all b-Brauer pairs before. Benson and Parker [2] suggested the possibility. Puig [22] analyzed virtual modules and gave a contruction of modules in Green rings.

2.9 It was a great success of the theory of pointed groups that with it, research on nilpotent blocks was developed deeply and reached its high peak. The nilpotent blocks are set in the new framework. The source algebra of a nilpoten block is given so satisfactorily and so precisely that all the structures of nilpotent blocks are known completely. Early in 1981 Puig [20] announced his results; his arguments at that time were dependent on the result [7]. Later Puig developed his ideas and techniques and understood more deeply what was going on; in [23] he completed the research on nilpotent blocks, and [7] became one of the consequences of the new theory.

2.10 Moreover, Külshammer and Puig [16] gave source algebras of extensions of nilpotent blocks.

2.11 The work [23] was done under the assumption that k is algebraically closed; though it was known that k need not be so large. The author discovered that the situation is somewhat different when the ground-field k is too small: two conditions which are equivalent to each other when k is large enough are no longer equivalent, see 6.8, 6.9 and 9.2.5 below; and local control (or, in other words, the local category of the block algebra is equivalent to the local category of its source algebra) is the essencial condition on the nilpotent blocks. And, based on [23], the source algebras of the nilpotent blocks over arbitrary ground-fields were given in [10]; see 9.2.4.

2.12 In [11] the so-called relative local control is considered and the source algebras in this case are described; as an application, it is shown how to understand the result in [28] from the point of view of Puig's theory.

2.13 What about the nilpotent condition in the sense of Frobenius (cf. 2.1 and 2.11) in small groud-fields? Such a consideration leads to a study of blocks which are decomposed into nipotent blocks in suitable coefficient extensions. It is finished in [12].

2.14 Let b and b' be two $\mathcal{O}$-blocks of G and G' resp. Assume that the block algebras of b and b' are Morita equivalent to each other, and assume that $\mathcal{O}$ is of characteristic zero and k is perfect. Puig [26] proved that if one of b and b' is nilpotent then so is the other. As a consequence, if $\mathcal{O}$ is as above and k is large enough, then an $\mathcal{O}$-block b of G with defect group P is nilpotent if and only if its block algebra is isomorphic to a full matrix algebra over $\mathcal{O}P$.

3. Interior Group Algebras

Part of the theory of pointed groups can be stated for G-algebras; however, many key constructions and results have been done only in interior G-algebras upto now.

Let A be an $\mathcal{O}$-algebra.

A is said to be a G-algebra if there is a group homomorphism $\psi : G \to \text{Aut}(A)$

where $\mathrm{Aut}(A)$ denotes the $\mathcal{O}$-algebra automorphism group of A; if this is the case we also say that G acts on A, and write $\psi(x^{-1}) \cdot a = a^x$ for $x \in G$ and $a \in A$.

3.1 Definition. A is said to be an interior G-algebra if there is a group homomorphism $\psi : G \to A^*$ where A^* denotes the multiplicative group of A (equivalently, a unitary algebra homomorphism $\psi : \mathcal{O}G \to A$); then ψ is called the interior G-structure map. For convenience we denote the element $\psi(x)$ in A for $x \in G$ also by $x \in A$ (in fact $x \in A^*$).

An interior G-algebra A is of course a G-algebra in a natural way: $a^x = x^{-1}ax$. It is just in this sense if we write A^G etc. for an interior G-algebra. However, an interior G-algebra A has a much richer structures than a G-algebra; A is a two-sided $\mathcal{O}G$-module in a natural way, and in fact we have ([21, 1.1])

(3.1.1). *An $\mathcal{O}$-algebra A is an interior G-algebra if and only if A is a two-sided $\mathcal{O}G$-module and $(a \cdot x)a' = a(x \cdot a')$ for all $a, a' \in A$ and $x \in G$ where $a \cdot x$ and $x \cdot a$ mean the module operation. If this is the case, the interior G-structure map is $\psi : G \to A^*, x \mapsto x \cdot 1_A$ (note that $x \cdot 1_A = 1_A \cdot x$ by the condition).*

(3.1.2) Remark. From a G-algebra A we can construct an interior G-algebra. Let AG be the free A-module of basis G, and define a distributive product on AG as follows:

$$(a_1 x_1)(a_2 x_2) = a_1 a_2^{x_1^{-1}} x_1 x_2 \quad \text{for} \quad a_1, a_2 \in A \quad \text{and} \quad x_1, x_2 \in G$$

Then AG is an interior G-algebra with structural map: $x \mapsto 1_A x$. We call AG the semidirect product of the G-algebra A and the group G. Note that, if A is itself an interior G-algebra, then $AG \cong A \otimes_{\mathcal{O}} \mathcal{O}G$ as interior G-algebras, where the latter has the diagonal map $x \mapsto x \otimes x$ as its structural map.

(3.1.3)Remaark. Let A be a G-algebra. Assume that A is isomorphic to a matrix algebra $\mathcal{M}_n(\mathcal{O})$ (A is said to be $\mathcal{O}$-simple in this case).

(3.1.3.1) If $(n, |G|) = 1$, then A can be uniquely lifted to an interior G-algebra (i.e. $x^{-1}ax = a^x$ for $a \in A$ and $x \in G$ is just the original G-action on A).

(3.1.3.2) If k is algebraically closed, then there is a central extension $\widehat{G}$ of G by a finite p'-group C and A can be lifted to an interior $\widehat{G}$-algebra such that C is mapped to the center of A and the action of $\widehat{G}$ on A is the same as the original G-action on A.

3.2 Trace maps and Brauer maps (Though we state results for G-algebras, they are valid even for G-modules). Let A be a G-algebra and H be a subgroup of G. Then we have the trace map

$$\mathrm{Tr}_H^G : \qquad A^H \to A^G \qquad a \mapsto \sum_{x \in G/H} a^x$$

which has many nice properties, for example:

(3.2.1) Mackey decomposition. *For subgroups H and K of G, and $a \in A^H$*

$$\mathrm{Tr}_H^G(a) = \sum_{x \in H \backslash G / K} \mathrm{Tr}_{H^x \cap K}^K(a^x)$$

Denote $\mathrm{Tr}_H^G(A^H) = A_H^G$, and for $Q \leq G$ denote

$$A(Q) = A^Q / (\sum_{R \lneq Q} A_R^Q + J(\mathcal{O}) \cdot A^Q)$$

which is a quotien algebra and can be regarded as an $N_G(Q)/Q$-algebra.

It is clear that $A(Q) = 0$ if Q is not a p-subgroup since any p'-integer is invertible in $\mathcal{O}$.

Let Q be a p-subgroup of G. The natural homomorphism $\quad \mathrm{Br}_Q : A^Q \to A(Q)$ is called the Brauer map with respect to Q.

(3.2.2) Proposition. *If A has a G-stable $\mathcal{O}$-basis B, then*

$$A^Q = \mathcal{O}(B^Q) \oplus \sum_{R < Q} A_R^Q \qquad \text{(as $\mathcal{O}$-modules)}$$

and

$$\overline{A}^Q = k(B^Q) \oplus \sum_{R < Q} \overline{A}_R^Q \qquad \text{(as k-spaces)}$$

hence the Q-fixed point set B^Q is a k-basis of $A(Q)$. In particular. $A(Q) \doteq kC_G(Q)$ if $A = \mathcal{O}G$ is the group algebra.

3.3 It is well-known which are algebra homomorphisms and which are G-algebra homomorphisms.

(3.3.1) Definition. An interior G-algebra homomorphism f from an interior G-algebra A to an interior G-algebra A' is both an $\mathcal{O}$-algebra homomorphism and a two-sided $\mathcal{O}G$-module homomorphism.

3.4 Note that, however, in the theory of pointed groups an algebra homomorphism $f : A \to A'$ is not necessarily unitary. However, $f(1_A)$ is always an idempotent of A', and $f(1_A) \cdot A' \cdot f(1_A)$ is an subalgebra of A' with unity $f(1_A)$. We have seen in 1.3.1 the module-theoretic interest of the following concept.

(3.4.1) Definition. f is said to be an embedding if $\mathrm{Im}(f) = f(1_A) \cdot A' \cdot f(1_A)$ and $\mathrm{Ker}(f) = \{0\}$; if f is an embedding, then we say that A is an embedded subalgebra of A'.

(3.4.2) Remark. Let $f : A \to A'$ be an embedding of algebras. Then $J(f(A)) = f(A) \cap J(A')$, and $i \in A$ is a primitive idempotent if and only if $f(i) \in A'$ is a primitive idempotent; further, if f is an embedding of G-algebras and $H \leq G$, then $f^H : A^H \to A'^H$ is still an embedding and $f(A_H^G) = f(A) \cap A_H'^G$

As in 2.2, we are concerned with the exomorphisms between G-algebras or between interior G-algebras.

3.5 Definition. Let $f : A \to A'$ be an interior G-algebra homomorphism. Then the exomorphism $\tilde{f}$ including f is the set of all compositions of f with an interior automorphism of A and an interior automorphism of A'. But note that an interior automorphism of an interior G-algebra A is a conjugation by an element in $(A^G)^*$.

If f is an embedding, then $\tilde{f}$ is called an exterior embedding.

It is easy to see that the composition $\tilde{f}\tilde{g}$ of two exomorphisms $\tilde{f}$ and $\tilde{g}$ of interior G-algebras is still an exomorphism of interior G-algebras ([22, 2.3]) and $\tilde{f}\tilde{g} = \widetilde{fg}$.

Thus we have a category of interior G-algebras (or G-algebras) with exomorphisms as morphisms. A commutative diagram in this category is said to be exterior commutative or said to be a commutative diagram of exomorphisms.

An easy but useful fact is ([21,1.5]):

(3.5.1) Lemma. Let the following be an exterior commutative diagram of (G- or interior G-) algebras with $\tilde{g}$ being an embedding. Then $\tilde{f}$ is determined by $\tilde{h}$ and $\tilde{f}$ is an embedding if and only if $\tilde{h}$ is an embedding.

$$
\begin{array}{ccc}
A & \xrightarrow{\tilde{h}} & A'' \\
\tilde{f}\downarrow & & \| \\
A' & \xrightarrow{\tilde{g}} & A''
\end{array}
$$

3.6   Restricted algebras.

Let A be a G-algebra with structure map $\psi : G \to \mathrm{Aut}(A)$. For any subgroup H of G we have the restricted H-algebra $\mathrm{Res}_H^G(A)$. In fact, for any group homomorphism $\varphi : H \to G$ we have the restricted H-algebra $\mathrm{Res}_\varphi(A)$ with the composition map $H \xrightarrow{\varphi} G \xrightarrow{\psi} \mathrm{Aut}(A)$ as its structure map. If A is an interior G-algebra with interior G-structure map $\psi : G \to A^*$, in the same way we define the restricted algebras. For a group homomorphism $\varphi : H \to G$, the restricted interior H-algebra $\mathrm{Res}_\varphi(A)$ has the composition map $H \xrightarrow{\varphi} G \xrightarrow{\psi} A^*$ as its structure map.

If $\tilde{f}$ is an exomorphism of interior G-algebras, then any two homomorphisms of $\tilde{f}$ are clearly located in one and the same exomorphism of the restricted interior H-algebras. Let $\mathrm{Res}_\varphi(\tilde{f})$ denote the exomorphism of the restricted interior H-algebras containing $\tilde{f}$. In other words, we get a functor Res_φ from the category of interior G-algebras to the category of interior H-algebras. It is clear that $\mathrm{Res}_{\psi\varphi} = \mathrm{Res}_\varphi \circ \mathrm{Res}_\psi$. These are of course still valid for G-algebras. But one of the distinguished features for interior G-algebras is ([21, 1.3]):

(3.6.1) Lemma. Let $\varphi : H \to G$ be a group homomorphism, let $\tilde{f}$ and $\tilde{g}$ be two interior G-algebra exomorphisms from an interior G-algebra to another one. Then $\tilde{f} = \tilde{g}$ if and only if $\mathrm{Res}_\varphi(\tilde{f}) = \mathrm{Res}_\varphi(\tilde{g})$.

3.7 Induced algebras.

A natural question is how to construct induced algebras. We can follow the structure of the $\mathcal{O}$-endomorphism algebra of an induced module. However the induced algebras can be constructed only for interior G-algebras.

Let H be a subgroup of G and B be an interior H-algebra. The induced interior G-algebra $\mathrm{Ind}_H^G(B)$ is defined to be the two-sided $\mathcal{O}G$-module $\mathcal{O}G \otimes_{\mathcal{O}H} B \otimes_{\mathcal{O}H} \mathcal{O}G$ endowed with the distributive product:

$$(x \otimes b \otimes y)(x' \otimes b' \otimes y') = \begin{cases} x \otimes byx'b' \otimes y' & \text{if } yx' \in H \\ 0 & \text{if } yx' \notin H \end{cases}$$

for $x, y, x', y' \in G$ and $b, b' \in B$; the interior G-structure is

$$z \longmapsto \sum_{x \in G/H} zx \otimes 1 \otimes x^{-1} \qquad \text{for} \qquad z \in G$$

It is can be checked by 3.1.1 that this is an interior G-algebra. Notice that the unity of $A = \mathrm{Ind}_H^G(B)$ is

$$1_A = \sum_{x \in G/H} x \otimes 1_B \otimes x^{-1} = \mathrm{Tr}_H^G(1_G \otimes 1_B \otimes 1_G)$$

One can see that it is as described in 3.1.1 that the structural map is $z \mapsto z \cdot 1_A$.

It is easy to prove that

(3.7.1) Proposition. *Let $H \leq G$ and N be an $\mathcal{O}H$-module. Then*

$$\mathrm{Ind}_H^G(\mathrm{End}_{\mathcal{O}}(N)) \cong \mathrm{End}_{\mathcal{O}}(\mathrm{Ind}_H^G(N))$$

as interior G-algebras.

3.8 Can the induced algebra $\mathrm{Ind}_\varphi(B)$ be constructed from an interior H-algebra B for any group homomorphism $\varphi : H \to G$? The answer is yes, but the construction is a little more complicated. The work [26] on Morita equivalences for nilpotent blocks realized necessary conditions for such construction. Puig discussed related questions. The statement similar to 3.7.1 is still true.

(3.8.1) Proposition. *Let $\varphi : H \to G$ be a group homomorphism and N be an $\mathcal{O}H$-module (remember that N is $\mathcal{O}$-free). Then*

$$\mathrm{Ind}_\varphi(\mathrm{End}_{\mathcal{O}}(N)) \cong \mathrm{End}_{\mathcal{O}}(\mathrm{Ind}_\varphi(N))$$

as interior G-algebras.

3.9 Now let $H \leq G$ and B be an interior H-algebra. There is a natural interior H-algebra embedding

$$d_H^G(B) : B \to \mathrm{Res}_H^G \mathrm{Ind}_H^G(B) \qquad b \mapsto 1 \otimes b \otimes 1$$

In fact it is a special case of

(3.9.1) Mackey decomposition. *For subgroups H, K of G and an interior H-algebra B, the unity of $\operatorname{Res}_K^G \operatorname{Ind}_H^G(B)$ has the following orthogonal (not primitive in general) idempotent decomposition:*

$$\operatorname{Tr}_H^G(1 \otimes 1 \otimes 1) = \sum_{x \in H \backslash G / K} \operatorname{Tr}_{H^x \cap K}^K(x^{-1} \otimes 1 \otimes x)$$

The Mackey decomposition of modules can be deduced from it.

Let $f : B' \to B$ be a homomorphism of interior H-algebras. It is clear that

$$\operatorname{Ind}_H^G(f) = 1 \otimes f \otimes 1 : \operatorname{Ind}_H^G(B') \longrightarrow \operatorname{Ind}_H^G(B)$$

is a homomorphism of interior G-algebras and the following diagram

$$
\begin{array}{ccc}
\operatorname{Ind}_H^G(B') & \xrightarrow{\operatorname{Ind}_H^G(f)} & \operatorname{Ind}_H^G(B) \\
{\scriptstyle d_H^G(B')} \uparrow & & \uparrow {\scriptstyle d_H^G(B)} \\
B' & \xrightarrow{\quad f \quad} & B
\end{array}
$$

is commutative. If $v \in B^*$, then $\operatorname{Tr}_H^G(v) = \sum_{x \in G/H} x \otimes v \otimes x^{-1}$ is invertible in $\operatorname{Ind}_H^G(B)$. With these observations we can see that if f_1 and f_2 are in one and the same exomorphism $\widetilde{f} : B' \to B$, then $\operatorname{Ind}_H^G(f_1)$ and $\operatorname{Ind}_H^G(f_2)$ are in one and the same exomorphism, denoted by $\operatorname{Ind}_H^G(\widetilde{f})$, of the induced algebras which contains the all $\operatorname{Ind}_H^G(f)$ with $f \in \widetilde{f}$.

In other words, we get a functor Ind_H^G from the category of interior H-algebras to the category of interior G-algebras.

4.   Pointed Groups

Let A be a G-algebra over $\mathcal{O}$ in this section except for further assumptions stated.

4.1   If H is a subgroup of G and β is an $(A^H)^*$-conjugacy class of primitive idempotents in A^H, then the pair (H, β), denoted by H_β in short, is called a pointed group on A; and β is called a point of H on A (cf. 1.2.3 and 1.4). The set of all points of H on A is denoted by $\mathcal{P}(A^H)$ or $\mathcal{P}_A(H)$. It is clear that $(H_\beta)^x = (H^x)_{\beta^x}$ for $x \in G$ is also a pointed group.

In a natural way (cf. 1.3) a relation of inclusion of pointed groups can be defined. Let K_γ and H_β be two pointed groups on A. We say that K_γ is contained in H_β, denoted by $K_\gamma \subset H_\beta$, if $K \subseteq H$ and there is $j \in \gamma$ and $i \in \beta$ such that $ij = j = ji$.

Since the β in a pointed group H_β is a conjugacy class of primitive idempotents in A^H, it corresponds to a unique simple factor in $A^H / J(A^H)$, denote the simple factor by $A(H_\beta)$; and the image of β in $A(H_\beta)$ are just the all primitive idempotents of $A(H_\beta)$. Then $A(H_\beta)$ is a matrix algebra (maybe over a division k-algebra); denote the order of the matrix by m_β, which is just equal to the number of the summands contained in β of an orthogonal primitive idempotent decomposition of 1_A in A^H.

So $A(H_\beta)$ is called the multiplicity algebra of H_β and m_β is called the multiplicity of β ([22, 2.10]). Now denote by $s_\beta : A^H \to A(H_\beta)$ the natural homomorphism. It is clear that $K_\gamma \subset H_\beta$ if and only if $s_\gamma(\beta) \neq \{0\}$.

Notice that A^H may be an $N_G(H)$-algebra while $A(H_\beta)$ may be an $N_G(H_\beta)$-algebra; since H acts trivially on them, they can be regarded as an $N_G(H)/H$-algebra and an $N_G(H_\beta)/H$-algebra resp.

However, if A is an interior G-algebra, the A^H and $A(H_\beta)$ can be regarded only as interior $C_G(H)$-algebras.

4.2 The behavior of pointed groups under homomorphisms is somewhat complicated; For it Puig developed a multiplicity technique in [19]. The situation becomes simple in two cases.

Let A and A' be two G-algebras and $\tilde{f} : A \to A'$ be an exterior embedding. Then by 3.4.2 for any pointed group H_β on A there is a unique pointed group $H_{\beta'}$ on A' such that $f(\beta) \subseteq \beta'$ (it is independent of the choice of $f \in \tilde{f}$) and $\tilde{f}$ induces an interior $C_G(H)$-algebra embedding $\tilde{f} : A(H_\beta) \to A'(H_{\beta'})$, hence we can denote β' and β by the one and the same symbol. In other words, $\tilde{f}$ induces a map from the pointed groups on A to the pointed groups on A'. Further discussion is continued in 6.4.

Another case is the so-called coverings which play an important role in research on nilpotent blocks; see Section 7.

4.3 Embedded algebras of pointed groups.

Associated with every pointed group H_β on A there is an H-algebra $A_\beta = iAi$ with $i \in \beta$. Since β is an $(A^H)^*$-conjugacy class of primitive idempotents, A_β is well-defined up to the exterior H-algebra embedding $\tilde{f_\beta} : A_\beta \to \mathrm{Res}_H^G(A)$ such that $f(1_{A_\beta}) \in \beta$. The pair $(A_\beta, \tilde{f_\beta})$ is called the embedded algerba of H_β.

Note that for pointed group H_β the $(A_\beta)^H$ is a local algebra and i is the unique non-zero idempotent of $iA^H i$. Such an interior H-algebra is said to be primitive.

With the embedded algerbas we can characterize the inclusion relationship of pointed groups as follows([21, 1.8 and 1.9]).

(4.3.1) Lemma. *If $K \subseteq H$, then $K_\gamma \subset H_\beta$ if and only if there is an exomorphism $\tilde{f_\gamma^\beta} : A_\gamma \to \mathrm{Res}_K^H(A_\beta)$ such that the following diagram is exterior commutative.*

$$
\begin{array}{ccc}
A_\gamma & \xrightarrow{\ \tilde{f_\gamma}\ } & \mathrm{Res}_K^G(A) \\[4pt]
{\scriptstyle \tilde{f_\gamma^\beta}}\big\downarrow & & \big\| \\[4pt]
\mathrm{Res}_K^H(A_\beta) & \xrightarrow[\ \tilde{f_\beta}\ ]{} & \mathrm{Res}_K^G(A)
\end{array}
$$

and $\tilde{f_\gamma^\beta}$ is an embedding uniqurly determined if this is the case.

Let $K_\gamma \subset H_\beta \subset G_\alpha$ be pointed groups on A. It is clear that

$$\tilde{f_\gamma^\alpha} = \mathrm{Res}_K^H(\tilde{f_\beta^\alpha}) \circ \tilde{f_\gamma^\beta}$$

because the compositions of both sides with $\widetilde{f_\alpha}$ are equal to the same embedding $\widetilde{f_\gamma}$ (cf. 3.5.1).

(4.3.2) Remark. Assume further that A is an interior G-algebra. Then the embedded algebra $A_\beta = iAi$ of a pointed group H_β is an interior H-algebra with the structure map $h \mapsto ihi = ih$ for $h \in H$; and the $\widetilde{f_\beta}$ is an interior H-algebra exterior embedding. And the $\widetilde{f_\gamma^\beta}$ in the above 4.3.1 is also an interior K-algebra embedding. Hence, by 3.6.1, the above lemma can be restated as

$$K_\gamma \subset H_\beta \quad \text{if and only if} \quad \operatorname{Res}_1^K(\widetilde{f_\gamma}) = \operatorname{Res}_1^H(\widetilde{f_\beta}) \circ \operatorname{Res}_1^K(\widetilde{f_\gamma^\beta})$$

We will consider several further connections between pointed groups with embedded algerbas like the above in section 6.

The following proposition ([19, 3.6]) is related to the relative projectivity of modules, compare with 1.3. By the way, for a pointed group H_β on A a two sided ideal $A^H \beta A^H$ of A^H is uniquely determined.

(4.3.3) Proposition. *Let $K_\gamma \subset H_\beta$ be two pointed groups on A. Then $\beta \subseteq \operatorname{Tr}_K^H(A^K \gamma A^K)$ if and only if there is an exomorphism $A_\beta \to \operatorname{Ind}_K^H(A_\gamma)$ such that the following diagram is exterior commutative*

$$
\begin{array}{ccc}
A_\gamma & \!\!\!\!=\!\!\!\!=\!\!\!\!=\!\!\!\! & A_\gamma \\
\widetilde{f_\gamma^\beta} \downarrow & & \downarrow d_K^H(A_\gamma) \\
\operatorname{Res}_K^H(A_\beta) & \longrightarrow & \operatorname{Res}_K^H \operatorname{Ind}_K^H(A_\gamma)
\end{array}
$$

and the exomorphism $A_\beta \to \operatorname{Ind}_K^H(A_\gamma)$ is an embedding uniquely determined if it exists.

Further developement will be given in 4.9.

4.4 Definition. Let A be a G-algebra over $\mathcal{O}$, and Q_δ be a pointed group on A. If $\operatorname{Br}_Q(\delta) \neq \{0\}$, then we say that Q_δ is local and δ is a local point of Q on A. By $\mathcal{LP}_A(Q)$ we denote the set of all local points of Q on A.

The Q must be a p-subgroup of G if Q_δ is a local pointed group, see 3.2.

The next theorem ([19, 1.2]) describes the structure of all local pointed groups just like the Sylow Theorem does, and it covers the idea of the vertices of modules, Green's idea on defect groups and the idea of the Brauer pairs as well.

4.5 Theorem. *Let G_α be a pointed group on a G-algebra A. Then*

(4.5.1). All maximal local pointed groups contained in G_α are conjugate to each other by G.

(4.5.2). A pointed group P_γ is minimal such that $\alpha \subseteq \operatorname{Tr}_P^G(A^P \gamma A^P)$ if and only if it is a maximal local pointed group contained in G_α.

(4.5.3) Definition. By defect pointed groups of G_α we mean the maximal local pointed groups contained in G_α. If P_γ is a defect pointed group of G_α, we also say that P_γ is a defect pointed group of the interior G-algebra A_α.

The following statement ([19, 1.3]) is of interests:

4.6 Lemma. *Let* Q_δ *be a local pointed group contained in* G_α. *Set* $N = N_G(Q_\delta)/Q$. *Then*

$$s_\delta(\mathrm{Tr}_Q^G(a)) = \mathrm{Tr}_1^N(s_\delta(q)) \qquad for \quad a \in A^Q \delta A^Q$$

$$s_\delta(A_Q^G) = A(Q_\delta)_1^N$$

One of its corollaries is ([19, 1.4]):

(4.6.1) Proposition. *Let* $P_\gamma \subset H_\beta \subset G_\alpha$. *If* $H \supseteq N_G(P_\gamma)$, *then* P_γ *is a defect pointed group of* G_α *if and only if* P_γ *is a defect pointed group of* H_β.

In fact, there is only one β satisfying the inclusions in the proposition.

One of the module-theoretic corollaries is the Green's correspondence and its partial inverse: "Assume M is an indecomposable $\mathcal{O}G$-module and $N_G(P) \leq H \leq G$. If one of the components of $\mathrm{Res}_H^G(M)$ is of vertex P, then so is M." This result was proved in [8] independentlly later.

(4.6.2) Remark. With 4.6 Puig gave an easy proof of the main result of [15], see [19, 1.6].

4.7 Connections of local pointed groups with Brauer pairs.

Let $A = \mathcal{O}G$ be the group algebra and $\alpha = \{b\}$ with b an $\mathcal{O}G$-block be a point of G on A.

Let Q_δ be a local pointed group contained in G_α. Since $\{0\} \neq \mathrm{Br}_Q(\delta) \subseteq A(Q) = kC_G(Q)$ and since $\mathrm{Br}_Q : A^Q \to A(Q)$ is surjective, $\mathrm{Br}_Q(\delta)$ is exactly a point on $kC_G(Q)$, cf. 1.2.5; in other words, $(Z(Q))_{\mathrm{Br}_Q(\delta)}$ is a pointed group on $kC_G(Q)$, therefore there is an unique b-Brauer pair (Q, b_δ) such that the pointed group $(C_G(Q))_{\{\overline{b_\delta}\}}$ contains $(Z(Q))_{\mathrm{Br}_Q(\delta)}$ on $kC_G(Q)$.

Conversely for each b-Brauer pair (Q, b_Q), lifting a point on $kC_G(Q)$ contained in $(C_G(Q))_{\{\overline{b_Q}\}}$ to A^Q, we see that there is at least one pointed group Q_δ such that the b-Brauer pair (Q, b_δ) obtained as above coincides with (Q, b_Q), i.e. $b_\delta = b_Q$.

And, from the Brauer First Main Theorem, the defect pointed groups (P_γ) are one-to-one corresponding to the Sylow b-Brauer pairs (P, b_γ).

About the inclusion of b-Brauer pairs (see 2.3), for b-Brauer pairs (Q, b_Q) and (R, b_R), we can show that $(Q, b_Q) \subset (R, b_R)$ if and only if there are local pointed groups Q_δ and R_ϵ contained in G_α corresponding to (Q, b_Q) and (R, b_R) resp. (i.e. $b_\delta = b_Q$ and $b_\epsilon = b_R$) such that $Q_\delta \subset R_\epsilon$, see [19, §1]. Thus, as a consequence of 4.5.1, we reprove that the all maximal b-Brauer pairs are G-conjugate to each other.

4.8 Definition. Let A be an interior G-algebra and G_α be a pointed group on A, and let A_α be the corresponding embedded interior G-algebra of A (see 4.3). Let P_γ be a defect pointed group of G_α, then the embedded interior P-algebra A_γ of P_γ is said to be the source algebra of G_α (and of A_α).

Since all the defect pointed groups form a G-conjugacy class, the source algebra is well-defined up to conjugation; if P is fixed, then it is well-defined up to exerior isomorphism.

For fixed P_γ, we have an interior P-algebra exterior embedding (see 4.3.1):

$$\widetilde{f_\gamma^\alpha} : A_\gamma \to \mathrm{Res}_P^G(A_\alpha)$$

And we have ([19, 3.4]):

4.9 Theorem. *Notations as above. Then*

(4.9.1). There is a unique exterior embedding $A_\alpha \to \mathrm{Ind}_P^G(A_\gamma)$ such that the following diagram is exterior commutative

$$
\begin{array}{ccc}
\mathrm{Res}_P^G(A_\alpha) & \longrightarrow & \mathrm{Res}_P^G \mathrm{Ind}_P^G(A_\gamma) \\
\widetilde{f_\gamma^\alpha} \uparrow & & \uparrow d_P^G(A_\gamma) \\
A_\gamma & =\!=\!=\!= & A_\gamma
\end{array}
$$

(4.9.2). Let $H \leq G$ and B be an interior H-algebra. If $A_\alpha \to \mathrm{Ind}_H^G(B)$ is an exterior embedding, then there is a G-exomorphism $\widetilde{\psi} : P \to H$ and an exterior embedding $\widetilde{f_\psi} : A_\gamma \to \mathrm{Res}_\psi(B)$ such that the following diagram is exterior commutative

$$
\begin{array}{ccc}
\mathrm{Res}_P^G(A_\alpha) & \longrightarrow & \mathrm{Res}_P^G \mathrm{Ind}_H^G(B) \\
\widetilde{f_\gamma^\alpha} \uparrow & & \uparrow \\
A_\gamma & \xrightarrow{\ \widetilde{f_\psi}\ } & \mathrm{Res}_\psi(B)
\end{array}
$$

Remark. It is clear that the first statement can be deduced from 4.3.3 and 4.5. The interest of the theorem is that the A_γ is the "minimal" one such that A_α can be naturally embedded in its induced algebra.

So the source algebra is of great importance; the following is an easy consequence ([19, 3.5]):

(4.9.3). A_α *and* A_γ *are Morita equivalent to each other and hence they have their centers isomorphic to each other.*

However, the role of the source algebra is far greater than this. When one requires only the Morita equivalence, the basic algebra of A_α is enough (take one idempotent from every points of A_α to form an orthogonal system; the embedded subalgebra associated with the idempotent which is the sum of the system is called the basic algebra of A_α). In contrast to this, the source algebra A_γ of A_α contains all the local information on A_α, see 6.2.2 and 6.6.2; the vertex and source of an A_α-module can be computed from the corresponding A_γ-module, etc. In the following we sketch how the characters of A_α are determined by its source algebra.

5. POINTED GROUPS AND CHARACTERS

5.1 Usually modular characters are defined on semisimple parts of an algebra. An $\mathcal{O}$-algebra is said to be $\mathcal{O}$-simple if it is isomorphic to a matrix algebra over $\mathcal{O}$. And a direct product of $\mathcal{O}$-simple algebras is called an $\mathcal{O}$-semisimple algebra. Using techniques for idempotents Puig proved ([19, 2.4]):

5.2 Theorem. *Let A be an $\mathcal{O}$-algebra (remember that it is $\mathcal{O}$-free of finite rank). Then*

(5.2.1). An $\mathcal{O}$-semisimple subalgebra S of A is maximal $\mathcal{O}$-semisimple if and only if $1_S = 1_A$ and an orthogonal primitive idempotent decomposition of 1_S in S is also an orthogonal primitive idempotent decomposition of 1_A in A.

(5.2.2). All maximal $\mathcal{O}$-semisimple subalgebras are A^*-conjugate to each other.

(5.2.3). If $T^a \subseteq S$ for an $\mathcal{O}$-semisimple subalgebra T of a maximal $\mathcal{O}$-semisimple subalgbra S and an $a \in A^*$, then there are $z \in C_A(T)^*$ and $s \in S$ such that $a = zs$.

Taking $\mathcal{O} = k$ to be a splitting field of A, one gets the Wedderburn-Malcev's Theorem from (5.2.1) and (5.2.2), while (5.2.3) is close to the Noether-Skolem Theorem.

The key point for the proof is the fact: If idempotents $i_1, \cdots, i_r$ of A are orthogonal to each other, and if they are conjugate to each other, i.e. there are $v_1, \cdots, v_r \in A^*$ such that $i_t = v_t^{-1} i_1 v_t$ for $t = 1, \cdots, r$, then the $\mathcal{O}$-submodule of A with basis $\{v_s^{-1} i_1 v_t \mid 1 \le s, t \le r\}$ is exactly a subalgebra of A which is isomorphic to the matrix algebra $\mathcal{M}_r(\mathcal{O})$.

The converse is also true. If $S \cong \mathcal{M}_r(\mathcal{O})$ is a subalgebra of A, then $1_S = \sum_{t=1}^{r} v_t^{-1} i v_t$ with $v_t \in S^*$ and $v_t^{-1} i v_t$, $t = 1, \cdots, r$ being orthogonal idempotents. Furthermore, if S is unitary (i.e. $1_S = 1_A$) and M is an A-module, then $A = \sum_{1 \le s, t \le r} i^{v_s} A i^{v_t}$ and $M = \sum_{1 \le t \le r} i^{v_t} M$. With this we can show ([19, 2.1]):

(5.2.4) Lemma. *If S is a unitary $\mathcal{O}$-simple subalgebra of A and M is an A-module, then*

$$A \cong S \otimes_{\mathcal{O}} C_A(S) \cong \mathcal{M}_r(C_A(S))$$

and

$$M \cong Si \otimes_{\mathcal{O}} iM \qquad as \qquad S \otimes_{\mathcal{O}} C_A(S)\text{-module}$$

With 5.2 the Brauer characters of A can be defined.

5.3 Definition. Let S be a maximal $\mathcal{O}$-semisimple subalgebra of an $\mathcal{O}$-algebra A and β be a point (a conjugacy class of primitive idempotents, see 1.2.3) on A. Take $j \in \beta \cap S$ (which is nonempty by 5.2.1). Then Sj is an S-module; its character, denoted by φ_β^A or φ_β for short, is said to be the Brauer character of A associated with the point β.

Remark. By 5.2, the character φ_β is defined on any $\mathcal{O}$-semisimple subalgebra T of A, and $\varphi_\beta(t^v) = \varphi_\beta(t)$ for any $t \in T$ and any $v \in A^*$. In particular, the above definition of φ_β is independent of the choice of S and $j \in \beta$.

Also, the Brauer character φ_β of A coincides with the modular character afforded by the simple A-module associated with the point β if $k = \mathcal{O}/J(\mathcal{O})$ is a splitting field for the simple A-module ([19, 4.2]).

5.4   From now on in this section 5 A is an interior G-algebra.

Let L_ε be a pointed group on A. One must note that the Brauer character φ_ε is a Brauer character of A^L (because ε is a point on A^L, not on A). Since A^L is an interior $C_G(L)$-algebra, the φ_ε is a modular character of $C_G(L)$ if k is large enough.

Let M be an A-module, and let χ be the $\mathcal{O}$-valued character of A afforded by M. For a pointed group H_β on A take an idempotent $j \in \beta$, then jM is an A_β-module where $A_\beta = jAj$ is the embedded interior H-algebra of H_β; hence jM can be regarded as an $\mathcal{O}H$-module whose character, denoted by χ^β, is an $\mathcal{O}$-valued character of H.

Now the Brauer Second Main Theorem can be extended as follows ([19, 4.3]).

5.5 Theorem. *Notations as above. Let u be a p-element of G and $U = \langle u \rangle$. Assume t is an element of an $\mathcal{O}$-semisimple subalgebra of A^U. Then*

$$\chi(ut) = \sum_{\varepsilon \in \mathcal{LP}_A(U)} \chi^\varepsilon(u)\varphi_\varepsilon(t)$$

If k is large enough, then φ_ε runs over the irreducible modular characters of A^U; moreover, in the special case that $A = \mathcal{O}G$ and $t \in C_G(U)$ is a p'-element, the image of t in A^U must be in an $\mathcal{O}$-semisimple subalgebra and $A(U) = kC_G(U)$, so we get the Brauer Second Main Theorem and $\chi^\varepsilon(u)$ is just the generalized decomposition number.

Sketch of the proof.   Take a maximal $\mathcal{O}$-semisimple subalgebra S of A^U such that $t \in S$. Decomposing S into $\mathcal{O}$-simple factors reduces it to the case that $t \in T$ where T is an $\mathcal{O}$-simple subalgebra of A^U associated with ε. Let e be the unity of T. Then eM is a $T \otimes_\mathcal{O} \mathcal{O}U$-module and $\chi(ut)$ is just the value of the character afforded by eM at $t \otimes u$. Take $j \in \varepsilon$ such taht $ej = j$. Then as $T \otimes_\mathcal{O} \mathcal{O}U$-module $eM \cong Tj \otimes_\mathcal{O} jM$ by 5.2.4. Therefore $\chi(ut) = \varphi_\varepsilon(t)\chi^\varepsilon(u)$.

5.6   Let A be an interior G-algebra over $\mathcal{O}$. Assume $\mathcal{O}$ is large enough. Let M be an A-module and χ be its $\mathcal{O}$-valued character. An observation is as follows. In the expression of $\chi(ut)$ in 5.5 φ_ε is determined by A itself, it is independent of χ. For a p'-element s of $C_G(u)$ the group algebra $\mathcal{O}\langle s \rangle$ is $\mathcal{O}$-semisimple and hence s is in an $\mathcal{O}$-semisimple subalgebra of A^U (we have used this fact in 5.5); so $\varphi_\varepsilon(s)$ is defined.

Return to the expression of $\chi(us)$, which is determined by $\chi^\varepsilon(u)$.

Definition. We call u_ε a local pointed element on A if $\langle u \rangle_\varepsilon$ is a local pointed group on A. If $\langle u \rangle_\varepsilon \subset G_\alpha$, we denote it by $u_\varepsilon \in G_\alpha$.

Let G_α be a pointed group on A and P_γ be its defect pointed group. From 5.5 we know that χ^α is an $\mathcal{O}$-valued character of G, and it is in fact determined on its values $\chi^\varepsilon(u)$ at all local pointed elements $u_\varepsilon \in G_\alpha$; but u_ε is conjugate to a local pointed element in P_γ; so the natural question is:

When can an $\mathcal{O}$-valued function λ defined on the local pointed elements of P_γ determine a virtual character λ^α of G ?

Of course, one necessary condition is: "If $u_\varepsilon \in P_\gamma$ and $x \in G$ such that $(u_\varepsilon)^x \in P_\gamma$, then $\lambda((u_\varepsilon)^x) = \lambda(u_\varepsilon)$". Such a function λ is said to be G-stable.

Now let λ be a G-stable $\mathcal{O}$-valued function defined on the local pointed elements of P_γ. For an element $x = us$ of G with p-part u and p'-part s, define

$$\lambda^\alpha(x) = \lambda^\alpha(us) = \sum_{\varepsilon \in \mathcal{LP}_A(u)} \lambda^\varepsilon(u)\varphi_\varepsilon(s)$$

where

$$\lambda^\varepsilon(u) = \begin{cases} \lambda((u_\varepsilon)^x) & \text{if} \quad (u_\varepsilon)^x \in P_\gamma \quad \text{for some} \quad x \in G \\ 0 & \text{otherwise} \end{cases}$$

It is clear that λ^α is an $\mathcal{O}$-valued class function on G. The following is [19, 5.2].

5.7 Theorem. *Notations as above. If $(\lambda^\alpha|_Q, \rho)_Q$ is a rational integer for every subgroup Q of P and every $\mathcal{O}$-valued linear character ρ of Q, then λ^α is a virtual character of G.*

Therefore the $*$-structure in 2.5.1 becomes a corollary [19, 5.3] of the above Theorem 5.7.

6.   Local Category and Local Fusion Category

In this section we present some deep theory on pointed groups inspired by research on the nilpotent blocks; then we give the definition of nilpotent blocks from the new point of view.

6.1 G-exomorphisms.

Let A be an interior G-algebra over $\mathcal{O}$, let K_γ and H_β be pointed groups on A.

Since pointed groups are not individual groups, group exomorphisms are not enough for them. It is reasonable to define:

(6.1.1) Definition. A G-exomorphism $\widetilde{\varphi}$ from K_γ to H_β is a group exomorphism $\widetilde{\varphi} : K \to H$ induced by an $x \in G$ (i.e. $\varphi(y) = y^x$ for all $y \in K$) such that $(K_\gamma)^x \subset H_\beta$. By $E_G(K_\gamma, H_\beta)$ we denote the set of all G-exomorphisms from K_γ to H_β.

The G-exomorphisms can be composed and can be onto-into decomposed ([21, 2.2 and 2.11]).

Just as in 4.3.1 and 4.3.2 we can characterize the G-exomorphisms by embedded algebras of the pointed groups. Note two facts. For any $K^x \leq L \leq G$ we have a group exomorphism $\widetilde{\varphi} : K \to L$ induced by x (cf. 2.2). For any embedded interior K-algebra B of A, $\widetilde{x} : B \to \mathrm{Res}_\varphi(B^x)$ is an exomorphism induced by conjugation with x.

(6.1.2) Lemma. *Let* $\widetilde{\varphi} : K \to H$ *be a group exomorphism induced by* $x \in G$. *Then* $\widetilde{\varphi} \in E_G(K_\gamma, H_\beta)$ *if and only if there is an interior K-algebra exomorphism* $\widetilde{f_\varphi} : A_\gamma \to Res_\varphi(A_\beta)$ *such that the diagram*

$$
\begin{array}{ccc}
A_\gamma & \xrightarrow{\;\widetilde{f_\varphi}\;} & Res_\varphi(A_\beta) \\
\widetilde{f_\gamma} \downarrow & & \downarrow {\scriptstyle Res_\varphi(\widetilde{f_\beta})} \\
Res_K^H(A) & \xrightarrow[\;\widetilde{x}\;]{} & Res_\varphi(A)
\end{array}
$$

is exterior commutative, if and only if (by 3.6.1)

(6.1.2.1) $$Res_1^K(\widetilde{f_\gamma}) = Res_1^H(\widetilde{f_\beta}) \circ Res_1^K(\widetilde{f_\varphi})$$

and $\widetilde{f_\varphi}$ *is a uniquely determined embedding if this is the case.*

Proof. Assume $(K_\gamma)^x \subset H_\beta$. Let $A_\gamma = jAj$ with $j \in \gamma$ and $A_\beta = iAi$ with $i \in \beta$ such that $j^x i = j^x = ij^x$. By 4.3.1, $\widetilde{f_{\gamma^x}} = Res_{K^x}^H(\widetilde{f_\beta}) \circ f_{\gamma^x}^\beta$. Applying the functor Res_φ, this becomes (note $Res_\varphi Res_{K^x}^G = Res_\varphi$)

$$Res_\varphi(\widetilde{f_{\gamma^x}}) = Res_\varphi(\widetilde{f_\beta}) \circ Res_\varphi(\widetilde{f_{\gamma^x}^\beta})$$

Composing both sides with $\widetilde{x}$ and noting that $Res_\varphi(\widetilde{f_{\gamma^x}}) \circ \widetilde{x} = \widetilde{x} \circ \widetilde{f_\gamma}$, we get the commutative diagram if we set $\widetilde{f_\varphi} = Res_\varphi(f_{\gamma^x}^\beta) \circ \widetilde{x}$.

Conversely, assume that $\widetilde{x} \circ \widetilde{f_\gamma} = Res_\varphi(\widetilde{f_\beta}) \circ \widetilde{f_\varphi}$. Let $\widetilde{\psi} : K^x \to K$ be the group exomorphism induced by conjugation by x^{-1}. Similarly to the above, applying the functor Res_ψ and setting $f_{\gamma^x}^\beta = Res_\psi(\widetilde{f_\varphi}) \circ \widetilde{x^{-1}}$, we get $\widetilde{f_{\gamma^x}} = Res_{K^x}^H(\widetilde{f_\beta}) \circ f_{\gamma^x}^\beta$. Hence $(K_\gamma)^x \subset H_\beta$ by 4.3.1.

6.2 *A*-fusions.

With the expression 6.1.2.1 Puig suppressed G to introduce the following concept on an interior G-algebra A.

(6.2.1) Definition. An *A*-fusion $\widetilde{\varphi}$ from pointed groups K_γ to H_β is an injective group exomorphism $\widetilde{\varphi} : K \to H$ for which there is an interior K-algebra exomorphism $\widetilde{f_\varphi} : A_\gamma \to Res_\varphi(A_\beta)$ such that

$$Res_1^K(\widetilde{f_\gamma}) = Res_1^H(\widetilde{f_\beta}) \circ Res_1^K(\widetilde{f_\varphi})$$

The *A*-fusions can be composed and can be onto-into decomposed ([21, 2.6, 2.11]).

From the 6.1.2 we have ([21; 2.10]):

(6.2.2) Proposition. $\quad E_G(K_\gamma, H_\beta) = E_G(K, H) \cap F_A(K_\gamma, H_\beta)$

(6.2.3) Remark. Let $\widetilde{\varphi} \in F_A(K_\gamma, H_\beta)$. Set $A_\gamma = jAj$ and $A_\beta = iAi$ just as in 6.1.2, and let $f_\gamma \in \widetilde{f_\gamma}$ and $f_\beta \in \widetilde{f_\beta}$ be the inclusion maps of A_γ and A_β respectively; choose $f_\varphi \in \widetilde{f_\varphi}$. Then the above definition implies that there is $u \in A^*$ such that

$$f_\varphi(a) = a^u \qquad \text{for all} \quad a \in A_\gamma$$

and j^u is a primitive idempotent in $(A_\beta)^{\varphi(K)}$ (in particular $\widetilde{\varphi}$ is an isomorphism if φ is surjective since in this case $(A_\beta)^{\varphi(K)} = (A_\beta)^H$ is a local algebra hence f_φ is an unitary embedding). Thus we have ([21, 2.12])

$$i(yj)^u = \varphi(y)j^u = j^u\varphi(y) = (yj)^u i \qquad \text{for all} \quad y \in K$$

and, conversely, this condition implies that $\widetilde{\varphi}$ is an A-fusion.

Intuitively speaking, an A-fusion is nearly induced by A^*. It is really induced by A^* if $\widetilde{\varphi}$ is surjective and K is into $(A_\gamma)^*$ ([21, 2.13]).

Two easy consequences:

(6.2.4). If $K \subseteq H$, then the inclusion map inc_K^H induces an A-fusion $\widetilde{inc_K^H} \in F_A(K_\gamma, H_\beta)$ if and only if $K_\gamma \subset H_\beta$.

(6.2.5). If $\widetilde{\varphi} \in F_A(K_\gamma, H_\beta)$ is surjective, then $\widetilde{\varphi^{-1}} \in F_A(K_\gamma, H_\beta)$.

Thus we have two categories for an interior G-algebra A.

6.3 Definition. The local category of an interior G-algebra A is the category whose objects are the local pointed groups on A and whose morphisms are the G-exomorphisms between local pointed groups.

The local fusion category of an interior G-algebra A is the category whose objects are the local pointed groups on A and whose morphisms are the A-fusions between local pointed groups.

The local category of an inerior G-algebra A is a subcategory of the local fusion category of A by 6.2.2.

Of course the above two categories can also be defined for G-algebras. But their behavior for interior G-algebras is better than for G-algberas. For example, much of 6.1 and 6.2 are no longer true for G-algebras.

6.4 Let $\widetilde{f} : A \rightarrow A'$ be an exterior embedding from an interior G-algebra A to an interior G-agebra A'.

As mentioned in 4.2, $\widetilde{f}$ induces a map from the set of pointed groups on A to the set of pointed groups on A'. Moreover, by 3.4.2, it is easy to see that the image of a local pointed group is again a local pointed group and vice versa; and, if $(A_\beta, \widetilde{f_\beta})$ is the embedded algebra of H_β on A, then the pair $(A_\beta, \widetilde{f} \circ \widetilde{f_\beta})$ is the embedded algebra of $H_{\beta'}$ on A' where $\widetilde{f}(\beta) \subseteq \beta'$, hence $\widetilde{f}$ induces a map from the G-exomorphisms (or A-fusions resp.) on A to the G-exomorphisms (or A'-fusions) on A'. In this way $\widetilde{f}$ induces a functor, denoted by $\widetilde{f_*}$, from the local category (or local fusion category resp.) of A to the local category (or local fusion category resp.) of A'. Since G-exomorphisms and A-fusions are characterized by the embedded algebras of the pointed groups, see 3.1.1, the following is clear ([21, 2.14]):

6.5 Proposition. *Let $\widetilde{f} : A \rightarrow A'$ be an embedding of interior G-algebras A and A', and let $\widetilde{f_*}$ be as above. Let H_β be a pointed group on A and $H_{\beta'} = \widetilde{f_*}(H_\beta)$ on A'.*

$(6.5.1)$. If $K_{\gamma'} \subset H_{\beta'}$ on A', then $K_{\gamma'} \in \mathrm{Im}(\widetilde{f}_*)$, i.e. there is K_γ on A such that $\widetilde{f}(\gamma) \subseteq \gamma'$.

$(6.5.2)$. $E_G(K_\gamma, H_\beta) = E_G(K_{\gamma'}, H_{\beta'})$, and $F_A(K_\gamma, H_\beta) = F_{A'}(K_{\gamma'}, H_{\beta'})$.

$(6.5.3)$. K_γ is local if and only if $K_{\gamma'}$ is local, hence K_γ is a defect pointed group of H_β if and only if $K_{\gamma'}$ is a defect pointed group of $H_{\beta'}$.

So the functor $\widetilde{f}_*$ will be an equivalence if it is onto the isomorphism classes of the objects of the codomain category.

We exhibit an important well-known example.

6.6   Let H be a subgroup of G and B be an interior H-algebra; then any local pointed group on $\mathrm{Ind}_H^G(B)$ is G-conjugate (hence isomorphic in the local category) to a local pointed group on B (by 3.9.1, see [21, 2.12]).

Thus we have ([21, 2.16]):

(6.6.1) Lemma. The canonical embedding $\widetilde{d_H^G(B)} : B \to \mathrm{Res}_H^G \mathrm{Ind}_H^G(B)$ induces an equivalence between the local fusion categories of B and $\mathrm{Ind}_H^G(B)$. Hence for any embedded algebra $\widetilde{g} : A \to \mathrm{Ind}_H^G(B)$ and exomorphism $\widetilde{f} : B \to \mathrm{Res}_H^G(A)$ such that $\mathrm{Res}_H^G(\widetilde{g}) \circ \widetilde{f} = \widetilde{d_H^G(B)}$, the map $\widetilde{f}$ induces an equivalence between the local fusion categories of B and A.

And from 4.9 we have:

(6.6.2) Corollary. Let G_α be a pointed group on an interior G-algebra A and P_γ be a defect pointed group of G_α. Then the canonical embedding $\widetilde{f_\gamma^\alpha} : A_\gamma \to \mathrm{Res}_P^G(A_\alpha)$ induces an equivalence between the local fusion categories of A_γ and A_α.

(6.6.3) Remark. The above conclusions are no longer true for local categories in general; G-exomorphisms are dependent on the global group G, but, e.g. the global groups of A_γ and A_α resp. are different in general. The special case that $\widetilde{f_\gamma^\alpha}$ induces an equivalence between the local categories of A_γ and A_α is exactly related to nilpotent blocks, see 6.10. And group algebras have their own features, see 8.6.4.

6.7   Let us consider again the embedding $\widetilde{d_H^G(B)} : B \to \mathrm{Res}_H^G \mathrm{Ind}_H^G(B)$ where H is a subgroup of G and B is an interior H-algebra. Just as pointed out in 6.6.3, the induced functor $\widetilde{d_H^G(B)}$ from the local category of B to the local category of $\mathrm{Ind}_H^G(B)$ is not an equivalence in general, i.e. for local pointed groups Q_δ and R_ε on B (hence also on $\mathrm{Ind}_H^G(B)$) we have

$$E_H(Q_\delta, R_\varepsilon) \subseteq E_G(Q_\delta, R_\varepsilon)$$

is a proper inclusion in general. Let us see what happens if it is an equality. Of course

$$E_G(Q_\delta, R_\varepsilon) = E_G(Q, R) \cap F_B(Q_\delta, R_\varepsilon)$$

by 6.2.2 and 6.6.1, on the right hand side $\mathrm{Ind}_H^G(B)$ has disappeared. We say that G is locally controlled by H on B in Puig's notation if

$$E_G(Q, R) \cap F_B(Q_\delta, R_\varepsilon) = E_H(Q_\delta, R_\varepsilon)$$

for all local pointed groups Q_δ and R_ϵ on B. In this case it is easy to see that, for any embedded algebra $\tilde{g} : A \to \mathrm{Ind}_H^G(B)$ and exomorphism $\tilde{f} : B \to \mathrm{Res}_H^G(A)$ such that $\mathrm{Res}_H^G(\tilde{g}) \circ \tilde{f} = \widetilde{d_H^G(B)}$, the map $\tilde{f}$ also induces an equivalence between the local categories of B and A. Hence, from 4.3.3, it is natural to define ([23, 3.5]):

(6.7.1) Definition. Let A be an interior G-algebra, G_α be a pointed group on A and H_β be a pointed group contained in G_α. We say that G_α is locally controlled by H_β if $\alpha \subseteq \mathrm{Tr}_H^G(A^H \beta A^H)$ and $\mathrm{E}_G(Q_\delta, R_\epsilon) = \mathrm{E}_H(Q_\delta, R_\epsilon)$ for any local pointed groups Q_δ and R_ϵ contained in H_β. In other words, there is an embadding $\tilde{f}_\beta^\alpha : A_\beta \to \mathrm{Res}_H^G(A_\alpha)$ (see 4.3.3) and $\tilde{f}_\beta^\alpha$ induces an equivalence from the local category of A_β (as an interior H-algebra) to the local category of A_α (as an interior G-algebra).

For block algebras Puig proved ([23, 3.8]):

6.8 Theorem. *Let* $A = \mathcal{O}G$. *Let* b *be an* $\mathcal{O}$-*block of* G *and* P_γ *be a defect pointed group of* G_α *where* $\alpha = \{b\}$. *If* $k = \mathcal{O}/J(\mathcal{O})$ *is algebraically closed, then the following three conditions are equivalent to each other:*

(6.8.1). G_α *is locally controlled by* P_γ.

(6.8.2). $\mathrm{E}_G(Q_\delta)$ *is a* p-*group for any local pointed group* Q_δ *of* G_α.

(6.8.3). $N_G(Q, b_Q)/C_G(Q)$ *is a* p-*group for any* b-*Brauer pair* (Q, b_Q).

However this is no longer true if k is not large enough. We showed ([10, 1.5]):

6.9 Theorem. *Let* $\alpha = \{b\}$ *and* P_γ *be as above. Then the following three conditions are equivalent to each other:*

(6.9.1). G_α *is locally controlled by* P_γ.

(6.9.2). $\mathrm{E}_G(Q_\delta)$ *is a* p-*group for any local pointed group* Q_δ *of* G_α *and* $\mathrm{E}_G(P_\gamma) = \{1\}$

(6.9.3). $N_G(Q, b_Q)/C_G(Q)$ *is a* p-*group for any* b-*Brauer pair* (Q, b_Q) *and* $N_G(P, b_\gamma)/(PC_G(P)) = \{1\}$ *for a Sylow* b-*Brauer pair* (P, b_γ).

Thus from the point of view of local categories it is natural to define:

6.10 Definition. An $\mathcal{O}$-block b of G is said to be nilpotent if G_α, where $\alpha = \{b\}$, is locally controlled by its defect pointed group P_γ.

7. COVERINGS

7.1 Coverings are very prominent in the theory of nilpotent blocks. Obviously an exomorphism $\tilde{f} : A' \to A$ of interior G-algebras A' and A over $\mathcal{O}$ induces an interior $C_G(H)$-exomorphism

$$\widetilde{f^H} : A'^H \longrightarrow A^H$$

for every subgroup H of G.

(7.7.1) Definition. $\widetilde{f}$ is said to be a covering of interior G-algebras if $A^H = \text{Im}(\widetilde{f^H}) + J(A^H)$ for every $H \leq G$; further, $\widetilde{f}$ is said to be strict if $\text{Ker}(\widetilde{f^H}) \subseteq J(A'^H)$ for every $H \leq G$.

The coverings are roughly dual to the embeddings in 6.5. For example, we have ([23, 4.18] and [11, 2.2]):

7.2 Proposition. *Let $\widetilde{f} : A' \longrightarrow A$ be a covering of interior G-algebras A' and A. Then $\widetilde{f}$ induces an injection $\widetilde{f}^*$ from the set of pointed groups on A to the set of pointed groups on A' such that $\widetilde{f}(\beta') \subseteq \beta$ for H_β on A and $H_{\beta'} = \widetilde{f}^*(H_\beta)$ on A', and*

(7.2.1). If $K_{\gamma'} \in \text{Im}(\widetilde{f}^*)$ and $K_{\gamma'} \subset H_{\beta'}$ on A', then $H_{\beta'} \in \text{Im}(\widetilde{f}^*)$.

(7.2.2). $E_G(K_{\gamma'}, H_{\beta'}) = E_G(K_\gamma, H_\beta)$ and $F_{A'}(K_{\gamma'}, H_{\beta'}) \subseteq F_A(K_\gamma, H_\beta)$.

(7.2.3). $K_{\gamma'}$ is local if and only if K_γ is local; hence $K_{\gamma'}$ is a defect pointed group of $H_{\beta'}$ if and only if K_γ is a defect pointed group of H_β .

(7.2.4). $\widetilde{f}$ induces an isomorphism $\widetilde{f}(H_\beta) : A'(H_{\beta'}) \longrightarrow A(H_\beta)$.

(7.2.5). $\widetilde{f}$ is strict if and only if $\widetilde{f}^*$ is bijective.

An important example of coverings will be given in 8.3.1 and 8.3.2.

In fact 7.2.4 is also a sufficient condition for $\widetilde{f} : A' \to A$ to be a covering, but we have a stronger criterion ([23, 4.22] and [11, 2.3]).

7.3 Theorem. *An exomorphism $\widetilde{f} : A' \to A$ of interior G-algebras A' and A is a covering if and only if for any local pointed group Q_δ on A there is a pointed group $Q_{\delta'}$ on A' such that $f(\delta') \subseteq \delta$ and $\widetilde{f}$ induces an isomorphism $\widetilde{f}(Q_\delta) : A'(Q_{\delta'}) \to A(Q_\delta)$. Further, $\widetilde{f}$ is strict if and only if the map $Q_\delta \mapsto Q_{\delta'}$ is onto the local pointed groups on A'.*

Note that $Q_{\delta'}$ in the statement must be local and unique, cf. 7.2.3.

The idea of the theorem is that: for any pointed group H_β on A, $A^H \beta A^H \subseteq \text{Tr}_P^G(A^P \gamma A^P)$ where P_γ is a defect pointed group of H_β; hence, by 1.2.4, A^H can be "covered" by local pointed group P_γ through the trace map; but the condition of the theorem implies

$$A^P = f(A'^P) + J(A^P) + \sum_{Q \lneq P} A_Q^P$$

thus it is enough to treat $\text{Tr}_P^H(J(A^P))$ carefully; Puig did this in [23, 4.21]. The argument is really independent (shown in [11, 2.3.1]) of the size of $\mathcal{O}$.

7.4   Another suggestion from 7.2.4 is that we can characterize the coverings by the multiplicities of pointed groups. In fact, for a general algebra homomorphism $f : A' \to A$ and points β' and β on A' and A resp. such that $f(\beta') \subseteq \beta$, it is clear that $m_{\beta'} \leq m_\beta$; and it is expected that f induces an isomorphism $f(\beta) : A'(\beta') \to A(\beta)$ if $m_{\beta'} = m_\beta$. However this is not enough for the expectation to be true when we consider arbitrary ground-fields. The following ([11, 2.4.1]) is useful to us.

(7.4.1) Lemma. Let $\widetilde{f} : A' \to A$ be an exomorphism of interior G-algebras A' and A, and let $H_{\beta'}$ and H_β be pointed groups on A' and A resp. such that $\widetilde{f}(\beta') \subseteq \beta$. Then the composition homomorphism $A'^H \to A^H \to A(H_\beta)$ induces an isomorphism $f(H_\beta) : A'(H_{\beta'}) \to A(H_\beta)$ if and only if $m_{\beta'} = m_\beta$ and the simple A'^H-module with β' and the simple A^H-module with β have their (division) endomorphism algebras isomorphic to each other.

7.5 Coverings of induced algebras.

Let H be a subgroup of G and B be an interior H-algebra. Let $A = \mathrm{Ind}_H^G(B)$ be the induced interior G-algebra.

In 6.6 we have seen that any local pointed group Q_δ on B can be regarded as a local pointed group on A, and any local pointed group on A is G-conjugate to such a local pointed group. However Q_δ has different multiplicities $m_\delta(B)$ on B and $m_\delta(A)$ on A ([24, 4.8] and verified in [11, 2.5.1] for arbitrary ground-fields).

(7.5.1) Lemma. Notation as above. Then

$$m_\delta(A) = \sum_{\beta \in \mathcal{P}_B(H)} m_\beta(B) \sum_{\overline{\varphi} \in \mathrm{E}_G(Q_\delta, H_\beta)} |C_G(\varphi(Q)) : C_H(\varphi(Q))| \cdot m_{\varphi(\delta)}^\beta$$

With the above and 7.3 and 7.4.1, we can show the following [24, 4.6] holds in any case (proved in [11, 2.6]):

7.6 Theorem. Let H be a subgroup of G and $\widetilde{g} : B' \to B$ be a covering of interior H-algebras B' and B. For pointed group Q_δ on B, let $Q_{\delta'}$ denote the pointed group on B' such that $\widetilde{g}(\delta') \subseteq \delta$. The following two statements are equivalent to each other:

(7.6.1). The induced exomorphism $\mathrm{Ind}_H^G(\widetilde{g}) : \mathrm{Ind}_H^G(B') \to \mathrm{Ind}_H^G(B)$ is a covering.

(7.6.2). For any local pointed group Q_δ and pointed group H_β on B

$$E_G(Q, H) \cap F_B(Q_\delta, H_\beta) \subseteq F_{B'}(Q_{\delta'}, H_{\beta'})$$

Further, if it is the case, $\mathrm{Ind}_H^G(\widetilde{g})$ is strict provided $\widetilde{g}$ is strict.

One of the advantages of the coverings of induced algebras was exhibited in [23, 7.2] and discussed for arbitrary ground-fields as follows ([10, 2.1]).

7.7 Lemma. Let b be an $\mathcal{O}$-block of G and P_γ be a defect pointed group of G_α where $\alpha = \{b\}$, let $B = A_\gamma$ be the source algebra of b. Let $\widetilde{g} : B' \to B$ be an exomorphism of interior P-algebras. If the induced exomorphism

$$\mathrm{Ind}_H^G(\widetilde{g}) : \mathrm{Ind}_H^G(B') \to \mathrm{Ind}_H^G(B)$$

is a strict covering, then there is a strict covering $\widetilde{g'} : B \to B'$ of interior P-algebras such that $\widetilde{g} \circ \widetilde{g'} = \widetilde{1_B}$.

8. DADE INTERIOR GROUP ALGEBRAS

8.1 Definittion. (8.1.1). A G-algebra A is said to be a permutatiton G-algebra if it had a G-stable $\mathcal{O}$-basis W, which is then called a permutation basis.

(8.1.2). A permutation G-algebra A with G-stable basis W is said to be a Dade G-algebra if $|W \cap A^G| \neq \emptyset$.

(8.1.3). An interior G-algebra A is said to be biregular if it has an $\mathcal{O}$-basis W such that $|Gw| = |G| = |wG|$ for any $w \in W$.

Let P be a p-group and B be a P-algebra.

(8.1.4). B is said to be primitive if 1_B is the unique non-zero idempotent of B^P.

(8.1.5). B is said to be a local Dade P-algebra if B is a primitive Dade P-algebra.

(8.1.6) Remark. It is clear by 3.2.2 that a permutation P-algebra B is a Dade P-algebra if and only if $B(P) \neq 0$; further, if this is the case, then a local pointed group Q_δ on B exists for any subgroup Q of P.

8.2 To treat tensor products of interior group algebras, we need permutation G-algebras or Dade G-algebras in many cases.

Let A and A' be interior G-algebras over $\mathcal{O}$.

It is trivially that $A \otimes_{\mathcal{O}} A'$ is again an interior G-algebra with structural map $x \mapsto x \otimes x$ for $x \in G$.

For interior G-algebra homomorphisms $f : A \to B$ and $f' : A' \to B'$, obviously $f \otimes f' : A \otimes_{\mathcal{O}} A' \to B \otimes_{\mathcal{O}} B'$ is again an interior G-algebra homomorphism. And, if f and g are in the same exomorphism and f' and g' are in the same exomorphism, then $f \otimes f'$ and $g \otimes g'$ are clearly in one and the same exomorphism of the tensor products; hence the exomorphism $\widetilde{f} \otimes \widetilde{g}$ is well-defined.

It is natural to ask: how are the pointed groups on $A \otimes_{\mathcal{O}} A'$ related to the pointed groups on A and A' ? The answer is not easy. Puig showed the following result ([23, 5.6]).

(8.2.1) Proposition. *Assume one of A and A' is a permutation P-algebra where P is a p-subgroup. Then for any subgroup Q of P*

(1). $A(Q) \otimes_{\mathcal{O}} A'(Q) \cong (A \otimes_{\mathcal{O}} A')(Q)$.

(2). *if k is splitting for one of $A(Q)$ and $A'(Q)$, then the above isomorphism induces a bijection*

$$\mathcal{LP}_A(Q) \times \mathcal{LP}_{A'}(Q) \longrightarrow \mathcal{LP}_{A \otimes_{\mathcal{O}} A'}(Q) \qquad (\delta, \delta') \longmapsto \delta \times \delta'$$

such that $\mathrm{Br}_Q(\delta) \otimes \mathrm{Br}_Q(\delta') \subseteq \mathrm{Br}_Q(\delta \times \delta')$, and there is an embedding

$$\widetilde{g_{\delta \times \delta'}} : (A \otimes_{\mathcal{O}} A')_{\delta \times \delta'} \longrightarrow A_\delta \otimes_{\mathcal{O}} A'_{\delta'}$$

such that $(\widetilde{f_\delta} \otimes \widetilde{f_{\delta'}}) \circ \widetilde{g_{\delta \times \delta'}} = \widetilde{f_{\delta \times \delta'}}$.

8.3 It is much more diffecult to treat the G-exomorphisms and fusions of pointed groups on tensor products. Puig got some general results ([23, 5.3]).

A simple situation will be useful to us.

An ideal $I(A)$ of an interior G-algebra A is called the augmentation ideal if $A/I(A) \cong \mathcal{O}$ is a trivial interior G-algebra.

(8.3.1) Lemma. *Let A and A' be interior G-algebras. Assume that A' has an augmentation ideal $I(A')$. Then (where $\overline{a'} \in A'/I(A')$)*

$$f : A \otimes_{\mathcal{O}} A' \longrightarrow A \qquad a \otimes a' \longmapsto a\overline{a'}$$

is a covering of interior G-algebras. Further, if $I(A') \subseteq J(A')$, then f is strict; hence for pointed group H_β on A and $j \in \beta$ there is a unique point $\beta \times 1$ of H on $A \otimes_{\mathcal{O}} A'$ such that $j \otimes 1 \in \beta \times 1$, and

$$F_A(K_\gamma, H_\beta) \cap F_{A'}(K_{\{1\}}, H_{\{1\}}) \subseteq F_{A \otimes_{\mathcal{O}} A'}(K_{\gamma \times 1}, H_{\beta \times 1}) \subseteq F_A(K_\gamma, H_\beta)$$

Proof. For any $H \leq G$, $A^H \otimes_{\mathcal{O}} A'^H \subseteq (A \otimes_{\mathcal{O}} A')^H$ and $A^H \otimes_{\mathcal{O}} A'^H$ is clearly mapped onto A^H by f; hence $f^H : (A \otimes_{\mathcal{O}} A')^H \to A^H$ is onto. Assume $I(A') \subseteq J(A')$ further. Then $\operatorname{Ker}(f) = A \otimes_{\mathcal{O}} I(A') \subseteq J(A \otimes_{\mathcal{O}} A')$, hence f is strict; and consequently, $j \otimes 1$ is a primitive idempotent since $f(j \otimes 1) = j$.

For $\widetilde{\varphi} \in F_A(K_\gamma, H_\beta) \cap F_{A'}(K_{\{1\}}, H_{\{1\}})$, there are

$$\widetilde{f_\varphi} : A_\gamma \to \operatorname{Res}_\varphi(A_\beta) \qquad \text{and} \qquad \widetilde{f'_\varphi} : A'_{\{1\}} \to \operatorname{Res}_\varphi(A'_{\{1\}})$$

which give that $\widetilde{\varphi} \in F_{A \otimes_{\mathcal{O}} A'}(K_{\gamma \times 1}, H_{\beta \times 1})$. The last assertion is by 7.2.2.

(8.3.2) Example. Let $\overline{G}$ be a quotient of G, and let $A' = \mathcal{O}\overline{G}$. Then the lemma is applied to get a covering $A \otimes_{\mathcal{O}} \mathcal{O}\overline{G} \to A$; further, if $\overline{G}$ is a p-group, then the covering is strict. This is [23, 4.25].

8.4 $\mathcal{O}$-**simple Dade interior P-algebras.**

Let P be a p-group. The following is an example of an $\mathcal{O}$-simple permutation interior P-algebra.

(8.4.1) Example. Let M be a permutation $\mathcal{O}P$-module with permutation basis W. Then $E = \operatorname{End}_{\mathcal{O}}(M)$ is an $\mathcal{O}$-simple permutation interior P-algebra: let $f_{ww'}$, $w, w' \in W$, be the $\mathcal{O}$-linear map such that

$$f_{ww'}(w'') = \begin{cases} w' & \text{if } w'' = w \\ 0 & \text{otherwise} \end{cases}$$

then $\{f_{ww'} \mid w, w' \in W\}$ is a permutation basis of E; and for any subgroup Q of P, $f_{ww'}$ is Q-fixed if and only if both w and w' are Q-fixed. Thus from 3.2.2 we see that ([22, 2.9.1])

$$(\operatorname{End}_{\mathcal{O}}(M))(Q) \cong \operatorname{End}_k(M(Q))$$

as $(N_G(Q)/Q)$-algebras over k; and that E is a Dade algebra if the permutation basis W of M has a P-fixed element w_0; and in this case $f_{w_0 w_0}$ is an idempotent and $f_{w_0 w_0} E f_{w_0 w_0} \cong \mathcal{O}$ is a trivial interior P-algebra.

A remarkable fact for $\mathcal{O}$-simple permutation P-algebras is ([23, 5.7 and 5.8])

(8.4.2) Proposition. *Let S be an $\mathcal{O}$-simple permutation P-algebra and Q be a subgroup of P. Then the local point of Q on S is unique if it exists. In particular, $|\mathcal{LP}_S(Q)| = 1$ if S is an $\mathcal{O}$-simple Dade P-algebra.*

The key idea is that the multiplication induces an isomorphism of interior P-algebras

$$S \otimes_\mathcal{O} S^o \longrightarrow \mathrm{End}_\mathcal{O}(S)$$

where S^o is the opposite algebra with structural map $y \mapsto \varphi(y)^{-1}$ (where $y \mapsto \varphi(y)$ is the structural map of S); by 8.2.1 and 8.4.2 it induces an isomorphism

$$S(Q) \otimes_k S(Q)^o \longrightarrow \mathrm{End}_k(S(Q))$$

Thus $S(Q)$ is a central simple k-algebra. The last assertion is by 8.1.6.

(8.4.3) Corollary. *If S is an $\mathcal{O}$-simple Dade P-algebra, then there is an unique embedding from the trivial interior P-algebra $\mathcal{O}$ to $S \otimes_\mathcal{O} S^o$.*

Indeed, if we identify $S \otimes_\mathcal{O} S^o$ with $\mathrm{End}_\mathcal{O}(S)$, the existence follows from 8.4.1 and the uniqueness follows from the uniqueness of the local point of P on $S \otimes_\mathcal{O} S^o$.

(8.4.4) Remark. From the above observations we see that the embedded algebra of the unique local point of P on $S \otimes_\mathcal{O} S^o$ is $\mathcal{O}$, and the multiplicity of the unique point of P on $S \otimes_\mathcal{O} S^o$ is just $|W \cap S^P|$ where W is the P-stable $\mathcal{O}$-basis of S. In particular, $|W \cap S^P| = 1$ if S is a primitive P-algebra.

8.5 Example. Let P be a p-group and S be an $\mathcal{O}$-simple interior P-algebra. Let $B = S \otimes_\mathcal{O} \mathcal{O}P$. Assume that B is a local Dade P-algebra.

(8.5.1). Since $S \cong S \otimes 1$ is a direct summand of B as $\mathcal{O}P$-modules, S must be a permutation P-algebra, hence S is a local Dade P-algebra too (cf. 8.3.2 and 8.1.6); moreover, the P-stable $\mathcal{O}$-basis W of S has a unique P-fixed element (cf. 8.4.4), hence we can chose W so that the unique P-fixed element is just 1_S (cf. the argument in [21, 3.4]).

Remark. Recently Puig [27] developed a creterion for Dade P-algebras which contains the present case as a simple special case.

(8.5.2). From 8.3.1, 8.3.2 and 8.4.2, there is a unique local pointed group Q_δ and R_ϵ for subgroups Q and R of P on B, and if we let S denote the unique local point of Q on S (the same for R) again, then $\delta = S \times 1$ and $\epsilon = S \times 1$ on $S \otimes_\mathcal{O} \mathcal{O}P$. By the uniqueness of local pointed group on S (see 8.4.2) we have

$$\mathrm{E}_P(Q_\delta, R_\epsilon) = \mathrm{E}_P(Q, R) = \mathrm{E}_P(Q_S, R_S) \subseteq \mathrm{F}_S(Q_S, R_S)$$

On the other hand, for $\mathcal{O}P$ by the forthcoming 8.6.4 we have

$$\mathrm{F}_{\mathcal{O}P}(Q_{\{1\}}, R_{\{1\}}) = \mathrm{E}_P(Q_{\{1\}}, R_{\{1\}}) = \mathrm{E}_P(Q, R) = \mathrm{E}_P(Q_\delta, R_\epsilon)$$

Thus by 8.3.1 and 8.3.2

$$\mathrm{E}_P(Q_\delta, R_\epsilon) \subseteq \mathrm{F}_B(Q_\delta, R_\epsilon) \subseteq \mathrm{F}_S(Q_S, R_S)$$

Using the unique embedding in 8.4.3, we get an interior P-algebra embedding

$$\mathcal{O}P \longrightarrow S^o \otimes_\mathcal{O} S \otimes_\mathcal{O} \mathcal{O}P = S^o \otimes_\mathcal{O} B$$

and $S^o \times \delta = S^o \times S \times 1$ is the unique local point of Q on $S^o \otimes_\mathcal{O} B$ by 8.4.2 and 8.3.1. Hence, by 6.5.2 and 8.3.1

$$\mathrm{F}_{\mathcal{O}P}(Q_{\{1\}}, R_{\{1\}}) = \mathrm{F}_{S^o \otimes_\mathcal{O} B}(Q_{S^o \times \delta}, R_{S^o \times \varepsilon}) \supseteq \mathrm{F}_{S^o}(Q_{S^o}, R_{S^o}) \cap \mathrm{F}_B(Q_\delta, R_\varepsilon)$$

It is clear that $\mathrm{F}_{S^o}(Q_{S^o}, R_{S^o}) = \mathrm{F}_S(Q_S, R_S)$. Thus

$$\mathrm{F}_{\mathcal{O}P}(Q_{\{1\}}, R_{\{1\}}) \supseteq \mathrm{F}_B(Q_\delta, R_\varepsilon)$$

So on B we get

$$\mathrm{F}_B(Q_\delta, R_\varepsilon) = \mathrm{E}_P(Q_\delta, R_\varepsilon)$$

8.6 Consider the following case: $A = \mathcal{O}G$, G_α with $\alpha = \{b\}$ is a pointed group on A, i.e. b is a $\mathcal{O}$-block of G, P_γ is a defect pointed group of G_α; hence A_γ is the source algebra of G_α and of b.

A is obviously a biregular Dade interior G-algebra. Note that any embedded algebra of a Dade algebra is again a permutation algebra. So we can show

(8.6.1). A_γ is a local Dade interior P-algebra in the sence of 8.1.5 (cf. 4.3).

(8.6.2). A_γ is biregular ([21, 3.4]).

Thus A_γ is projective either as a left $\mathcal{O}P$-module or as a right $\mathcal{O}P$-module. Moreover, we have ([23, 7.7]):

(8.6.3). A_γ-module N is projective if and only if N is a projective $\mathcal{O}P$-module through the structural map $\mathcal{O}P \to A_\gamma$.

With these facts Puig showed ([21, 3.6])

(8.6.4) Theorem. *The local category and the local fusion category of A coincide.*

8.7 Notations are as above again. Now we show a condition for blocks such that 7.7 can be applied.

Recall that, if G_α is locally controlled by P_γ (see 6.7.1), then for any $\tilde{\varphi} \in \mathrm{E}_G(Q_\delta, P_\gamma)$ induced by $x \in G$ there are $z \in C_G(Q)$ and $u \in P$ such that $x = zu$.

For a normal subgroup D of P we denote

$$C_G(Q, D) = \{x \in G \mid [Q, x] \subseteq D\}$$

where $[-, -]$ denotes the commutator, and define

(8.7.1) Definition. We say that G_α is locally controlled by P_γ relative to D (or D-locally controlled by P_γ) if for any $\tilde{\varphi} \in \mathrm{E}_G(Q_\delta, P_\gamma)$ induced by $x \in G$ there are $z \in C_G(Q, D)$ and $u \in P$ such that $x = zu$.

It is obvious that 1-local controls are just the usual local controls in 6.7.1.

(8.7.2) Proposition. *Assume G_α is D-locally controlled by P_γ. Then for the interior P-algebra strict covering (see 8.3.2)*

$$\tilde{g} : A_\gamma \otimes_{\mathcal{O}} \mathcal{O}(P/D) \longrightarrow A_\gamma$$

there is an interior P-algebra strict covering

$$\tilde{g'} : A_\gamma \longrightarrow A_\gamma \otimes_{\mathcal{O}} \mathcal{O}(P/D)$$

such that $\tilde{g} \circ \tilde{g'} = \tilde{1}_{A_\gamma}$.

Indeed, we can check that 7.6.2 of the Theorem 7.6 holds in the present case, hence 7.7 can be applied to get the desired conclusion ([11, 1.5]).

9.   STRUCTURE THEOREM ON SOURCE ALGEBRAS OF NILPOTENT BLOCKS

Now we come to the Puig's Theorem on the source algebras of nilpotent blocks.

9.1 Theorem. *Let b be an $\mathcal{O}$-block of G and P_γ be a defect pointed group of G_α where $\alpha = \{b\}$, and let A_γ be the source algebra of G_α on $A = \mathcal{O}G$. If $k = \mathcal{O}/J(\mathcal{O})$ is algebraically closed, then the following two statements are equivalent to each other:*

(1). The block b is nilpotent.

(2). There is an $\mathcal{O}$-simple interior P-algebra S over $\mathcal{O}$ such that

$$A_\gamma \cong S \otimes_{\mathcal{O}} \mathcal{O}P$$

And if this is the case, the P-algebra S is unique upto isomorphism and has a P-stable $\mathcal{O}$-basis which contains the unity as the unique P-fixed element.

Remark. Denote $B = S \otimes_{\mathcal{O}} \mathcal{O}P$ for short. By 2.1.1 Remark, $B \cong SP$, where S is only a P-algebra. Puig in [23, 6.2] showed that the P-structrue can be lifted to a unique interior P-structure of S, see 3.1.3.1.

First we see that the last part "And $\cdots$" has been shown in 8.5.1.

The uniqueness is somewhat subtle. Puig showed that for any maximal $\mathcal{O}$-simple subalgebra S' of B there is a $w \in B^*$ which is fixed by P such that $S' = S^w$ ([23, 6.9]). In a technical way Puig proved that if an interior P-subalgebra C of B such that $C + J(B) = B$ then $C = B$. With this the implication (1) $\implies$ (2) can be sketched as folows.

From 8.7.2 we have a strict covering

$$\tilde{g} : A_\gamma \otimes_{\mathcal{O}} \mathcal{O}P \longrightarrow A_\gamma$$

and there is a strict covering

$$\tilde{g'} : A_\gamma \longrightarrow A_\gamma \otimes_{\mathcal{O}} \mathcal{O}P$$

such that $\tilde{g} \circ \tilde{g'} = \tilde{1}_{A_\gamma}$. If there is an $\mathcal{O}$-simple interior P-algebra S and a surjection $\tau : A_\gamma \to S$, we let $f = (\tau \otimes 1) \circ g'$ which is an interior P-homomorphism from A_γ

to $S \otimes_{\mathcal{O}} \mathcal{O}P$. Denote $h : S \otimes_{\mathcal{O}} \mathcal{O}P \to S$, $s \otimes u \mapsto s$ for $s \in S$ and $u \in P$. Then $\tilde{h} \circ \tilde{f} = \tilde{1}_S$, hence we can show that f is surjective and $S \otimes_{\mathcal{O}} \mathcal{O}P$ is $\mathcal{O}P$-projective. With these an easy argument ([23, 7.4] or [11, 3.1]) shows that f is an isomorphism. The above is a sketch of [23, 7.2]. Now if the characteristic of $\mathcal{O}$ is p, it is reduced to the case of k; and S is easy to find ([23, 7.9]). For the general case, with the help of 5.5 and the fact (which follows from the above case of k) that for any $Q \leq P$ there is only one local point of Q on B, Puig got an ordinary character χ which affords the unique irreducible Brauer character of B, hence the desired S exists.

The implication $(2) \Longrightarrow (1)$ follows from 8.5.2, since by 8.6.4 and 6.6.2

$$\mathrm{E}_G(Q_\delta, R_\varepsilon) = \mathrm{F}_B(Q_\delta, R_\varepsilon)$$

for any local pointed groups Q_δ and R_ε on B.

9.2 Related questions.

Let b be an $\mathcal{O}$-block of G and P_γ be its defect pointed group again.

If b is nilpotent, from Puig's Theorem 9.1 the source algebra of b is isomorphic to a full matrix algebra over $\mathcal{O}P$, hence the block algebra $\mathcal{O}Gb$ is isomorphic to a full matrix algebra over $\mathcal{O}P$ too, cf. 4.9.1.

(9.2.1) Question. If the block algebra $\mathcal{O}Gb$ is isomorphic to a full matrix algebra over $\mathcal{O}P$, is the $\mathcal{O}$-block b nilpotent ?

If the defect group is abelian, the answer is positive; it is proved in [17].

Just as mentioned in 2.14 recently Puig proved an interesting result ([26]) on Morita equivalence of nilpotent blocks; as an obvious consequence, the positive answer to the question is given for the case that $\mathcal{O}$ is of characteristic zero.

If b is nilpotent, from Theorem 9.1 and 8.4.2 and 8.5.1 we see that for any subgroup $Q \leq P$ there is a unique local point δ on A_γ; and, by 4.7, this is equivalent to the statement: for any b-Brauer pair (Q, b_Q) the $kC_G(Q)\overline{b_Q}$ has a unique irreducible modular character.

(9.2.2) Question. If $kC_G(Q)\overline{b_Q}$ has a unique irreducible modular character for any b-Brauer pair (Q, b_Q), is b nilpotent ?

Watanabe [29] showed that Alperin's weight conjecture implies the positive answer to the question.

Soon afterward; Puig proved the positive answer if the defect group is abelian, and Watanabe shortened his proof. This is [25].

The question is still open for the general case though it seems probably positive.

(9.2.3) Another natural question is what about the source algebras of nilpotent blocks when the ground-fields are not large enough.

Now assume that k is arbitrary and b is an $\mathcal{O}$-block of G with defect pointed group P_γ and source algebra A_γ.

It is reasonable to consider the integral extension $\widehat{\mathcal{O}}$ of $\mathcal{O}$ by adding all the $\varphi(x')$ where φ runs over the absolutely irreducible modular characters belonging to b and x' runs over the p'-elements of G. Then we have ([10, 1.3]):

(9.2.4) Theorem. *The following two statements are equivalent to each other:*

(1). The block b is nilpotent.

(2). There is an $\widehat{\mathcal{O}}$-simple interior P-algebra S over $\widehat{\mathcal{O}}$ such that as interior P-algebras over $\mathcal{O}$ (not over $\widehat{\mathcal{O}}$!)

$$A_\gamma \cong S \otimes_{\mathcal{O}} \mathcal{O}P$$

And if this is the case, the P-algebra S is unique up to isomorphism and has a P-stable $\widehat{\mathcal{O}}$-basis which contains the unity as the unique P-fixed element.

As a corollary we show that ([10, 1.7]):

(9.2.5). *The three condition (6.8.1) (6.8.2) (6.8.3) of Theorem 6.8 are equivalent to each other whenever k is a splitting field for $kC_G(P)\overline{b_\gamma}$ where (P, b_γ) is a Sylow b-Brauer pair (see 4.7); and $\mathcal{O} = \widehat{\mathcal{O}}$ in this case.*

REFERENCES

1. J.L. Alperin and M. Broué, *Local methods in block theory*, Ann. Math. **110** (1979), 143–157.

2. D. Benson and R.A. Parker, *The Green ring of a finite group*, J. of Algebra **87** (1984), 290–331.

3. R. Brauer, *Some applications of the theory of blocks of characters of finite groups*, J. of Algebra **1** (1964), 152–167.

4. R. Brauer, *On blocks and sections in finite groups, II*, Amer. J. Math. **90** (1968), 895–925.

5. M. Broué, *Radical, hauteus, p-sections et blocs*, Ann. of Math. **107** (1978), 89–107.

6. M. Broué and L. Puig, *Characters and local structures in G-algebras*, J. of Algebra **63** (1980), 306–317.

7. M. Broué and L. Puig, *A Frobenius theorem for blocks*, Invent. Math. **56** (1980), 117–126.

8. D.W. Berry and J.F. Carlson, *Restriction of module to local subgroups*, Proc. Amer. Math. Soc. **84** (1982), 181–184.

9. C.W. Curtis and I. Reiner, *Methods of Representation Theory* Vol I, Wiley, New York, 1981.

10. Y. Fan, *The source algebras of nilpotent blocks over arbitrary ground-fields*, J. of Algebra **168** (1994), 606–632.

11. Y. Fan, *Relative local controls and the block source algebras*, Preprint (1995).

12. Y. Fan and L. Puig, *On blocks with nilpotent coefficient extensions*, In preparation.

13. W. Feit, *The Representation Theory of Finite Groups*, North–Holland, Amsterdam, 1982.

14. J.A. Green, *Some remarks on defect groups*, Math. Z. **107** (1968), 133–150.

15. R. Knörr, *On the vertices of irreducible modules,*, Ann. Math. **110** (1979), 487–499.

16. B. Külshammer and L. Puig, *Extensions of nilpotent blocks*, Invent. Math. **102** (1990), 17–71.

17. T. Okuyama and Y. Tsushima, *Local properties of p-block algebras of finite groups*, Osaka J. Math. **20** (1983), 33–41.

18. L. Puig, *Sur un Théorème de Green*, Math. Z. **168** (1979), 487–499.

19. L. Puig, *Pointed groups and construction of characters*, Math. Z. **176** (1981), 265–292.

20. L. Puig, *The source algebra of a nilpotent block*, Preprint (1981).

21. L. Puig, *Local fusions in block source algebras*, J. of Algebra **104** (1986), 358–369.

22. L. Puig, *Pointed groups and construction of modules*, J. of Algebra **116** (1988), 7–129.

23. L. Puig, *Nilpotent blocks and their source algebras*, Invent. Math. **93** (1988), 77–116.

24. L. Puig, *Une correspondence de modules pour les blocs à groupes de défaut abéliens*, Geometriae Dedicata **37** (1991), 9–43.

25. L. Puig and A. watanabe, *On blocks with one simple module in any Brauer correspondant*, J. of Algebra **163** (1994), 135–138.

26. L. Puig, *From two-sided modules to interior group algebras*, Lecture Notes in Nankai Math. Institute, 1994. (Chinese)

27. L. Puig, *A sufficient condition for Dade P-interior algebras*, Lecture Notes in Nankai Math. Institute, 1994.

28. A. Watanabe, *Some studies on p-blocks with abelian defect groups*, Kumamoto J. Sci. Math. **16** (1985), 49–67.

29. A. Watanabe, *On nilpotent blocks of finite groups*, J. of Algebra **163** (1994), 128–134.

Near-Rings in China: Past and Present

Yuen Fong

Department of Mathematics, National Cheng Kung University
Tainan, Taiwan 701

Dedicate to my beloved teachers Prof. I. H. Lin and Dr. J. D. P. Meldrum,
and in homage of the late Mr. Chao-Hsueh Hsu

§1. Introduction

A *near-ring* is a set N together with two binary operations "+" and "·" such that (i) $(N, +)$ is a not necessarily abelian group, (ii) $(N, \cdot)$ is a semigroup, and (iii) $x \cdot (y + z) = x \cdot y + x \cdot z$ for all $x, y, z \in N$. To be more precise, this is a *left near-ring* because it satisfies only the left distributive law. If we replace (iii) by (iii') $(x + y) \cdot z = x \cdot z + y \cdot z$ for all $x, y, z \in N$, we then have a *right near-ring*. In each concept, or theorem about left near-rings, there is an analogous one for right near-rings, and vice versa. Throughout this paper, we will use the word near-ring to mean "left near-ring". The aim of this survey article is to give a detailed account of the recent development of near-ring theory and its applications, especially the recent research on this topic done in China, Hong Kong and Taiwan.

Although near-rings were defined in 1905, rigorous research on the theory was not started until 1950. The study of near-ring theory in fact began rather late in China and the surrounding areas. Before 1990, near-rings in China and Hong Kong received little attention. Totally there were not more than twenty papers and most of them were written in Chinese and published in the local mathematical journals (see [97, 133, 134, 138, 139, 140, 141, 142, 143, 144, 145, 146, 147, 148]). They are hard to obtain.

In Taiwan, the systematic investigation of the theory of near-rings was started by the school of Y. Fong, and his former students W. F. Ke, C. S. Wang, F. K. Huang and S. Y. Liang since 1980 (see [4, 5, 6, 7, 8, 9, 21, 35, 39, 46, 47, 48, 49, 50, 51, 52, 53, 54, 55, 56, 57, 58, 59, 60, 61, 62, 63, 64, 65, 66, 67, 68, 69, 70, 71, 72, 73, 74, 75]), but the first near-ring article is due to B. S. Du [42].

Near-rings are generalized rings. Roughly speaking, as one sees from the definition, near-rings satisfy all the axioms of a ring except for two: commutativity of addition and one of the two distributive laws. They are algebraic structures which arise very naturally in the study of group mappings. If $(G, +)$ is a not necessarily abelian group, then $M(G) = \{f \mid f : G \to G\}$, the set of all group mappings from G into itself, will form an honest near-ring under the usual pointwise addition and composition of mappings whenever the cardinality of G is greater than one. Examples of near-rings are abundant.

1.1. Examples Let $(G, +)$ be a group written additively and 0 the identity of G.
 (a) The following sets of mappings from G into itself are near-rings under the usual pointwise addition and composition of group mappings.

    (i) $M(G) = \{f \mid f : G \to G\}$.

    (ii) $M_0(G) = \{f \in M(G) \mid 0f = 0\}$.

    (iii) $M_C(G) = \{f \in M(G) \mid f \text{ is a constant mapping}\}$.

    (iv) $M_C^0(G) = \{f_y \in M(G) \mid y \in G, xf_y = \begin{cases} 0, & \text{if } x = 0; \\ y, & \text{if } x \neq 0 \end{cases}\}$.

Note that $(G, +) \cong (M_c(G), +) \cong (M_c^0(G), +)$.

(b) The following triples are near-rings on G:

    (i) $(G, +, \cdot)$ with $x \cdot y = 0$ for all $x, y \in G$.

    (ii) $(G, +, *)$ with $x * y = y$ for all $x, y \in G$.

    (iii) Take any $0 \notin A \subset G$ and define

$$x \cdot_A y = \begin{cases} y, & \text{if } x \in A; \\ 0, & \text{if } x \notin A. \end{cases}$$

Then $(G, +, \cdot_A)$ forms a near-ring.

The above three types of near-ring multiplications are called the *trivial multiplications* and were introduced by J. J. Malone [114] in 1967.

(c) All rings are near-rings.

Here we mention a key result (the embedding theorem) of near-ring theory in the following:

1.2. Theorem (H. E. Heatherly and J. J. Malone [90]) *Every near-ring N can be embedded into the near-ring $M(G)$ for some suitable group G.*

One realizes immediately from the above theorem that every near-ring can be embedded into a near-ring with identity. In the same spirit as the Cayley's embedding theorem for groups, the study of various classes of subnear-rings of $M(G)$ is thus of utmost importance. This article is mainly focused on various classes of near-rings of group mappings done by Chinese mathematicians.

The reader is referred to the books by Clay [33], Meldrum [121] and Pilz [129] for terminology, definitions and and basic facts of near-rings not mentioned in this paper.

§2. Endomorphism near-rings

As we have seen from the embedding theorem of (1.2), we know that the motivation for the theory of distributively generated near-rings (d. g. near-rings) arises from the study of the algebraic system which is additively generated by all the endomorphisms of a nonabelian group. Once the notion of d. g. near-rings is formulated, the question arises of whether an arbitrary d. g. near-ring can be embedded into a suitable endomorphism near-ring (a prototype d. g. near-ring) or not. In ring theory, it is well-known that every ring can be embedded into an endomorphism ring of a suitable abelian group. However, instead of the embedding theorem (1.2), we also obtain a useful theorem that a d. g. near-ring with a left identity can be embedded into an endomorphism near-ring of a suitable not necessarily abelian group. Thus, the study of endomorphism near-rings is one of the main streams in study of near-rings of group mappings.

One sees from the definition of near-ring that only the left distributive law is available. If we induce the other distributive law in our near-rings, we then have *distributive near-rings*. Distributive near-rings are very close to rings and of course carry a lot more good properties compared to that of near-rings in general. In this section, we consider a large class of near-rings which has a great deal of distributivity built in.

Between 1958 and 1962, A. Fröhlich published a series of papers on d. g. near-rings [77, 78, 79, 80, 81, 82, 83, 84]. These lay a solid foundation stone for the beginning of this subject although H. Neumann [126, 127] had used the same ideas in the early fifties.

Though the right distributive law is not available in a left near-ring in general, there are elements that can be distributed on both sides (e.g., the additive zero element of a near-ring is definitely a two-sided distributive element) over an arbitrary sum of elements. Such elements have a special name.

2.1. Definition Let $(N, +, \cdot)$ be a near-ring. An element $d \in N$ is said to be distributive if $(x + y) \cdot d = x \cdot d + y \cdot d$ for all $x, y \in N$.

2.2. Examples (a) Let $(N, +, \cdot)$ be a near-ring. Then 0_N is distributive and the identity 1_N, if exists, is also distributive.

(b) Let $(G, +)$ be a group. Then $End\,G = \{f \in M(G) \mid (x + y)f = xf + yf \text{ for all } x, y \in G\}$, the set of all endomorphisms of G, consists of all the distributive elements of $M(G)$.

Here we notice that the set of all distributive elements of a near-ring N forms a semigroup under multiplication. In the rest of this section, N_d denotes the multiplicative subsemigroup of all distributive elements of the near-ring N.

Thus we have

2.3. Definition A near-ring $(N, +, \cdot)$ is called a *distributively generated near-ring* if N_d generates the whole N additively.

From the above definition, we immediately notice that every d. g. near-ring is zero-symmetric (a near-ring N is said to be *zero-symmetric* if $0 \cdot x = 0$ for all $x \in N$).

In the sequel, we emphasize only those results of endomorphism near-rings, a special class of d. g. near-rings, which were done in Taiwan.

Here we give a definition for endomorphism near-rings.

2.4. Definition Let $(G, +)$ be a group and S a subsemigroup of $End\,G$. Then (N, S), the sub-d. g. near-ring of $M_0(G)$ generated by S is called an *endomorphism near-ring* on G.

For each group G, there are three endomorphism near-rings which are of particular significance. They are

(i) The inner automorphism near-ring of G: The sub-d. g. near-ring of $M_0(G)$ which is generated by $Inn\,G$, the group of all inner automorphisms of G, is denoted by $I(G)$.

(ii) The automorphism near-ring of G: The sub-d. g. near-ring of $M_0(G)$ which is generated by $Aut\,G$, the group of all automorphisms of G, is denoted by $A(G)$.

(iii) The endomorphism near-ring of G: The sub-d. g. near-ring of $M_0(G)$ which is generated by $End\,G$, the semigroup of all endomorphisms of G, is denoted by $E(G)$.

Here we remark that

$$I(G) \leq A(G) \leq E(G),$$

where the symbol "$\leq$" denotes the phrase "is a subnear-ring of".

The detailed structure of near-rings of the form $I(G)$, $A(G)$ or $E(G)$ for various groups, say, finite simple nonabelian group, Engel 2-group, dihedral group, infinite dihedral group, generalized odd dihedral group, generalized quaternion group of order 2^n for $n \geq 3$, finite general linear group, symmetric group and direct sum of n copies of isomorphic finite simple nonabelian groups, were studied by several authors (see [67, 121]).

The following results obtained by Fong and Meldrum [65] outline structures of $E(S_n)$.

2.5. Theorem *Let S_n be the symmetric group of degree n, $n \geq 5$, and A_n the alternating group of degree n. Then*

$$E(S_n) = A(S_n) = I(S_n) = N$$

and N has an ideal I such that

$$I^2 = \{0\},$$
$$N/I \cong M_0(A_n) \oplus \mathbb{Z}_2,$$
$$I = \{\alpha \in M_0(S_n) \mid S_n\alpha \subseteq A_n \text{ and } A_n\alpha = 0\}.$$

Here $\oplus$ indicates the direct sum and $\mathbb{Z}_2$ the ring of integers modulo 2. Furthermore,

$$N = I + M_0'(A_n) + Z_2 \tag{A}$$

where $M_0'(A_n)$ is a subnear-ring of N which is isomorphic to $M_0(A_n)$, and acts in the natural way on A_n, annihilating $S_n \setminus A_n$ and $Z_2 = \{0, \theta\}$ is a subnear-ring, where $\theta \in \operatorname{End} S_n$ maps $S_n \setminus A_n$ to (12) and A_n to 0.

From (A) we can represent an arbitrary element of N as

$$\nu + \beta + \alpha \text{ or } (\nu, \beta, \alpha)$$

where $\nu \in I$, $\beta \in M_0'(A_n)$ and $\alpha \in Z_2$.

A cumbersome calculation shows that every ν can be written in the following two forms

$$\nu = \sum_{i=1}^{m+1} \delta_i \zeta_i$$

or

$$\nu = \sum_{i=1}^{m+1} \nu_{ij(i)}.$$

Here we denote $A_n = \{0 = g_0, g_1, \ldots, g_m\}$, $S_n \setminus A_n = \{g_{m+1}, \ldots, g_{2m+1}\}$, $m = (n!)/2 - 1$, δ_i maps g_{i+m} to g_1 and $S_n \setminus \{g_{i+m}\}$ to 0, ζ_i maps g_i to itself and $A_n \setminus \{g_i\}$ to 0, and ν_{ij} maps g_{m+i} to g_j and all other elements to 0. Thus, we can now give the addition and multiplication rules for $E(S_n)$.

2.6. Theorem *Let $N = E(S_n) = \{(\nu, \beta, \alpha) \mid \nu \in I, \beta \in M_0'(A_n), \alpha \in Z_2\}$. Then*
(a) $(\nu, \beta, \alpha) + (\nu', \beta', \alpha') = (\nu + \nu'^\alpha, \beta + \beta', \alpha + \alpha')$,
(b) $(\sum_{i=1}^{m+1} \delta_i \zeta_i, \beta, 0)(\nu', \beta', \alpha') = (\sum_{i=1}^{m+1} \delta_i \zeta_i \beta', \beta\beta', 0)$,
(c) $(\sum_{i=1}^{m+1} \nu_{ij(i)}, \beta, \theta)(\sum_{i=1}^{m+1} \delta_i \zeta_i, \beta', \alpha) = (\sum_{i=1}^{m+1} \delta_i \zeta_{j(i)+1}, \beta\beta', \alpha)$.

Here we present some results for ideals.

2.7. Theorem *The following is a complete list of right ideals of $E(S_n)$, $n \geq 5$:*

$$\sum_{j \in J} \delta_j M_0'(A_n) + \sum_{i \in K} \epsilon_i M_0'(A_n), \sum_{i \in K} \epsilon_i M_0'(A_n) + I + Z_2,$$

where $J \subseteq \{1, 2, \ldots, m+1\}$, $K \subseteq \{1, 2, \ldots, m\}$.

2.8. Theorem *The following is a complete list of left ideals of $E(S_n)$, $n \geq 5$ and is at the same time a complete list of ideals of $E(S_n)$, $n \geq 5$:*

$$\{0\}, I, I + Z_2, M_0'(A_n) + I, N.$$

The following theorem is a combination of the results due to Fong and Meldrum [66], and Fong and Kaarli [59].

2.9. Theorem *Let S_4 be the symmetric group of degree 4, A_4 the alternating group of degree 4 and $V_4 = \{0, (12) + (34), (13) + (24), (14) + (23)\}$ the Vier-gruppe. Then*

$$E(S_4) = A(S_4) = I(S_4)$$

and

$$E(S_4)/N \cong E(S_3) \oplus M_2(\mathbb{Z}_2)$$

where $N = \{\alpha \in E(S_4) \mid S_4 \alpha \subseteq V_4, V_4 \alpha = 0\}$, the ideal of $E(S_4)$, and $|N| = 4^{15}$, $M_2(\mathbb{Z}_2)$ is the ring of 2×2 matrices over the Galois field of order 2, and $E(S_3)$ is the endomorphism near-ring of S_3, the result of Lyons and Malone [115].

A canonical form for the elements of $E(S_4)$ and the rules for their addition and multiplication are given in Fong and Meldrum [66] or Fong's Ph. D. thesis [50], but are too complicated to reproduce here. However, we do present some more facts for $E(S_4)$:

(i) The size of $E(S_4)$ is $|E(S_4)| = 2^{35} \cdot 3^3 = 927{,}712{,}935{,}936$.

(ii) All the radicals J_0, $J_{1/2}$, J_1, J_2, N and P coincide and the radical J is nilpotent of nilpotency class 3. (For the definition of the above radicals, please refer to the book of Meldrum [121]). Also,

$$E(S_4)/J \cong \mathbb{Z}_2 \oplus \mathbb{Z}_3 \oplus M_2(\mathbb{Z}_2).$$

A complete list of maximal right ideals of $E(S_4)$ is also available in Fong and Meldrum [66].

Now we turn our attention to endomorphism near-rings of a direct sum of isomorphic finite simple non-abelian groups.

Again, the following theorems are due to Fong and Meldrum [67].

2.10. Theorem *Let G be an arbitrary finite simple nonabelian group and $H = \oplus^n G$, the external direct sum of n copies of G. Then*

(i) *$End H = \{[\alpha_{ij}] \mid \alpha_{ij} \in End G, 1 \leq i, j \leq n\}$ is the set of all $n \times n$ matrices which have at most one nonzero entry in each column;*

(ii) *$E(H) \supseteq \oplus^n M_0(G)$;*

(iii) *H is an $E(H)$-module of type 2 and $E(H)$ is 2-primitive.*

We close this section with some nice theorems on the minimal generating sets of $E(D_n)$, the endomorphism near-ring of the dihedral group D_n of order $2n$.

In 1963 [85], L. Fuchs had raised the following famous question: *For which abelian group G does the automorphism group of G generate the endomorphism ring $End G$?* R. S. Pierce [128], R. W. Stringall [132], H. Freedman [76] and F. Castagna [23] gave certain results on both the positive and negative sides in the subsequent five years.

Here one can immediately notice that $I(G) \subseteq A(G) \subseteq E(G)$. Somewhat parallel to the spirit of Fuchs' question, A. Fröhlich [79] (C. G. Lyons and J. J. Malone [115], [116]) in his (their) early work showed that $I(G) = A(G) = E(G)$ if G is a finite simple group (finite dihedral group of order $2n$, n odd). Here one immediately realizes that the above results are far better than that of Castagna and the others since the generating set of $E(G)$ is $Inn\, G$ which is obviously more economical and applicable than that of $Aut\, G$.

Now we arrive at an even more difficult and hence more economical and applicable question for the minimal generating sets of the above mentioned near-rings. Fong and Ke [63] at the 1989 Oberwolfach Near-rings and Near-fields conference (Germany), presented a very satisfactory result: Only three *fixed elements* from the generating set $End\, D_n$ are needed to generate additively the whole $E(D_n)$ whenever $n \geq 3$ and n is odd. Unlike the odd case, if $n > 3$ and n is even, we have $I(D_n) \subset A(D_n) \subset E(D_n)$. So we need to discuss the problem in three different cases. These results are due to Fong, Huang and Ke [54].

We end this section with the following four theorems on the minimal generating sets of $E(D_n)$, $A(D_n)$ and $E(D_n)$. (See [62, 54]).

2.11. Theorem   *Let $D_n = \langle a, b \mid na = 2b = 0, a + b = b - a \rangle$ be the dihedral group of order $2n$, $n \geq 3$. If n is odd and $i, j \in \{0, 1, 2, \ldots, n-1\}$ and $(i - j, n) = 1$, then $\{\rho_0, \varphi_{ia+b}, \varphi_{ja+b}\}$ is a minimal generating set of $E(D_n)$. Here, $\rho_0 = id_{D_n}$ is the identity automorphism of D_n, and $\varphi_x \in End\, D_n$ sends the normal subgroup $\langle a \rangle$ to the additive identity 0 and the coset $\langle a \rangle + b$ to the assigned element x where $x \in \{ia + b \mid i = 0, 1, \ldots, n-1\}$.*

2.12. Theorem   *Let $n \geq 4$ be even and $D_n = \langle a, b \mid na = 0 = 2b, a + b = b - a \rangle$. Then*

$$I(D_n) = \langle \{\rho_0, \rho_a, \rho_b\}, + \rangle,$$

where ρ_x is the inner automorphism induced by x.

In order to determine some minimal generating sets for $A(D_n)$ and $E(D_n)$ with even n, we need to introduce a few mappings from $M(D_n)$, the near-ring of all mappings from D_n into itself. Let $A_0 = \langle a \rangle$, $A_1 = A_0 \setminus \langle 2a \rangle$, $B_0 = A_0 + b$, $B_1 = A_1 + b$, and $B_2 = B_0 \setminus B_1$. For any $x \in D_n$, define the mappings $\lambda_{0,x}, \lambda_{1,x}$, and $\lambda_{2,x}$ from D_n into D_n via

$$y\lambda_{0,x} = \begin{cases} x, & \text{if } y \in B_0; \\ 0, & \text{otherwise}; \end{cases}$$

$$y\lambda_{1,x} = \begin{cases} x, & \text{if } y \in A_1 \cup B_1; \\ 0, & \text{otherwise}; \end{cases}$$

$$y\lambda_{2,x} = \begin{cases} x, & \text{if } y \in A_1 \cup B_2; \\ 0, & \text{otherwise}; \end{cases}$$

A routine check shows that the above three mappings are endomorphisms of D_n. Moreover, for $j \in \{0, 1, 2, \ldots, n-1\}$, define $\Omega_j : D_n \to D_n$ via

$$(sa + \epsilon b)\Omega_j = jsa,$$

for $s \in \{0, 1, \ldots, n-1\}$ and $\epsilon \in \{0, 1\}$.

2.13. Theorem   *$A(D_n) = \langle \{\rho_0, \rho_b, \alpha\}, + \rangle$, where α is an outer automorphism of D_n and, in fact, $\alpha = \Omega_1 + \lambda_{0,a+b}$.*

2.14. Theorem   *The set $\{id_{D_n}, \lambda_{0,b}, \lambda_{0,a+b}, \lambda_{1,b}, \lambda_{1,a+b}\}$ is a minimal generating set of $E(D_n)$ for even n with $n \geq 4$.*

2.15. Remark In the same spirit, readers are invited to investigate all the algebraic structures of $I(G)$, $A(G)$ and $E(G)$ for various classes of groups G and at the same time their minimal generating sets.

§3. Syntactic near-rings

Automata theory gives rise to a lot of interesting and good mathematics, as well as applications. Techniques of obtaining near-rings from automata were first initiated by M. Holcombe [94]. For certain semiautomata, one obtains near-rings in a natural way, and these near-rings can be related and analogous to d. g. near-rings. We start with

3.1. Definition A *semiautomaton* is a triple $\mathcal{S} = (G, I, \delta)$ where G is the *state set*, I is the *input set* and $\delta : G \times I \to G$ is the *transition function*. If we impose a group structure on G, we then call $\mathcal{S}$ a *group-semiautomaton*, in short *GSA*.

Given any GSA $\mathcal{S} = (G, I, \delta)$, the transition function δ defines a family of mappings $M_\delta = \{f_a : G \to G \mid a \in I\}$ where $x f_a = \delta(x, a)$. The group $(G, +)$ has a near-ring $(M(G), +, \circ)$ and $M_\delta \subseteq M(G)$. So $M_\delta \cup \{id_G\}$ generates a subnear-ring of $M(G)$, denoted by $N(\mathcal{S})$ and called the *syntactic near-ring* of $\mathcal{S}$. Thus, $N(\mathcal{S})$ is a near-ring with identity. If G is finite, then $N(\mathcal{S})$ is finite too.

The following two theorems were first presented by Fong and Ke at the General Algebra conference in Krems (Austria), 1988.

3.2. Theorem *If $(G, +)$ is a group with finite exponent $n > 0$ and $\mathcal{S} = (G, G, \delta)$ a GSA with $\delta(x, y) = x + y$ for all $x, y \in G$ then*

$$N(\mathcal{S}) = \left\{ \sum_{i=1}^{m} f_{a_i} \,\middle|\, a_i \in G, m \geq 1 \right\}.$$

3.3. Theorem *If $(G, +)$ is a group with finite exponent $n > 0$ and $\mathcal{S} = (G, G, \delta)$ a GSA with $\delta(x, y) = x + y$ for all $x, y \in G$ then*

$$N(\mathcal{S}) = I(G) + M_C(G),$$

where $I(G)$ is the inner automorphism near-ring of G and $M_C(G)$ is the constant subnear-ring of $M(G)$.

Let $(G, +)$ be a finite group and $\sigma, \tau \in M(G)$. If we define a mapping $\delta_{(\sigma, \tau)} : G \times G \to G$ via $\delta_{(\sigma, \tau)}(x, y) = x\sigma + y\tau$ for all $x, y \in G$, then $\mathcal{S}_{(\sigma, \tau)} = (G, G, \delta_{(\sigma, \tau)})$ is a GSA. The the corresponding syntactic near-ring is denoted by $N_{(\sigma, \tau)}$ and is generated by $M_{\delta_{(\sigma, \tau)}} = \{f_a \mid a \in G\} \cup \{id_G\}$ where $x f_a = \delta_{(\sigma, \tau)}(x, a) = x\sigma + a\tau$, and id_G is the identity mapping of G. Thus we have the following results (see Y. Fong, W.-F. Ke, C.-S. Wang [64]).

3.4. Theorem *Let $(G, +)$ be a finite group and $\sigma \in M(G)$ be fixed. If $\rho_a, \rho_b \in \operatorname{Inn} G$, then $N_{(\sigma, \rho_a)} = N_{(\sigma, \rho_b)}$.*

3.5. Theorem *Let $(G, +)$ be a finite group and $\tau \in M(G)$. If $m, n \in \mathbb{Z}$ and $\bar{m}, \bar{n} \in M(G)$ such that $\bar{m} : G \to G$ is a mapping that sends x into mx for all $x \in G$, then $N_{(\bar{m}, \tau)} = N_{(\bar{n}, \tau)}$.*

The following theorem is just an application of (3.5) together with a cumbersome calculation.

3.6. Theorem   Let $G = S_3$, the symmetric group of degree 3. Then there are exactly three different syntactic near-rings of $\mathcal{S} = (G, G, \delta_{(\bar{m},\bar{k})})$ no matter what the values of $m, k \in \mathbb{Z}$ and their sizes are:

$$|N_{(\bar{m},\bar{k})}| = \begin{cases} 324, & \text{if } k \equiv 1 \pmod 2; \\ 54, & \text{if } k \equiv 2 \text{ or } 4 \pmod 6; \\ 6, & \text{if } k \equiv 0 \pmod 6. \end{cases}$$

Now we turn our attention to the following cases:

(i) $G = (\mathbb{Z}_n, +)$, the cyclic group of order n;
(ii) G is an abelian group.

3.7. Theorem (Y. Fong and J. R. Clay [53])   Let $\mathcal{S} = (\mathbb{Z}_n, \mathbb{Z}_n, \delta)$ be a GSA with $\delta : \mathbb{Z}_n \times \mathbb{Z}_n \to \mathbb{Z}_n$ via $\delta(x,y) = x\varphi_m + y\varphi_k = mx + ky$ for all $x, y \in \mathbb{Z}_n$, where $\varphi_m, \varphi_k \in End\,\mathbb{Z}_n$. Then

(i) $N(\mathcal{S}) = End\,\mathbb{Z}_n + M_C(\mathbb{Z}_n \mid \langle k \rangle)$;
(ii) $N(\mathcal{S})/M_C(\mathbb{Z}_n \mid \langle k \rangle) \cong (\mathbb{Z}_n, +, \cdot)$;
(iii) If $(n, k) = 1$, then $N(\mathcal{S}) = End\,\mathbb{Z}_n + M_C(\mathbb{Z}_n)$. Here $End\,\mathbb{Z}_n$ is the endomorphism ring of $\mathbb{Z}_n$, $(\mathbb{Z}_n, +, \cdot)$ is the ring of integers modulo n and $M_C(\mathbb{Z}_n \mid \langle k \rangle) = \{\theta_x \in M_C(\mathbb{Z}_n) \mid x \in \langle k \rangle\}$ is a subnear-ring of the constant near-ring $M_C(\mathbb{Z}_n)$ where $y\theta_x = x$ for all $y \in \mathbb{Z}_n$ and $\langle k \rangle$ is the subgroup of $(\mathbb{Z}_n, +)$ generated by k.

3.8. Theorem (Y. Fong [52])   Let $N(\mathcal{S}) = \{\varphi_t + \theta_x \mid \varphi_t \in End\,\mathbb{Z}_n, \theta_x \in M_C(\mathbb{Z}_n \mid \langle k \rangle)\}$ be as mentioned above. Then

$$(\varphi_t + \theta_x) + (\varphi_s + \theta_w) = \varphi_{t+s} + \theta_{x+w}$$

and

$$(\varphi_t + \theta_x)(\varphi_s + \theta_w) = \varphi_{ts} + \theta_{xs+w}.$$

Furthermore, $|N(\mathcal{S})| = n \cdot |\langle k \rangle|$. In particular, if $(n, k) = 1$, then $|N(\mathcal{S})| = n^2$.

The above theorems are generalized results of Fong and Clay [53].

3.9. Theorem (Y. Fong and J. R. Clay [53])   Let A be an abelian group of order m and exponent n. If $\mathcal{S} = (A, A, \delta)$ with $\delta(x, y) = x + y$ for all $x, y \in A$, then

(i) $|N(\mathcal{S})| = mn$;
(ii) $N(\mathcal{S}) \cong \mathbb{Z}_n \oplus A$ as abstract affine near-rings and A is an $\mathbb{Z}_n$-module.

3.10. Theorem (Y. Fong and J. R. Clay [53])   Let $\mathcal{S} = (\mathbb{Z}, \mathbb{Z}, \delta)$ with $\delta(x, y) = x + y$ for all $x, y \in \mathbb{Z}$. Then

$$N(\mathcal{S}) = End\,\mathbb{Z} + M_C(\mathbb{Z}_n) \quad \text{and} \quad |N(\mathcal{S})| = \aleph_0.$$

Before we close this section, we wish to point out that structure theorems for syntactic near-rings for nonabelian groups are rather complicated and are omitted here. Readers are encouraged to make a detailed study of the recent paper by Clay and Fong [35] on dihedral groups.

§4. Semi-endomorphism near-rings and infra-endomorphism near-rings

It is a well-known fact that $End\,G$ is a ring with identity under the usual pointwise addition and composition of mappings if G is an abelian group. If G is not abelian, then the sum of two arbitrary endomorphisms is not necessarily an endomorphism. However, as shown in §2, the set of all endomorphisms of a nonabelian group does generate additively a near-ring with identity. Such a near-ring is called the endomorphism near-ring of G. For four decades, $E(G)$ was studied by several authors for different classes of groups (see Meldrum [121]). With a slight different concept than that of an endomorphism of a group, Herstein [91] in 1968 called a group mapping $f : G \to G$ a semi-endomorphism if $(x+y+x)f = xf+yf+xf$ for all $x, y \in G$. He was interested only in developing some theorems for when those semi-endomorphisms were endomorphisms.

In this section we first present some very interesting results for those near-rings which are generated by $SEnd\,G$, the set of all semi-endomorphisms of G. We denote these near-rings by $S(G)$, and call them the semi-endomorphism near-rings of G. All those works on semi-endomorphism presented here are due to Fong and Pilz [68], and Fong, Huang, Ke and Yeh [56].

In the same spirit of pursuing new classes of near-rings of group mappings and from a slight different point of view, we call a group mapping $f : G \to G$ an infra-endomorphism if $(x + x)f = xf + xf$ for all $x \in G$. Thus, the last part of this section is dedicated to those near-rings of group mappings that are generated by $IEnd\,G$, the set of all infra-endomorphisms. We denote these near-rings by $\mathcal{I}(G)$ and call them the infra-endomorphism near-rings of G. The work is due to Fong and Yeh [74], and Bouchard, Fong, Ke and Yeh [21].

Before we proceed any further, we first want to point out that all endomorphisms and anti-endomorphisms are both semi-endomorphisms and infra-endomorphisms but the converses are not true in general.

4.1. Remark   Here one can easily notice that the two subsets $SEnd\,G$ and $IEnd\,G$ of $M(G)$ are not comparable. For if we let $G = (S_3, +)$, where $S_3 = \{0 = (1), a = (12), b = (13), c = (23), d = (123), e = (132)\}$ and write maps from the right, we get the addition table

+	0	a	b	c	d	e
0	0	a	b	c	d	e
a	a	0	d	e	b	c
b	b	e	0	d	c	a
c	c	d	e	0	a	b
d	d	c	a	b	c	0
e	e	b	c	a	0	d

Again, if we take $f, g \in M(S_3)$ where

$$xf = \begin{cases} x, & \text{if } x \neq b; \\ a, & \text{if } x = b, \end{cases}$$

and

$$xg = a \quad \text{for all } x \in S_3,$$

then a routine check shows that $f \in IEnd\,S_3$ but $f \notin SEnd\,S_3$ and $g \in SEnd\,S_3$ but $g \notin IEnd\,S_3$.

Now we divide this section into the following two subsections.

A. Semi-endomorphism near-rings

We start with the following results.

4.2. Lemma (I. N. Herstein [91]) If $f \in SEnd\,G$, then $(-x)f = -(xf)$ holds for all $x \in G$.

4.3. Lemma ([70]) If $f \in SEnd\,G$ and G is an abelian group containing no nonzero elements of order 2, then $f \in End\,G$ and hence $SEnd\,G = End\,G$.

4.4. Lemma ([68]) $SEnd\,G$ contains all constant mappings with (constant) values of order 2.

Now we present some basic properties of $S(G)$ in the following:

4.5. Theorem (i) $(End\,G, \circ) \leq (SEnd\,G, \circ)$ as subsemigroups.
(ii) $(E(G), +, \circ) \leq (S(G), +, \circ)$ as subnear-rings.

In [52], it was shown that $E(G) + M_C(G) \leq (M(G), +, \circ)$. Thus we can write down a Hasse diagram as follows:

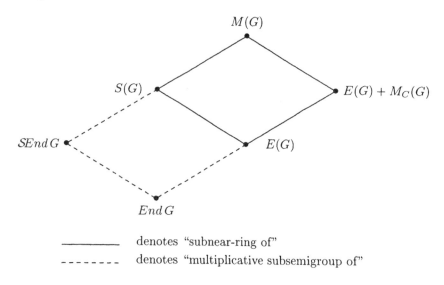

_____ denotes "subnear-ring of"

- - - - - - denotes "multiplicative subsemigroup of"

Figure A

The rest of the results in this subsection can be found in Fong and Pilz [68] and Fong, Huang, Ke and Yeh [56].

4.6. Lemma Let H be an abelian subgroup of an arbitrary group G and let $SEnd\,_H G = \{f \in SEnd\,G \mid Gf \subseteq H\}$. Then $(SEnd\,_H G, +, \circ)$ is an abelian subnear-ring of $S(G)$.

4.7. Corollary If G is an abelian group, then $S(G) = SEnd\,G$ is an abelian near-ring with identities.

4.8. Corollary If every element of G is of order 2 then $SEnd\,G = S(G) = M(G)$.

From (4.4) we get

4.9. Lemma *If the elements of order 2 generate G, then $M_C(G) \subseteq S(G)$.*

4.10. Theorem *If for some $n \in \mathbb{N}$, $n \geq 2$, the set $\{x \in G \mid nx = 0\}$ forms a normal subgroup N of G, then $S(G) \leq \{f \in M_0(G) \mid Nf \subseteq N\} + M_C(G)$. In particular, $S(G)$ is then strictly smaller than $M(G)$ unless $N = G$.*

4.11. Theorem *If $N = \{x \in G \mid nx = 0\}$ as in (4.10) is cyclic and normal, then*

$$S(G) \leq \{f \in M_0(G) \mid f|_N \in End\,(N, +)\} + M_C(G).$$

4.12. Theorem *If G is an abelian group, then*

$$SEnd\,_0 G = \{f \in SEnd\,G \mid 0f = 0\} \subseteq M_{\mathbb{Z}}(G),$$

where $M_{\mathbb{Z}}(G) = \{f \in M(G) \mid f(ng) = nf(g) \text{ for all } g \in G, n \in \mathbb{Z}\}$.

4.13. Theorem *If every nonzero semi-endomorphism of G is either an anti-automorphism or an automorphism (Herstein [91] described such a case), then*

$$S(G) \leq \{f \in M_0(G) \mid Z(G)f \subseteq Z(G)\} + M_C(G),$$

where $Z(G)$ is the center of the group G.

We close this subsection with the following theorems for the semi-endomorphisms for special classes of groups.

4.14. Theorem *Let $(\mathbb{Z}, +)$ be the group of integers and $(\mathbb{Z}_n, +)$ the group of integers modulo n. Then we have*
   (i) *$S(\mathbb{Z}) \cong (\mathbb{Z}, +, \cdot)$, the ring of integers.*
   (ii) *$S(\mathbb{Z}_{2n+1}) \cong (\mathbb{Z}_{2n+1}, +, \cdot)$, the ring of integers modulo n.*
   (iii) *$S(\mathbb{Z}_{2n}) \cong \mathbb{Z}_{2n} \oplus \mathbb{Z}_2$ as groups; $S(\mathbb{Z}_{2n})$ is an abstract affine near-ring.*
   (iv) *If $G = \oplus^n \mathbb{Z}_2$, then $S(G) = M(G)$.*
   (v) *$S(\mathbb{Z}_2) \times S(\mathbb{Z}_2) = M(\mathbb{Z}_2) \times M(\mathbb{Z}_2) \ncong M(\mathbb{Z}_2 \times \mathbb{Z}_2)$. So we see that $S(G \times H) \ncong S(G) \times S(H)$ in general.*
   (vi) *If $G = \oplus^n \mathbb{Z}_p$, where p is an odd prime, then $S(G) \leq M_0(G)$.*

4.15. Theorem *Let $G = \mathbb{Z}_m \oplus \mathbb{Z}_n$. Then*

$$|S(G)| = \begin{cases} mn \cdot (m, n)^2, & \text{if } 2 \nmid mn; \\ 16 \cdot mn \cdot (m, n)^2, & \text{if } 2 \mid m \text{ and } 2 \mid n; \\ 2 \cdot mn \cdot (m, n)^2, & \text{otherwise.} \end{cases}$$

4.16. Theorem *If G is a finite simple nonabelian group, then $S(G) = M(G)$.*

4.17. Theorem *Let $\mathbb{Z}_p^\infty$ be the quasicyclic groups. Then $S(\mathbb{Z}_p^\infty) \leq \{f \in M_0(\mathbb{Z}_p^\infty) \mid f|_H \in End\,H, \text{ for all } H \leq \mathbb{Z}_p^\infty\} + M_C(\mathbb{Z}_p^\infty)$.*

4.18. Theorem *If G is a finite nilpotent group and S_p (p a prime) is the p-Sylow-subgroup of G, then*

$$S(G) \leq \cap_{p\in\mathbf{P}}\{f \in M_0(G) \mid S_p f \subseteq S_p\} + M_C(G),$$

where $\mathbf{P}$ denotes the set of all primes. Hence if G is not a p-group, then $S(G) < M(G)$.

4.19. Theorem *Let S_n be the symmetric group of degree n. Then $M_C(S_n) \subseteq S(S_n)$.*

4.20. Theorem *The following near-rings are equal:*
(i) $S(S_3)$,
(ii) $E(S_3) + M_C(S_3)$,
(iii) $N(\mathcal{S})$, *the syntactic near-ring of GSA $\mathcal{S} = (S_3, S_3, \delta)$ with $\delta(x,y) = x + y$.*
(iv) $P(S_3)$, *the near-ring of all polynomial functions on S_3.*

Conjecture: $S(S_n) = E(S_n) + M_C(S_n)$ holds for all $n \in \mathbf{N}$.

Here we wish to point out that by (4.19), $E(S_n) + M_C(S_n) \subseteq S(S_n)$, but the other inclusion is still open.

B. Infra-endomorphism near-rings

The following theorem is immediate.

4.21. Theorem *Let G be a group. Then*
(i) $(End\,G, \circ) \leq (IEnd\,G, \circ) \leq (M_0(G), \circ)$,
(ii) $(E(G), +, \circ) \leq (\mathcal{I}(G), +, \circ) \leq (M_0(G), +, \circ)$.

Thus we can generalize the diagram of Figure A into the following Figure B:

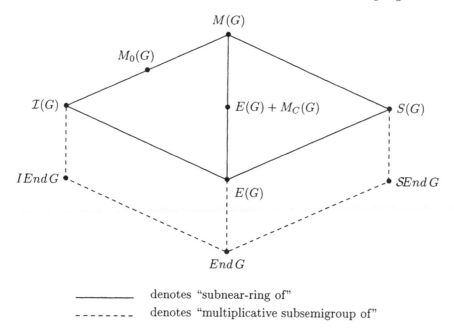

_____ denotes "subnear-ring of"

- - - - - - denotes "multiplicative subsemigroup of"

Figure B

4.22. Lemma *Let H be an abelian subgroup of a group G and $IEnd_H G = \{f \in IEnd\, G \mid Gf \subseteq H\}$. Then $(IEnd_H G, +, \circ)$ is an abelian subnear-ring of $\mathcal{I}(G)$.*

4.23. Corollary *If G is an abelian group, then $\mathcal{I}(G) = IEnd\, G$ is an abelian near-ring with identity.*

4.24. Corollary *If every element of a group is of order 2, then $IEnd\, G = \mathcal{I}(G) = M_0(G)$.*

4.25. Theorem *If G is a finite simple nonabelian group, then $\mathcal{I}(G) = I(G) = A(G) = E(G) = M_0(G)$.*

4.26. Definition For any $x, m \in \mathbb{Z}$, we define $\mathrm{Ord}_m(x) = k$ if

$$|\{1, x, x^2, \ldots \ (\mathrm{mod}\ m)\}| = k.$$

4.27. Theorem *Let $p \in \mathbf{P} \setminus \{2\}$. Then*

(i)
$$|\mathcal{I}(\mathbb{Z}_{2^m})| = 2^{2^m - 1}\ \text{for all}\ m \in \mathbb{N}$$

(ii)
$$|\mathcal{I}(\mathbb{Z}_{p^m})| = \prod_{i=0}^{m-k-1} (p^{m-i})^{p^{m-i-1}(p-1)/p^{m-k-i}\mathrm{Ord}_p(2)}$$
$$\times \prod_{i=m-k}^{m-1} (p^k)^{p^{m-i-1}(p-1)/\mathrm{Ord}_p(2)}.$$

4.28. Corollary *$\mathcal{I}(\mathbb{Z}_m) = End\, \mathbb{Z}_m$ if and only if 2 is a strictly primitive root over $\mathbb{Z}_m$.*

§5. Centralizer Near-rings.

It is well-known that primitive rings play an important role in the structure theory of rings. The celebrated Jacobson Density Theorem states that these rings are dense subrings of complete rings of linear transformations of linear spaces over skew fields. A nonlinear analog of a linear transformation ring is a centralizer near-ring. Let G be an additively written (not necessarily abelian) group and A a group of its automorphism. The set

$$\mathcal{C}_G(A) = \{f : G \to G \mid 0f = 0 \quad \text{and} \quad fa = af \quad \text{for all} \quad a \in A\}$$

endowed with pointwise addition and composition is a left near-ring and is called the *centralizer near-ring* on G determined by A. Analogously to that of the ring case, the structure theory of 2-primitive near-rings is closely connected with centralizer near-rings (see [12]).

The ideal structure of linear transformation rings is well-known and easily determined. In contrast with this, the ideal structure of centralizer near-rings is very complicated and has, in fact, not been completely described up to now, although these near-rings play an important role in the structure theory of near-rings (see [121] and [129]). The study of the ideal structure of centralizer near-rings was carried forward in the last fifteen years mainly in the papers of C. J. Maxson, J. D. P. Meldrum, A. Oswald, K. C. Smith and M. Zeller (see [87],[119], [120], [122], [123]). Starting from the celebrated theorem of G. Berman and R. J. Silverman [11] which established the simplicity of $M_0(G)$ (provided $|G| > 2$), the research in this area was mainly concentrated on finding sufficient conditions for simplicity of $C_G(A)$. All results (for infinite G) were obtained under the assumption that the stabilizers $St_A(g) = \{a \in A \mid ga = g\}$ are conjugate. For finite group G Maxson and Smith [120] had proved that the simplicity of $C_G(A)$ is equivalent to all stabilizers $St_A(g)$, $g \in G$, are conjugate. We note that in this case they are maximal also. The following question was raised by Meldrum and Zeller [123] (see also [129], Problem 22).

Problem: *Does $C_G(A)$ simple imply all stabilizers are maximal and conjugate?*

This problem was recently completely solved in a joint paper by K. I. Beidar, Y. Fong and W.-F. Ke [5]. Namely the following results were proved.

5.1. Theorem *If A is a nilpotent group and $C_G(A)$ is simple, then:*
(i) *All stabilizers are maximal and conjugate;*
(ii) *$C_G(A)$ is a von Neumann regular near-ring.*

5.2. Theorem *Let F be a countable field and V a linear space over F of countable dimension. Then there exists a solvable subgroup A of the automorphism group $\mathrm{GL}_F(V)$ of class 2 such that:*
(i) *$C_G(A)$ is simple;*
(ii) *$C_G(A)$ is not regular*
(iii) *There is no nonzero element of V with maximal stabilizer;*
(iv) *Not all stabilizers are conjugate.*

§6. Near-ring Multiplications on Groups

In this section, we use right near-rings instead of left near-rings for a certain reason of convenience.

In a number of papers the problem of definition of nontrivial near-ring multiplication on an arbitrary or given group was investigated (see [25, 26, 27, 28, 34, 37, 43, 44, 99, 110, 111, 112, 117, 118] and also [129] for the survey on this topic). H. E. Heatherly [89] and R. Dover [41] had considered a question of existence of distributive near-ring multiplication on groups. In a series of papers, J. R. Clay had investigated the existence of nontrivial near-ring multiplication in terms of coupling maps.

It is well-known that any ring R whose additive group is torsion divisible has a trivial multiplication (i.e., $R^2 = 0$). Such an abelian group $(A, +)$ possessing only trivial ring structure, is called a nil group [33, p. 15]. It is natural to ask what would be an analogous concept for right near-rings? If $(G, +)$ is any group, and $T \subseteq G \setminus \{0\}$, define $\cdot_T$ by the rule

$$a \cdot_T b = \begin{cases} a, & \text{if } b \in T; \\ 0, & \text{if } b \notin T. \end{cases}$$

Then $(G, +, \cdot_T)$ is a right zero-symmetric near-ring. This multiplication $\cdot_T$ is called *trivial*, as is the resulting near-ring. The following problem was posed in [33, Problem 2.16, p. 16]. Other than the groups of order 1 and 2, does there exist a group $(G, +)$ having the property that if $(G, +, \cdot)$ is a right near-ring, then $\cdot$ is a trivial multiplication?

Recently this problem was investigated in a joint paper written by K. I. Beidar, Y. Fong, W.-F. Ke and S.-Y. Liang [7]. Now we proceed to describe the results obtained. Before we go into a detailed discussion, we recall some definitions. Given any semigroup S with 0, a set G is called a *right $S, 0$-act* if there exists a fixed element $0 \in G$ and a mapping $G \times S \to G, (g, s) \mapsto gs$ such that the following conditions are satisfied:

(1) $0s = 0$ for all $s \in S$;
(2) $g0 = 0$ for all $g \in G$;
(3) $g(st) = (gs)t$ for all $s, t \in S$ and $g \in G$.

6.1. Example (1) Let $(G, +)$ be a group and let S be a subsemigroup containing 0 of the semigroup $End\,(G)$ of all endomorphisms of G. Then G and $End\,(G)$ are right $S, 0$-acts canonically.

(2) Let G be a right near-ring and $S = (G, \cdot)$ its multiplicative semigroup. Then G is a right $S, 0$-act canonically.

Let G and H be two right $S, 0$-acts. A mapping $f : G \to H$ is said to be a *homomorphism* of the $S, 0$-acts if $f(gs) = f(g)s$ for any $s \in S$ and $g \in G$.

6.2. Example Let N be a right near-ring and let $S = \{R_n \mid n \in N\} \subseteq End\,(G)$ be the subsemigroup of right multiplications (i.e., $xR_n = xn$ for all $x, n \in N$). Then N is a right $S, 0$-act and the mapping $f : N \to End\,(N)$ given by $f(n) = R_n$ is a homomorphism of right $S, 0$-acts.

The following result is well-known in different terminology and notations (see [129, § 9d]).

6.3. Proposition *Let $(G, +)$ be a group and $f : G \to End(G)$ a mapping such that the following conditions are fulfilled:*

(a) *$S = f(G)$ is a subsemigroup of $End\,(G)$;*
(b) *$0 \in S$;*
(c) *$f : G \to End\,(G)$ is a homomorphism of right $S, 0$-acts.*

Define a multiplication $\cdot = \cdot_f$ on G by the rule $a \cdot b = af(b)$ for all $a, b \in G$.

Then:

(1) *$(G, 0, +, \cdot)$ is a (right zero-symmetric) near-ring;*
(2) *the multiplication $\cdot$ is trivial if and only if $S \subseteq \{0, id_G\}$;*
(3) *$(G, 0, +, \cdot)$ is a ring if and only if (i) G is an abelian group, (ii) S is a subring of the ring $End\,(G)$, and (iii) f is a homomorphism of right S-modules.*

6.4. Example (1) Keep the notations of (6.2). Then the multiplication $\cdot_f$ coincides with the ordinary multiplication on the near-ring N. Thus any near-ring may be obtained as a result of the above construction.

(2) Let $(G, +)$ be a group and let $T \subseteq G \setminus \{0\}$. Define the mapping $f : G \to End\,(G)$ via

$$f(t) = \begin{cases} id_G, & \text{if } t \in T; \\ 0, & \text{if } t \notin T. \end{cases}$$

Then $S = f(G) \subseteq \{0, id_G\}$ (in fact, $S = \{0, id_G\}$ if and only if $T \neq \varnothing$) is a subsemigroup of $End\,(G)$ and f is a homomorphism of $S, 0$-acts. Clearly $\cdot_f = \cdot_T$.

(3) ([44]; [129, 1.4(b)]) Let H be a fixed point free automorphism group on $(G, +)$ (i.e., if $g \in G \setminus \{0\}$ and $h \in H \setminus \{id_G\}$, then $gh \neq g$.) Choose any subset $\mathcal{C} = \{U_i \mid i \in I\}$

of the set of all orbits of H and let $U = \{u_i \in U_i \mid i \in I\}$ be a set of representatives of the orbits in $\mathcal{C}$. Define a multiplication $*_U$ by

$$a *_U b = \begin{cases} 0, & \text{if } b \notin \bigcup_{i \in I} U_i; \\ ah_b, & \text{if } b \in U_i \text{ for some } i \in I \text{ and } u_i h_b = b. \end{cases}$$

We show that $*_U = \cdot_f$ for some mapping $f : G \to End(G)$.

Set $S = H \cup \{0\} \subseteq End(G)$. Clearly, S is a subsemigroup of $End(G)$. Define the mapping $f : G \to End(G)$ via

$$f(b) = \begin{cases} 0, & \text{if } b \notin u; \\ h_b, & \text{if } b \in U_i, \text{ and } u_i h_b = b. \end{cases}$$

Then $f(G) = S$. It is easy to check that f is a homomorphism of right $S, 0$-acts (also see (4.19) below), and that $*_U = \cdot_f$.

6.5. Theorem   *Let $(G, +)$ be a group and S a subsemigroup of $End(G)$ such that $\{0, id_G\} \subseteq S$, and let $a \in G$. Suppose for any $b \in G$ and $r \in S$, the relation $br \in aS$ implies the existence of a unique element $t = t(b, r) \in S$ satisfying $br = atr$. Assume further that $rt \neq 0$ and $rS \cap tS \neq \{0\}$ for all $r, t \in S \setminus \{0\}$. We set $H = \bigcup_{s \in S \setminus \{0\}} Ker(s)$. Then:*

(1) *The mapping $f : G \to S$ given by*

$$f(b) = \begin{cases} 0, & \text{if } bS \cap aS = \{0\}; \\ t, & \text{if } br = atr \neq 0 \text{ for some } r, t \in S \end{cases}$$

   is a homomorphism of right $S, 0$-acts.
(2) *H is an ideal of the near-ring $N = (G, 0, +, \cdot_f)$, $aS \cap H = 0$ and $b \cdot_f (c + g) = b \cdot_f g$ for all $b, g \in G$, $c \in H$.*
(3) *If a is a sum of distributive elements of the near-ring N, then $G = H + aS$.*
(4) *The following conditions are equivalent:*
   (i) *$(G, 0, +, \cdot_f)$ is a distributive near-ring;*
   (ii) *$(G, 0, +, \cdot_f)$ is a ring;*
   (iii) *G is an abelian group, S is a subring of the ring $End(G)$, and f is a homomorphism of right S-modules.*

6.6. Theorem   *Let $(G, +)$ be a group and $s \in End(G)$ an injective endomorphism of G. Let $a \in G$. Assume that (a) $as^t = a$ if and only if $s^t = id_G$ and (b) if s is not an automorphism, then $a \notin Gs$. Set $S = \{s^n \mid n = 1, 2, \ldots\} \cup \{0, id_G\}$ if s is not an automorphism, and $S = \{s^n \mid n \in \mathbb{Z}\} \cup \{0, id_G\}$ if s is an automorphism. Define a mapping $f : G \to S$ via*

$$f(b) = \begin{cases} 0, & \text{if } b \notin aS; \\ s^t, & \text{if } b = as^t \text{ for some } t \in \mathbb{Z}. \end{cases}$$

Then f is a homomorphism of $S, 0$-acts. Moreover, if we set N to be the near-ring $(G, 0, +, \cdot_f)$, then the following conditions are equivalent:

   (i) *a is the sum of some distributive elements of the near-ring N;*
   (ii) *N is a distributively generated near-ring;*
   (iii) *N is a ring;*
   (iv) *N is a finite field;*
   (v) *$G = aS$.*

Furthermore, any of the conditions (i)–(v) *implies that s is an automorphism of order* $p^n - 1$ *for some positive integers p and n, p a prime, and the group G is isomorphic to the direct sum of n copies of cyclic groups of order p.*

Using (6.6), we obtain as corollaries some conditions which, when imposed on a group G, ensure the existence of a nontrivial right near-ring multiplication $\cdot$ on G such that $(G, 0, +, \cdot)$ is not distributively generated.

6.7. Corollary   *Let $(G, +)$ be a group, and $s \in End\,(G) \setminus \{0\}$ an injective endomorphism of G which is not surjective. Then the group G has a nontrivial near-ring multiplication $\cdot$ such that $(G, 0, +, \cdot)$ is not a distributively generated near-ring.* ∎

6.8. Corollary   *Let $(G, +)$ be a group, $s \in Aut(G)$, $s \neq id_G$, and $a \in G$. Suppose that $o(s) = \infty$ and $as^n \neq a$ for all $n \neq 0$. Then the group G has a nontrivial near-ring multiplication $\cdot$ such that $(G, 0, +, \cdot)$ is not a distributively generated near-ring.* ∎

6.9. Corollary   *Let $(G, +)$ be any group which is not a cyclic group of prime order and let s be an automorphism of G of a finite order n, $n \geq 2$. Then G has a nontrivial near-ring multiplication $\cdot$ such that $(G, 0, +, \cdot)$ is not a distributively generated near-ring.*

6.10. Corollary   *Any abelian group G of order greater than 2 has a nontrivial near-ring multiplication $\cdot$ such that $(G, 0, +, \cdot)$ is not a distributively generated near-ring.*

6.11. Corollary   *Any finitely generated group G of order greater than 2 has a nontrivial near-ring multiplication $\cdot$ such that $(G, 0, +, \cdot)$ is not a distributively generated near-ring. In particular, the statement holds for any finite groups of order greater than 2.*

6.12. Proposition   *Let $(G, +)$ be a group and let $e \in End\,(G)$ be an idempotent endomorphism of G which is different from 0 and id_G. Then G has a nontrivial right near-ring multiplication $\cdot$ such that $N = (G, 0, +, \cdot)$ is not a distributively generated near-ring.*

6.13. Corollary   *If a group G is isomorphic to a semidirect product of two nonzero groups, then it admits a nontrivial right near-ring multiplication $\cdot$ such that the near-ring $N = (G, 0, +, \cdot)$ is not distributively generated.* ∎

For any group G, let $s \in End\,(G)$ and set $K(s) = \{g \in G \mid s^m \neq 0 \text{ and } gs^m = 0 \text{ for some } m \geq 0\}$. It is easy to check that $K(s)$ is a normal subgroup of G. With the help of $K(s)$, we have the following theorem.

6.14. Theorem   *Let $(G, +)$ be a group and let $s \in End\,(G) \setminus \{0\}$. Assume that $K(s) + \operatorname{Im} s \neq G$. Then the group G has a nontrivial near-ring multiplication.*

6.15. Corollary   *Let $(G, +)$ be a group and let $s \in End\,(G) \setminus \{0\}$ be a nilpotent endomorphism of G. Then the group G has a nontrivial near-ring multiplication.*

Having results on groups which admit nontrivial right near-ring multiplications, we now summarize the necessary conditions for a group to admit only trivial right near-ring multiplications. Here we recall that an endomorphism s of a group G is called *locally finite* if for any $a \in G$ there exists a number $m = m(a) > 0$ such that $as^m = a$.

6.16. Theorem   *Let $(G, +)$ be a group of order greater than 2. Suppose that G has no nontrivial right near-ring multiplication. Then:*

(1) G is not an abelian group.

(2) G is not a finitely generated group.

(3) Any nonidentical injective endomorphism of G is a locally finite automorphism of infinite order.

(4) Any element of G of finite order belongs to its center $Z(G)$.

(5) For any $x, y \in G$ there exists a positive integer $n = n(x, y)$ such that $(nx) + y = y + (nx)$.

(6) Let $x \in G$ and m be an integer. Then $mx \in Z(G)$ implies that either $x \in Z(G)$ or $m = 0$.

(7) G is not a semidirect product of two nonzero subgroups.

(8) G has no nonzero nilpotent endomorphism.

(9) If $s \in End(G)$, then $G = K(s) + \operatorname{Im} s$.

Coming back to the problem of Clay, we state the following

6.17. Proposition Let $(G, +)$ be a group of order greater than 2 such that

(1) it has no nonzero nilpotent endomorphisms;

(2) it has no nontrivial idempotent endomorphisms;

(3) it has no nonidentical automorphisms of finite order;

(4) all of its nontrivial endomorphisms are locally finite.

Then G has no nontrivial right near-ring multiplications.

We can now put down a necessary and sufficient condition for a simple group to possess only trivial right near-ring multiplications.

6.18. Theorem Let $(G, +)$ be a simple group. Then the following conditions are equivalent:

(1) G has no nontrivial right near-ring multiplication;

(2) any nonzero endomorphism of G is an automorphism and any nonidentical automorphism is locally finite and has an infinite order.

Now we proceed to discuss some model theoretical aspects of the above problem. The readers are referred to [22] and [96] for basic notions of model theory.

6.19. Theorem Let $\mathcal{C}_0$ be the class of all groups having no nontrivial right near-ring multiplications and $\mathcal{C}_1$ the class of all groups having a nontrivial near-ring multiplication. Then the following conditions are equivalent:

(1) The class $\mathcal{C}_0$ is axiomatizable.

(2) The class $\mathcal{C}_1$ is axiomatizable.

(3) Any group of order greater then 2 has a nontrivial near-ring multiplication. ∎

6.20. Remark The last result shows that it is impossible to describe the class of all groups having no nontrivial right near-ring multiplications in terms of properties of elements of groups. Therefore, the only possible ways to characterize it are the descriptions of the type of (6.18) since they involve the "second order properties" that are expressible in terms of endomorphisms and subgroups.

Note that it was shown in [2], that the class of multiplicative semigroups of rings is not axiomatizable.

§7. Planar near-rings

Soon after Dickson [40] discovered near-fields, Veblen and Wedderburn [135] used them to create a plane having unusual properties. In order to extend the geometrical applications from near-fields to a much bigger class of near-rings, J. R. Clay introduced a new class of planar near-rings in 1967. It turned out that this class of near-rings has an important applications to coding theory and cryptography (M. Modisett [125] for the circular near-rings, P. Fuchs, G. Hofer and G. Pilz [86] for the general case of planar near-rings), and combinatorics (G. Ferrero [45]).

Recently, under the supervision of T.-Y. Huang, I-Hsing Chen [24] found some interesting applications of planar near-rings to geometry and combinatorics. They constructed nets from arbitrary finite planar near rings, which automatically leads to the construction of transversal designs, families of mutually orthogonal Latin squares, and to strongly regular graphs. They also constructed a family of association schemes and partially balanced incomplete block designs.

For any given near-ring N, define a relation $=_m$ on N, and define $a =_m b$ if and only if $ax = bx$ for all $x \in N$. One can readily checks that $=_m$ is an equivalence relation. Following J. R. Clay, a near-ring N is called *planar* if $N/=_m$ has at least three equivalence classes, and for all $a, b, c \in N$ with $a \neq_m b$ there exists a unique $x \in N$ such that $ax = bx+c$ (see [33]).

To any given planar near-ring N there corresponds a uniquely determined *Ferrero pair*, i.e., a pair (N, Φ), where N is a group and $\Phi \subseteq Aut(N)$ such that if $\phi \in \Phi$ is not the identity mapping, then $-\phi + id_N$ is bijective. Actually, numerous planar near-rings can be constructed from any Ferrero (N, Φ).

Given a finite Ferrero pair (N, Φ), i.e., $|N| \leq \aleph_0$, the semidirect product of N and Φ is a Frobenius group, and, in fact, any Frobenius group is of this form. Thus finite planar near-rings are closely connected with Frobenius groups, which are of great importance in group theory.

Let N be a finite planar near-ring with corresponding Ferrero pair (N, Φ) and set

$$N^* = N \setminus \{x \in N \mid x \neq_m 0\}$$

and

$$B^* = \{N^* a + b \mid a, b \in N, a \neq 0\} = \{\Phi(a) + b \mid a, b \in N, \ a \neq 0\},$$

then the incidence structure $(N, B^*, \in)$ is a *balanced incomplete block design*, which is to say that there are positive integers k, r and λ so that (i) each *block* $B \in B^*$ contains exactly k elements, (ii) any given element $x \in N$ belongs to exactly r blocks from B^*, and (iii) any given pair $y, z \in X$ of distinct elements belongs to exactly λ subsets from B^*. The term "balanced incomplete block design" is usually abbreviated as "BIBD". (See [33]).

Next, a planar near-ring N is called *circular* (and the corresponding Ferrero pair (N, Φ) and BIBD $(N, B^*, \in)$ as well) if any three distinct points $x, y, z \in N$ belong to at most one block $B \in B^*$. In this case, a block $N^* a + b$ is referred to as a *circle* with center b and radius a. The class of circular planar near-rings is a proper subclass of the class of all planar near-rings. The circularity property provides some important and useful additional properties of geometries, combinatorial objects (for example, BIBD's), codes and cryptography systems arising from planar near-rings as well as related Frobenius groups.

Ke and Wang [105] investigated the Frobenius groups having kernels of order 64 since 64 is the smallest possible order for a nonabelian kernel to exist (cf. Adams [1]). Among the 267 nonisomorphic groups of order 64 (cf. Hall and Senior [113]), only three nonabelian ones can be the kernels of a Frobenius group. These three nonabelian kernels of order 64 are listed below.

(i) Let $F = GF(8)$, the Galois field of order 8, and let θ be a nonidentity automorphism of F. Define, for $a, b \in F$,

$$u(b, a) = \begin{pmatrix} 1 & 0 & 0 \\ a & 1 & 0 \\ b & \theta(a) & 1 \end{pmatrix}.$$

Then the set $A(3, \theta)$ of all $u(b, a)$ is a nonabelian Frobenius kernel with $|A(3, \theta)| = 64$.

(ii) Let S be the set of all matrices of the form

$$Q(a, b) = \begin{pmatrix} 1 & a & b \\ 0 & 1 & a^4 \\ 0 & 0 & 1 \end{pmatrix},$$

where $a, b \in GF(2^4)$ and $b + b^4 + a^5 = 0$. Then S is a nonabelian Frobenius kernel of order 64.

(iii) Consider the finite field $GF(2^2)$. Let T be the set of all matrices of the form

$$B(t, r, s) = \begin{pmatrix} 1 & 0 & 0 \\ r & 1 & 0 \\ t & s & 1 \end{pmatrix},$$

where r, s, t are elements of $GF(2^2)$. Then T is a nonabelian Frobenius kernel of order 64.

The following two problems were posed by J. R. Clay in [33, Problems 5.47 and 5.48].

Problem 1. Does there exist a circular Ferrero pair (N, Φ) with Φ nonabelian?

Problem 2. Does there exist a circular Ferrero pair (N, Φ) with N nonabelian and $|\Phi| > 3$?

It is also mentioned in [33, Page 81] that in all the known examples of circular Ferrero pairs (N, Φ) the group Φ is cyclic. We shall see that this is "almost" the general case.

These problems were recently studied by Beidar, Fong and Ke [4]. The following results were obtained.

7.1. Theorem   *If (N, Φ) is a circular Ferrero pair, then all Sylow p-subgroups of Φ are cyclic. In particular Φ is metacyclic.*

In the connection with the above theorem, the following question arises naturally.

Problem 3. Given any finite group Φ all of whose Sylow p-subgroups are cyclic and whose subgroups of order a product of two distinct primes are cyclic, does there exist a circular Ferrero pair (N, Φ)?

In fact, (7.1) is a corollary of the following results on Frobenius groups.

7.2. Theorem   *Let $(F, +)$ be an additively written Frobenius group with Frobenius kernel M and let H be a complement of M. For $m \in M$ and $h \in H$ we set $m^h = -h + m + h$ and $m^H = \{m^h \mid h \in H\}$. Suppose that for all $a, b, c, d \in M$, the relation $|(a + b^H) \cap (c + d^H)| \geq 3$ implies that $a = c$. Then all Sylow p-subgroup of H are cyclic.*

7.3. Theorem *Given any finite circular Ferrero pair* (N, Φ) *such that* Φ *has odd order, there exists a nonabelian nilpotent group* N' *such that* (N', Φ) *is a circular Ferrero pair. In particular, for any cyclic group* Ψ *of odd order there exists a nonabelian nilpotent group* M *such that* (M, Ψ) *is a circular Ferrero pair.*

7.4. Theorem *Let* G_1 *and* G_2 *be groups defined by*

$$G_1 = \langle x, y \mid x^3 wxyz = e = y^4, \ y^{-1}xy = x^2 \rangle$$

and

$$G_2 = \langle u, v \mid x^7 = e = y^9, \ y^{-1}xy = x^2 \rangle.$$

Then there exists an abelian group N *and a nonabelian nilpotent group* M *such that* (N, G_1) *and* (M, G_2) *are circular Ferrero pairs.*

In his fundamental paper [3], Amitsur had completed the description of the finite groups embeddable into a multiplicative group of division rings that was first initiated by Herstein. In particular, it was shown that both groups G_1 and G_2 are embeddable. In the connection with the above Problem 3, we make the following

Conjecture 1. Let Φ be a finite group such that all of its Sylow p-subgroups are cyclic. Suppose that Φ is embeddable into a division ring. Then there exists a circular Ferrero pair (N, Φ).

Conjecture 2. Let (N, Φ) be a finite circular Ferrero pair. Then Φ is embeddable into a multiplicative group of a division ring.

Near-rings are known to lack symmetry, i.e., one of the distributive law is absent from the axioms. However, circular planar near-rings show a kind of symmetry when suitably displayed on a plane. We take the following materials from W.-F.Ke's Ph. D. thesis written under the supervision of J. R. Clay. A general version of the results also can be found in a paper by Ke and Kiechle [107].

Let p be a prime and $k \geq 3$ such that $k \mid (p - 1)$. Let $\Phi = \Phi_k$ be the unique multiplicative subgroup of $\mathbb{Z}_p^* = \mathbb{Z}_p \setminus \{0\}$ of order k. Thus, we have a BIBD $(\mathbb{Z}_p, \mathcal{B}^*, \in)$ with $\mathcal{B}^* = \{\Phi a + b \mid a, b \in \mathbb{Z}_p, a \neq 0\}$. Assume that $(\mathbb{Z}_p, \mathcal{B}^*, \in)$ is circular throughout the rest of this section.

For any $c, r \in \mathbb{Z}_p^*$, define $E_c^r = \{\Phi r + b \mid b \in \Phi c\}$. Intuitively, E_c^r is the collection of circles with radius r and center on the circle Φc. Although they are not really circles in a complex plane, one still can image the "circles" of an E_c^r in the complex plane. In fact, this definition came from the model of the real circles.

Each E_c^r of $(\mathbb{Z}_p, \mathcal{B}^*, \in)$ yields a graph $\Gamma(E_c^r)$ (cf. Clay [32], and Clay and Yeh [38]): The vertices $\Gamma(E_c^r)$ are the centers of the circles of E_c^r, i.e., the points of Φc. An edge exists between two vertices a and b, $a \neq b$, if $(\Phi r + a) \cap (\Phi r + b) \neq \varnothing$. The edge is *even* if $|(\Phi r + a) \cap (\Phi r + b)| = 2$, and the edge is *odd* if $|(\Phi r + a) \cap (\Phi r + b)| = 1$. Therefore, $\Gamma(E_c^r)$ is a weighted graph.

A suitably displayed graph $\Gamma(E_{12}^1)$ in $(\mathbb{Z}_{73}, \mathcal{B}_8^*, \in)$ looks like the following picture.

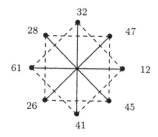

The graph of E_{12}^1 in $(\mathbb{Z}_{73}, \mathcal{B}_8^*, \in)$
with $\Phi = \langle 10 \rangle$

There corresponds with each $\Gamma(E_c^r)$ a sequence: Let $a = \varphi^l c \in \Phi c$, where $\varphi \in \Phi$ is a generator of Φ. Define a sequence $s_l = (i_1 i_2 \ldots i_{k-1})$ by

$$i_j = |(\Phi r + \varphi^l c) \cap (\Phi r + \varphi^{l+j} c)|,$$

where $j = 1, 2, \ldots, k-1$ and $|\Phi| = k$. So, each entry i_j of s_l is one of 0, 1, or 2.

Conversely, a graph $\Gamma(E_c^r)$ can be recovered from its corresponding sequence $(i_1 i_2 \ldots i_{k-1})$. First, take a regular k-gon and fix an orientation (clockwise or counterclockwise). Choose a vertex, label it by 0, the next vertex by 1, and the next vertex by 2, etc. Now, for each j, $j \in \{1, 2, \ldots, k-1\}$, draw a solid line segment from vertex 0 to vertex j if $i_j = 2$, and a broken line segment if $i_j = 1$ (Of course, nothing is drawn if $i_j = 0$ because this means that there is no edge between vertex 0 and vertex j). This is one step. Repeat this procedure for every vertex of the k-gon. Then a diagram of the graph of E_c^r is obtained.

Following this path, the concept of basic graphs, the building blocks of the graphs, is introduced.

7.5. Definition Let $k \geq 3$ and let $1 \leq j \leq k-1$. A sequence $(e_1 e_2 \ldots e_{k-1})$ is the jth *even basic k-sequence* if $e_j = e_{k-j} = 2$, and $e_s = 0$ for all $s \notin \{j, k-j\}$. Similarly, $(o_1 o_2 \ldots o_{k-1})$ is the jth *odd basic k-sequence* if $o_j = o_{k-j} = 1$, and $o_s = 0$ for all $s \notin \{j, k-j\}$. These basic k-sequences will also be simply referred to as *basic sequences*.

7.6. Definition Let $k \geq 3$ and let $1 \leq j \leq k-1$. The jth *even basic k-graph*, denoted by Π_j^k, is the graph associated to the jth even basic k-sequence; the jth *odd basic k-graph*, denoted by Γ_j^k, is the graph associated to the jth odd basic k-sequence. We will also refer to these j^{th} even or odd basic k-graphs as *basic k-graphs*, or simply *basic graphs*.

With these basic graphs, one can see more clearly the structure of a graph $\Gamma(E_c^r)$.

7.7. Theorem *If $\Gamma(E_c^r)$ is not null, then it is a union of spanning subgraphs (i.e., subgaphs whose vertex sets are the whole vertex set of the original graph), and each of these spanning subgraphs is isomorphic to an even basic k-graph or an odd basic k-graph.*

The following example of $\Gamma(E_6^1)$ in $(\mathbb{Z}_{229}, \mathcal{B}_{12}^*, \in)$ illustrates (7.7).

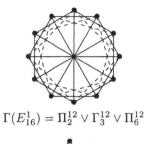

$$\Gamma(E_{16}^1) = \Pi_2^{12} \vee \Gamma_3^{12} \vee \Pi_6^{12}$$

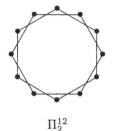

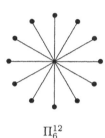

$$\Pi_2^{12} \qquad\qquad \Gamma_3^{12} \qquad\qquad \Pi_6^{12}$$

Let $j \in \{1, 2, \ldots, k-1\}$ and let $r \in \mathbb{Z}_p^*$. Define

$$\gamma_j(r) = |\{E_c^r \mid \Gamma_j^k \text{ is a subgraph of } \Gamma(E_c^r)\}|$$

and

$$\pi_j(r) = |\{E_c^r \mid \Pi_j^k \text{ is a subgraph of } \Gamma(E_c^r)\}|.$$

7.8. Theorem *Suppose $2|k$. Let $c = (\varphi^j - 1)^{-1}(\psi - 1)r$, where $1 \le j \le k-1$, $\psi \in \Phi \setminus \{1\}$, and $r \in \mathbb{Z}_p^*$. Then $|(\Phi r + c) \cap (\Phi r + \varphi^j c)| = 2$ if and only if $\psi \ne -1$. Moreover, if $\psi \ne -1$, then $(\Phi r + c) \cap (\Phi r + \varphi^j c) = \{-r + c, \psi r + c\}$.*

7.9. Corollary *Suppose $2 \mid k$. Then for a fixed $r \in \mathbb{Z}_p^*$, the only c for which Γ_j^k is a subgraph of $\Gamma(E_c^r)$ is $c = -2r(\varphi^j - 1)^{-1}$. Hence, $\gamma_j(r) = 1$.*

7.10. Theorem *Let $2 \big| |\Phi|$ and let $1 \le j \le k-1$. For a fixed $r \in \mathbb{Z}_p^*$, Π_j^k is a subgraph of the graph of $E_{c_i}^r$, where $c_i = (\varphi^j - 1)^{-1}(\varphi^i - 1)r$ for $1 \le i \le k/2 - 1$. Also, if $1 \le i' \le k/2 - 1$ and $i \ne i'$, then $\Phi c_i \ne \Phi c_{i'}$. Therefore, $\pi_j(r) = k/2 - 1$.*

7.11. Theorem *Suppose $2 \mid k$. If $r \in \mathbb{Z}_p^*$, then the total number of appearances of the basic graphs as subgraphs of the graphs of E_c^r's with $c \in \mathbb{Z}_p^*$, including even ones as well as odd ones, is $(k/2)^2$.*

7.12. Theorem *Suppose that $(\mathbb{Z}_p, \mathcal{B}_k^*, \in)$ and $(\mathbb{Z}_p, \mathcal{B}_{2k}^*, \in)$ are both circular, where k is odd. Then the total number of the appearances of the basic graphs as subgraphs of the graphs of the E_c^r's in $(\mathbb{Z}_p, \mathcal{B}_k^*, \in)$ with a fixed $r \in \mathbb{Z}_p^*$, is $(k-1)^2/2$.*

As has been mentioned before, the above results have been generalized. The generalized version also helps in finding the solutions of the equation $x^m + y^m - z^m = 1$ in a finite field. (See Ke and Kiechle [104]).

Let $(F, +, \cdot)$ be the Galois field of order $q = p^s$, where p is a prime and $s > 0$. Let $k \ge 2$ such that $k \big| (q-1)$. Suppose that with the subgroup $\Phi \le F^*$ of order k, the Ferrero pair (F, Φ) is circular, and put $m = (q - 1)/k$. Denote the number of solutions of the equation

$$x^m + y^m - z^m = 1$$

in F by N. Also, let N' be the number of solutions with $xyz \neq 0$.

7.13. Theorem   (1) *If k is even, then*

$$N = \begin{cases} 3(k - 1)m^3 + 6m^2 + 3m & \textit{if } 6 \mid k; \\ 3(k - 1)m^3 + 3m^2 + 3m & \textit{if } p = 3; \\ 3(k - 1)m^3 + 3m & \textit{otherwise,} \end{cases}$$

and $N' = 3(k - 1)m^3$.

(2) *If k is odd, and if $(q, 2k)$ is also circular, then $N = (2k - 1)m^3 + 2m$ and $N' = (2k - 1)m^3$.*

§8. Miscellany

This section contains those results which for one reason or another do not fit into any of earlier sections, but which are worth including. The first result which we present here concerning derivations in 3-prime near-rings is due to X. K. Wang [139], the first Chinese near-ringer in Mainland China who has published the first near-ring paper in an international level journal.

Historically speaking, the study of derivations of near-rings was initiated by H. E. Bell and G. Mason in 1987 [10], but up to now only a few papers on this subject were published because the additive groups of the near-rings are not abelian in general and there is only one distributive law.

Recall that a near-ring N is called 0-prime if the product of any two nonzero ideals is nonzero. Further, a near-ring N is said to be 3-prime if for any nonzero $x, y \in N$, $xNy \neq 0$ (see [137], [88] for further information). An additive mapping $d : N \to N$ is said to be a derivation of N if $(xy)^d = xy^d + x^d y$ for all $x, y \in N$.

With an additional assumption, X. K. Wang proved for 2-torsion free 3-prime near-rings a weaker version of the famous Posner's theorem on a composition of two derivations of a prime ring.

8.1. Theorem (X. K. Wang [139])   *Let N be a 2-torsion free 3-prime near-ring, and d_1 and d_2 are derivations on N such that $d_1 d_2$ is also a derivation. Then the following two conditions are equivalent:*
  (i) *Either $d_1 = 0$ or $d_2 = 0$;*
  (ii) *$[x^{d_1}, y^{d_2}] = 0$ for all $x, y \in N$.*

In order to prove the equality $Z^d = 0$ for a nilpotent derivation d of a 3-prime near-ring N where Z is its center, Wang [139] extended Leibniz's rule for derivations of rings to near-rings.

8.2. Proposition [Leibniz's rule for near-rings]   *For any $n \geq 2$ and any $x, y \in N$, we have*

$$(xy)^{d^n} = x^{d^n} y + \binom{n}{1} x^{d^{n-1}} y^d + \ldots + \binom{n}{i} x^{d^{n-1}} y^{d^i} + \ldots$$

$$+ \binom{n}{n-1} x^d y^{d^{n-1}} + xy^{d^n}.$$

For nilpotent derivations of near-rings without divisors of zero he proved the following stronger result.

8.3. Theorem *Let n be a natural number and N an $n!$-torsion free near-ring with no divisors of zero. Then N admits no nonzero derivation d with $d^n = 0$.*

Recently the analog of Posner's theorem on composition of two derivations in prime rings was proved for a 3-prime near-ring in a forthcoming paper jointly written by K. I. Beidar, Y. Fong and X. K. Wang [9]. Besides, some other results on derivations of Posner and Herstein were generalized to near-rings (see [92], [93] and [130]). The following 5 results are due to Beidar, Fong and Wang [9].

8.4. Theorem *Let N be a zero-symmetric 3-prime near-ring with derivations d_1 and d_2. Suppose that $2N \neq 0$. Then $d_1 d_2$ is a derivation if and only if either $d_1 = 0$, or $d_2 = 0$.*

8.5. Theorem *Let N be a zero-symmetric 3-prime near-ring with a derivation d and $a \in N$. Suppose that $d^2 \neq 0$ and $x^d a = a x^d$ for all $x \in N$. Then a belongs to the center*

$$Z(N) = \{b \in N \mid bx = xb \quad \text{for all} \quad x \in N\}$$

of N.

8.6. Corollary *Let N be a zero-symmetric 3-prime near-ring with a nonzero derivation d and $a \in N$. Suppose that $2N \neq 0$ and $x^d a = a x^d$ for all $x \in N$. Then a belongs to the center $Z(N)$ of N.*

8.7. Theorem *Let N be a zero-symmetric 3-prime near-ring with derivations d_1 and d_2. Suppose that $d_1^2 \neq 0 \neq d_2^2$ and $x^{d_1} y^{d_2} = y^{d_2} x^{d_1}$ for all $x, y \in N$. Then N is a commutative ring without nontrivial zero divisors.*

8.8. Corollary *Let N be a zero-symmetric 3-prime near-ring with nonzero derivations d_1 and d_2. Suppose that $2N \neq 0$ and $x^{d_1} y^{d_2} = y^{d_2} x^{d_1}$ for all $x, y \in N$. Then N is a commutative ring without nontrivial zero divisors.*

We close the discussion of derivations on near-rings with the following interesting example which was constructed in [9] and which shows that (8.4), (8.5) and (8.7) do not hold even for simple 0-prime near-rings with a left identity. We also note that the idea of this example was suggested by the results of Beidar, Fong, Ke and Liang on near-ring multiplication on groups listed in the previous section.

8.9. Example Let V be a linear space with a basis $e_1, e_2, \ldots, e_n$ over a field F of characteristic different from 2. Define a multiplication $\cdot : V \times V \to V$ by the rule $vw = 0$ for all $v, w \in V$ with $v \notin \{e_1, -e_1\}$ and $e_1 w = w$, $(-e_1)w = -w$. One can easily check that V is a left zero-symmetric near-ring with respective to this multiplication (see [5]).

We claim that V is a simple near-ring with the left identity e_1. Indeed, let I be a nonzero proper ideal of V. If $e_1 \in I$, then $V = e_1 V \subseteq I$, a contradiction. Hence $e_1 \notin I$. Let $0 \neq v \in I$. Then either $e_1 + v \neq -e_1$, or $-e_1 + v \neq e_1$. Without loss of generality we consider the first case only. Clearly $e_1 + v \neq e_1$ as well. Hence $(e_1 + v)w = 0$ for all $w \in V$. Since I is an ideal, we have $-w = (e_1 + v)w - e_1 w \in I$ for all $w \in V$, which implies that $V \subseteq I$, a contradiction. Thus V is simple. Obviously V is not a ring.

One can easily check that any linear transformation $d : V \to V$ such that $V^d \cap \{e_1, -e_1\} = \varnothing$ is a derivation of the near-ring V. Suppose now that $n = 3$. Define linear transformations d_1, d_2 as follows:

$$e_1^{d_1} = 0 = e_1^{d_2}, \; e_2^{d_1} = e_2, \; e_3^{d_1} = 0, \; e_2^{d_2} = 0, \; e_3^{d_2} = e_3.$$

Clearly $d_1 d_2 = 0$ is a derivation. Further $x^{d_1} y^{d_2} = y^{d_2} x^{d_1}$ for all $x, y \in V$. Thus (8.4), (8.5) and (8.7) do not hold for V.

It is well-known that a heart of a subdirectly irreducible ring is either a simple ring or square zero. However the analogous result does not hold for zero-symmetric near-rings. On this aspect, Kaarli constructed an example of a finite abelian zero-symmetric near-ring whose heart is neither a simple near-ring nor square zero in [102]. Up to now no other examples are know.

The study of hearts of subdirectly irreducible near-rings is equivalent to that of minimal ideals (see [18, Proposition 2]). Works on minimal ideals in near-rings go back (at least) to D.W. Blackett [19] and [20]. Since the first paper devoted to minimal ideals was written by S.D. Scott [131] in 1974, this subject was investigated in a number papers [14],[15],[16], [17], [18], [98] and [100],

The example given by Kaarli is a subnear-ring N of a zero-preserving transformation near-ring of a cyclic group of order 8. The heart K of N is a subdirectly irreducible near-ring with the heart L which is a nilpotent ideal of K. Moreover, the factor near-ring K/L is simple. Thus the following two questions arise naturally.

Suppose that a heart K' of a subdirectly irreducible near-ring N' is again subdirectly irreducible with a heart L', then:

(1) Is L' a nilpotent ideal of K'?
(2) Is L' a prime ideal of K'?

Using some results and constructions given in [5], K. I. Beidar, Y. Fong and K. P. Shum recently constructed a series of abelian subdirectly irreducible near-rings N' whose heart K' is also a subdirectly irreducible near-ring with a nontrivial minimal ideal L' which is neither nilpotent nor a prime ideal of K'. Thus, the work of Kaarli was amplified and the above two questions are therefore answered in a negative way.

8.10. Theorem   *Given any prime number p there exists a zero-symmetric near-ring N with a chain of subnear-rings $N = N_0 \supset N_1 \supset \ldots \supset N_t \supset \ldots$ such that:*

(a) *$(N, +)$ is a direct sum of countably many cyclic groups of order p;*
(b) *$N_i = N^{2^i}$ for all $i = 1, 2, \ldots$;*
(c) *N_i is a subdirectly irreducible irreducible near-ring with a heart N_{i+1} for all $i = 0, 1, \ldots$;*
(d) *$\cap_{i=0}^{\infty} N_i = 0$.*

Now we proceed to discuss the results obtained by Y. Fong and K. Kaarli in [59]. From the universal algebra viewpoint elements of near-ring $I(G)$ are exactly the zero-preserving unary polynomial functions. In order to describe the properties of $I(G)$ it is important to have a characterization of elements of $I(G)$. In [101] Kaarli proved that in the case G is a strictly non-abelian group (i.e., commutator subgroup $[H, H]$ of any normal subgroup $H \subseteq G$ coincides with H) the near-ring $I(G)$ is dense in the near-ring $C_0(G)$ of all 0-preserving compatible functions on G. The compatibility of f means here that for any normal subgroup $H \subseteq G$, $a - b \in H$ implies $af - bf \in H$. If G is finite then the equality $I(G) = C_0(G)$ holds and this is actually the best one could want.

In fact more was proved in [101]. If A is a subset of G and $f : A \to G$ then we call f a *partial function* on G and A is called a *domain* of f. If f is a restriction of some unary polynomial function of G then we call f a *partial polynomial function* of G. It was proved in [101] that a function $f : A \to G$, where G is strictly non-abelian and A is a finite subset containing 0, is a polynomial if and only if it is compatible.

Let f be a partial function on G with a finite domain A and let H be a normal subgroup of G. Suppose we are looking for conditions for f to be polynomial. Then the first necessary condition is that f must be compatible with H and hence it must induce the function $\overline{f} : \overline{A} \to \overline{G}$ where $\overline{A}$ is the canonical image of A in $\overline{G} = G/H$. The second necessary condition is that $\overline{f}$ must be a polynomial function of $\overline{G}$. Clearly any polynomial function of a quotient group is induced by some polynomial function of the original group. Thus there must exist a polynomial function p of G inducing on $\overline{A}$ exactly $\overline{f}$. Then for function $f_1 = f - p$ we have $Af_1 \subseteq H$. We see that the problem we started with splits into two new problems:

(1) *Describe partial polynomial functions of G/H with domain $\overline{A}$;*
(2) *Describe partial polynomial functions $A \to H$ of G.*

Obviously at least in the case of finite G both problems are easier than the original one.

The concrete aim of the paper [62] was to solve problem **(2)** in the case when $0 \in A$, H is the only minimal normal subgroup of G, H is abelian and the following non-commutativity condition holds: if two nonzero normal subgroups of G centralize each other, then they are both equal to H. Although these restrictions are definitely rather strong, the results obtained have several interesting applications. In particular, the results on endomorphism near-rings of symmetric groups obtained originally by lengthy calculations follow easily from them.

We fix the following notations. In what follows R is an unital left d.g. near-ring with the fixed set D of distributive generators and G_R is an R-group such that all $d \in D$ act on G by endomorphisms and all inner automorphisms of G are induced by the elements of D. Our general problem may be restated now as follows. Given any finite subset $A \subseteq G$ and a function $f : A \to G$, can we find $r \in R$ such that $af = ar$ for all $a \in A$? If the answer is *yes* that we say that f admits the *interpolation property*. Let H be an R-ideal of G. Then a finite subset $A \subseteq G$ is called *independent* with respect to H if any function $f : A \to H$ admits the interpolation property. Here we note that if H is a minimal abelian ideal of G, then H is an irreducible module over the ring $R/(0 : H)_R$. In this case the centralizer Γ of this module is a skew field and H is a linear space over Γ.

The following 5 results were proved by Fong and Kaarli in [62].

8.11. Theorem *If H is an abelian minimal ideal of G and a subset $A \subseteq G$ is independent with respect to H, then:*

(i) *$A \cap H$ is a linearly independent subset of the linear space $_\Gamma H$;*
(ii) *For any $a \in A \setminus H$, the set $\{a - x \mid x \in (a + H) \cap S\}$ is linearly independent in $_\Gamma H$.*

8.12. Theorem *Let H be a minimal abelian ideal of G such that for any two non-zero ideals F_1 and F_2 of G, $[F_1, F_2] = 0$ implies $F_1 = H = F_2$. Then a subset $A \subseteq G$ is independent with respect to H if and only if it satisfies the conditions (i) and (ii) of (8.11)*

The following results are obtained as applications of the above theorems.

8.13. Theorem *Let G and H be as in (8.12) and R satisfies the descending chain condition for right ideals. Then:*
  (i) *H is finite dimensional over Γ, let $n = \dim_\Gamma H$;*
  (ii) *There exists an idempotent $e \in R$ such that $Ge = H$ and $(G \setminus H)e = 0$;*
  (iii) *$(0 : e)_R = (0 : H)_R$ is an ideal of R;*
  (iv) *$eR = (0 : G \setminus H)_R$ is a right ideal of R isomorphic to the matrix ring $M_n(\Gamma)$;*
  (v) *R is a direct sum of eR and $(0 : e)_R$.*

8.14. Theorem *Let G and H be as in (8.12) and $|G| < \infty$. If $|G/H| = m$ and $\dim_\Gamma H = n$, then*

$$|R/(0 : G)_R| = |R/(H : G)_R| \cdot |H|^{mn+m-1}.$$

8.15. Corollary *Let G be a finite group with a unique minimal normal subgroup H and let H be the only nonzero normal subgroup of G having a nonzero centralizer. Then*

$$|I(G)| = |I(G/H)| \cdot |H|^{mn+m-1}$$

where $m = |G/H|$ and n is the dimension of H over $\Gamma = \mathrm{End}(H_{I(G)})$.

Furthermore it was shown that if G is a nonabelian group of order pq where p and q are primes and $p < q$, then any near-ring R of transformations on G generated by the set of endomorphisms of G containing all inner automorphisms of G has the size pq^{2p-1}. In particular

$$|I(S_3)| = |A(S_3)| = |E(S_3)| = 2 \cdot 3^3.$$

Further the old result of Fong and Meldrum that

$$|A(S_4)| = |E(S_4)| = |I(S_4)| = 2^{35} \cdot 3^3$$

was obtained as an easy corollary of the above results. Also it was shown that

$$|A(A_4)| = |E(A_4)| = 3 \cdot 4^8, \quad |I(A_4)| = 3 \cdot 4^5.$$

Note that G. Mason reached the same result using a computer program.

The concept of semi-homomorphism of near-rings was discussed by Y. Fong and L. van Wyk in [70]. The notion of semi-homomorphism of ring was introduced by S. A. Huq [95] as a generalization of Jordan homomorphisms of a ring and semi-homomorphisms of a groups. A mapping $f : N \to M$ of near-rings N and M is called a *semi-homomorphism* if

$$(a + b + a)f = af + bf + af \quad \text{and} \quad (aba)f = afbfaf$$

for all $a, b \in N$. In particular it was shown that every semi-automorphism of the smallest Dickson near-field is an automorphism and its automorphism group is isomorphic to S_3 (the last result is a special case of the famous Zassenhaus's theorem on finite Dickson near-fields).

Now we proceed to discuss results of Y. Fong, K. Kaarli and W.-F. Ke on varieties of near-rings [60] and [61].

Recall that a near-ring is called *arithmetical* if its ideal lattice is distributive. A variety of near-rings is called *arithmetical* if so are all its members. Discriminator varieties form an important class of arithmetical varieties. Recall that a (ternary) *discriminator* of a set A is a ternary function t on A defined by the rule

$$t(x, x, z) = z \quad \text{and} \quad t(x, y, z) = x \quad \text{if} \quad x \neq y.$$

A finite near-ring N is *quasiprimal* if the discriminator is among its term functions. A variety is called *discriminator* if it is generated by a class $\mathcal{K}$ of *jointly quasiprimal* algebras. The latter means that there is a ternary term which realizes the discriminator on all members of the class $\mathcal{K}$. Further, a variety is said to be *finitely generated* if it is generated by a single finite algebra. Here we call a near-ring N *strongly semiprime* (*strongly prime*) if it is semiprime (prime) and so are all its subnear-rings. The following 6 results were obtained by Fong, Kaarli and Ke in [61]

8.16. Theorem *A finite near-ring generates an arithmetical variety if and only if it is zero-symmetric and strongly semiprime.*

8.17. Proposition *A finite strongly semiprime zero-symmetric near-ring N is a direct sum of simple 2-primitive strongly semiprime near-rings. Every ideal of N has a left unity element.*

Let G be an additively written group, $\Gamma \subseteq \mathrm{Aut}(G)$ a subgroup and ρ an congruence relation of the Γ-act G. The latter means that ρ is an equivalence relation of G such that $g_1 \rho g_2$ implies $(\gamma g_1)\rho(\gamma g_2)$ for all $\gamma \in \Gamma$. The set $M(G, \Gamma, \rho)$ of all transformations f of the set G satisfying the following conditions:

(i) $0f = 0$;
(ii) $g_1 \rho g_2 \Rightarrow g_1 f = g_2 f$;
(iii) $(\gamma g)f = \gamma(gf)$ for all $\gamma \in \Gamma, g \in G$

forms a (Polin) near-ring with respect to pointwise addition and composition of mappings. We define the following equivalence relation τ of G:

$$\tau = \{(g_1, g_2) \in G \times G \mid (\gamma g_1)\rho g_2 \text{ for some } \gamma \in \Gamma\}.$$

8.18. Theorem *For a finite zero-symmetric near-ring N the following conditions are equivalent:*
(i) *N is simple and strongly semiprime;*
(ii) *N is strongly prime;*
(iii) *N is isomorphic either to a Galois field or to a near-ring $M(G, \Gamma, \rho)$ with the triple (G, Γ, ρ) satisfying the following conditions:*
   (a) *$(\gamma g)\rho g \Rightarrow \gamma = 1$ or $g\rho 0$;*
   (b) *every nonzero subgroup of G intersects every τ-class of G.*

8.19. Theorem *A finite near-ring is quasiprimal if and only if it is zero-symmetric and strongly semiprime.*

8.20. Corollary *All finitely generated arithmetical varieties of near-rings are discriminator.*

Given a variety V and a near-ring N, we set $(N)V$ to be equal to the intersection of all ideals I of N such that $N/I \in V$. The variety V is said to have *attainable identities* if $((N)V)V = (N)V$ for all near-rings N. The study of near-rings varieties having attainable identities was initiated by S. Veldsman [136]. In particular he proved that there exists an arithmetical variety of near-rings which has no attainable identities. In contrast with this we have the following

8.21. Theorem *Every finitely generated arithmetical variety of near-rings has attainable identities.*

A variety V of near-rings is called *locally finite* if all its finitely generated members are finite. A near-ring N is said to be *strictly simple* if it has no proper subnear-rings. It is well-known that every minimal variety of associative rings is generated by a finite ring of prime order and, in particular, is locally finite. In [61] Fong, Kaarli and Ke investigated locally finite minimal varieties of near-rings. It was shown that such varieties are exactly those which are generated by finite strictly simple near-rings. They proved that every finite strictly simple near-ring is either a near-ring with trivial multiplication on a group of prime order or a finite planar near-ring whose additive group is elementary abelian. They describe the multiplicative subgroups of Galois fields which lead to strictly simple planar near-rings and prove that they are exactly those strictly simple planar near-rings which satisfy the identity $xyz = yxz$.

It is well-known that not all the good properties of the Kurosh-Amitsur radical theory in the variety of associative rings are preserved in the bigger variety of near-rings. In the smaller and better behaved variety of zero-symmetric near-rings the radical theory is much more satisfactory. In [71] Y. Fong, S. Veldsman and R. Wiegandt showed that many of the results of radical theory of the zero-symmetric near-rings can be extended to a much bigger variety of near-rings which, amongst others, includes all zero-symmetric as well as constant near-rings. The varieties considered are varieties of near-rings, called Fuchs varieties, in which the constants form an ideal. The good arithmetic of such varieties makes it possible to derive more explicit conditions

(i) for the subvariety of constant near-ring to be a semisimple class (i.e, to have attainable identities);

(ii) for semisimple classes to be hereditary.

It is proved that the subvariety of zero-symmetric near-rings has attainable identities in a Fuchs variety. The theory of overnilpotent radicals of zero-symmetric near-rings is extended to the largest Fuchs variety $\mathcal{F}$. The following surprising fact was discovered. Let $\mathcal{R}$, $\mathcal{N}_0$ and $\mathcal{N}$ denote the varieties of all associative rings, all zero-symmetric near-rings and all near-rings respectively. Then obviously

$$\mathcal{R} \subset \mathcal{N}_0 \subseteq \mathcal{F} \subset \mathcal{N}.$$

According to the result of Kaarli and Betsch [13], Mlitz and Oswald [124], radical classes with hereditary semisimple classes in varieties $\mathcal{N}_0$ and $\mathcal{N}$ contain all nilpotent near-rings, but this is no longer true for varieties $\mathcal{R}$ and $\mathcal{F}$.

Finally, the following result was proved by Fong and Wiegandt [72].

8.22. Theorem *Let $\mathcal{W}$ be a universal class of associative rings, alternative Φ-algebras $(1/3 \in \Phi)$, Jordan Φ-algebras $(1/2 \in \Phi)$, or 0-symmetric near-rings. If $B \lhd A \in \mathcal{W}$ and B is Brown-McCoy semisimple, that is,*

$$B = \sum_{subdirect} (B/D_\alpha \mid B/D_\alpha \text{ is simple with unity for all } \alpha \in I),$$

then

(i) *for each $\alpha \in I$ there exists an ideal K_α of A such that $B/D_\alpha \cong (B + K_\alpha)/K_\alpha = A/K_\alpha$,*

(ii) *A has a subdirect decomposition $A = A/B \oplus_s A/K$ where $K = \cap(K_\alpha \mid \alpha \in I)$ such that $B \cong (B + K)/K$ and $\mathcal{G}(A) \subseteq K$.*

Here we finish our voyage to the wonderful ocean of near-rings. If anyone still wishes to keep in touch with these charming creatures, please do not hesitate to send your address to the author and you will get a copy of "Near-ring Newsletters" twice a year.

Acknowledgement

I would like to express my deepest thanks to Prof. K. I. Beidar and Prof. W.-F. Ke for their helpful suggestions and discussions.

References

[1] W. B. Adams. Near integral domains on nonabelian groups. *Monatsh. Math.* **81** (1976), 178–183.

[2] A. Adler. On the multiplicative semigroups of rings. *Communications in Algebra* **6** (1978), 1751–1753.

[3] S. A. Amitsur. Finite subgroups of division rings. *Trans. Amer. Math. Soc.* **80** (1955), 361–386.

[4] K. I. Beidar, Y. Fong, and W. F. Ke. On the circularity of finite Ferrero pairs. To appear.

[5] K. I. Beidar, Y. Fong, and W. F. Ke. On the simplicity of centralizer nearrings. To appear.

[6] K. I. Beidar, Y. Fong, and W. F. Ke. When are the endomorphism rings generated by automorphisms? To appear.

[7] K. I. Beidar, Y. Fong, W. F. Ke, and S. Y. Liang. Nearring multiplications on groups. *Comm. in Algebra* (1995), to appear.

[8] K. I. Beidar, Y. Fong, and K. P. Shum. On the hearts of subdirectly irreducible near-rings. To appear in SEAMS Math. Bulletin.

[9] K. I. Beidar, Y. Fong, and X. K. Wang. Posner and Herstein theorems for derivations of 3-prime nearrings. Submitted.

[10] H. E. Bell and G. Mason. On derivations in near-rings. In G. Betsch, editor, *Near-rings and Near-fields*. North-Holland/American, Elsevier, Amsterdam, 1987.

[11] G. Berman and R. J. Silverman. Simplicity of near-rings of transformations. *Proc. Amer. Math. Soc.* (1959), 456–459.

[12] G. Betsch. Some structure theorems on 2-primitive near-rings. *Colloquia Math. Societatis Janus Bolyai 6, Rings, modules and Radicals. Keszthely (Hungary), 1971* (1973), 73–102.

[13] G. Betsch and K. Kaarli. Supernilpotent radicals and hereditariness of semisimple classes of near-rings. *Coll. Math. Soc. J. Bolyai* (1985), 47–58.

[14] G. Birkenmeier and H. Heatherly. Minimal ideals in near-rings and localized distributivity conditions. submitted.

[15] G. Birkenmeier and H. Heatherly. Medial near-rings. *Monatsh. Math.* (1989).

[16] G. Birkenmeier and H. Heatherly. Left self distributive near-rings. *J. Austral. Math. Soc. Ser. A* (1990).

[17] G. Birkenmeier and H. Heatherly. Medial and distributively generated near-rings. *Monatsh. Math.* (1990).

[18] G. Birkenmeier and H. Heatherly. Minimal ideals in near-rings. *Comm. in Algebra* (1992).

[19] D. W. Blackett. *Simple and semisimple near-rings.* PhD thesis, Princeton Univ., 1950.

[20] D. W. Blackett. Simple and semisimple near-rings. *Proc. Amer. Math. Soc.* (1953).

[21] P. Bouchard, Y. Fong, W. F. Ke, and Y. N. Yeh. Counting f such that $f \circ g = g \circ f$. Submitted.

[22] S. Burris and H. P. Sankappanavar. *A course in universal algebra.* Springer-Verlag, New York, 1981.

[23] F. Castagna. Sums of automorphisms of a primary group. *Pacific J. Math* **27** (1968), 463–473.

[24] I. H. Chen. *Some combinatorial structures arising from finite planar near-rings.* PhD thesis, Nat'l Chiao Tung Univ., Hsinchu, 1991.

[25] J. R. Clay. *The near-rings definable on an arbitrary group and the group of left distributive multiplications definable on an abelian group.* Doctoral dissertation, University of Washington, 1966.

[26] J. R. Clay. The group of left distributive multiplications on an abelian group. *Acta Math. Sci. Hungar* **19** (1968), 221–227.

[27] J. R. Clay. The near-rings on groups of low order. *Math. Z.* **104** (1968).

[28] J. R. Clay. Research in near-rings using a digital computer. *Bit* **10** (1970), 249–265.

[29] J. R. Clay. Generating balanced incomplete block designs from planar near-rings. *J. Algebra* **22** (1972).

[30] J. R. Clay. Circular block designs from planar nearrings. *Ann. Discrete Math.* **37** (1988), 95–106.

[31] J. R. Clay. Geometric and combinatorial ideas related to circular planar nearrings. *Bulletin of the Institute of Mathematics Academia Sinica* **16** (1988), 275–283.

[32] J. R. Clay. Compound closed chains in circular planar nearrings. *Ann. Discrete Math.* **52** (1992), 93–106.

[33] J. R. Clay. *Nearrings: Geneses and Applications.* Oxford Univ. Press, Oxford, 1992.

[34] J. R. Clay and D. K. Doi. Near-rings with identity on alternating groups. *Math. Scand.* **23** (1968), 54–46.

[35] J. R. Clay and Y. Fong. On syntactic near-rings of even dihedral groups. *Results in Mathematics, Germany* **23** (1993), 23–44.

[36] J. R. Clay and H. Karzel. Tactical configurations derived from groups having a group of fixed-point-free automorphism. *J. Geom.* **27** (1986), 473–502.

[37] J. R. Clay and J. J. Malone. The near-rings with identities on certain finite groups. *Math. Scand.* **19** (1966), 146–150.

[38] J. R. Clay and Y. N. Yeh. On some geometry of mersenne primes. To appear.

[39] B. de la Rosa, Y. Fong, and R. Wiegandt. Complementary radicals revisited. *Acta Math. Hungar.* **65** (1994), 253–264.

[40] L. E. Dickson. Definitions of a group and a field by independent postulates. *Trans. Amer. Math. Soc.* **6** (1905), 198–204.

[41] R. Dover. Quasi-rings. Master's thesis, Univ. of Texas at Arlington, 1968.

[42] B. S. Du. On regular near-rings. Master's thesis, Nat'l Tsing Hua Univ., Hsinchu, 1974.

[43] S. Feigelstock. Generalized nil 2-groups and near-rings. *Indian J. Math.* **22** (1980), 99–103.

[44] G. Ferrero. Classificazione e costruzione degli stems p-singolari. *Istituto Lombardo Accad. Sci. Lett., Rend. A.* **102** (1968), 597–613.

[45] G. Ferrero. Stems planari e bib-disegni. *Riv. Mat. Univ. Parma (2)* **11** (1970), 79–96.

[46] Y. Fong. Endomorphism near-rings of symmetric groups. Conf. Edinburgh, Edinburgh, August 1978.

[47] Y. Fong. On strictly semi-perfect near-ring modules. Conf. Oberwolfach, April 1980.

[48] Y. Fong. Rings and near-rings generated by semi-endomorphisms of groups. Conference on General Algebra, Wien 1990.

[49] Y. Fong. A theorem on strictly semi-perfect near-ring modules. Symp. Summer Math. Res. Center Report, Taipei, pp. 339–344.

[50] Y. Fong. *The endomorphism near-rings of the symmetric groups.* PhD thesis, University of Edinburgh, 1979.

[51] Y. Fong. On near-rings and automata. *Proc. National Sci. Counc. R.O.C. (A)* **12** (1988), 240–246.

[52] Y. Fong. On the structure of abelian syntactic near-rings. *Proc. of the First Int'l Symposium of Algebraic Structures and Number Theory, 1988 Hong Kong* (1990), 114–123.

[53] Y. Fong and J. R. Clay. Computer programs fro investigating syntactic near-rings of finite group-semiautomata. *Bull. Inst. Math. Academia Sinica* **16** (1988), 295–304.

[54] Y. Fong, F. K. Huang, and W. F. Ke. On the minimal generating sets of the endomorphism near-rings of the dihedral groups D_n with even n. Submitted.

[55] Y. Fong, F. K. Huang, and W. F. Ke. Syntactic near-rings associated with group semiautomata. *PU. M. A. Ser. A* **2** (1992), 187–204.

[56] Y. Fong, F. K. Huang, W. F. Ke, and Y. N. Yeh. On semi-endomorphisms of finite groups and transformation near-rings. Submitted.

[57] Y. Fong, F. K. Huang, and C. S. Wang. Group semiautomata and their related topics. To appear in the proceedings of the conference on Words, Languages and Combinatorics.

[58] Y. Fong, F. K. Huang, and R. Wiegandt. Radical theory for group semiautomata. To appear in Acta Cynbernetica.

[59] Y. Fong and Kaarli. Unary polynomials on a class of groups. Submitted.

[60] Y. Fong, K. Kaarli, and W. F. Ke. On arithematical varieties of nearrings. Submitted.

[61] Y. Fong, K. Kaarli, and W. F. Ke. On minimal varieties of near-rings. To appear.

[62] Y. Fong and W. F. Ke. On the minimal generating sets of the endomorphism near-rings of the dihedral groups D_{2n} with odd n. To appear in the Proceedings of the 1989 Near-rings and Near-fields conference, Oberwolfach.

[63] Y. Fong and W. F. Ke. Syntactic near-rings of finite group-semiautomata. Int'l Conference on General Algebra, Krems/Donau, 8/21–8/27, 1988.

[64] Y. Fong, W. F. Ke, and C. S. Wang. Syntactic near-rings. To appear in the Proceedings of the 1992 Nearrings and Nearfields conference, New Brunswick.

[65] Y. Fong and J. D. P. Meldrum. The endomorphism near-ring of the symmetric group of degree at least five. *J. Australia Math. Soc. (A)* **30** (1980), 37–49.

[66] Y. Fong and J. D. P. Meldrum. The endomorphism near-ring of the symmetric group of degree four. *Tamkang Journal of Math.* **12** (1981), 193–203.

[67] Y. Fong and J. D. P. Meldrum. Endomorphism near-rings of a direct sum of isomorphic finite simple non-abelian groups. *Porc. of the 1985 Near-rings and Near-fields conference, Tübingen* (1987), 73–78.

[68] Y. Fong and G. Pilz. Near-rings generated by semi-endomorphisms of groups. *Contr. General Algebra* **7** (1991), 159–168.

[69] Y. Fong and L. van Wyk. A note on semi-homomorphisms of rings. *Bull. Austral. Math. Soc.* **40** (1989), 481–486.

[70] Y. Fong and L. van Wyk. Semi-homomorphisms of near-rings. *Mathematica Pannonica* **3** (1992), 3–17.

[71] Y. Fong, S. Veldsman, and R. Wiegandt. Radical theory in varieties of near-rings in which the constants form an ideal. *Comm. in Algebra* **21** (1993), 3369–3384.

[72] Y. Fong and R. Wiegandt. Subdirect irreducibility and radicals. *Quaestions Mathematicae* **16** (1993), 103–113.

[73] Y. Fong and Y. H. Xu. Non-associative and non-distributive rings. *Contr. General Algebra* **8** (1992).

[74] Y. Fong and Y. N. Yeh. A combinatorial approach on rings and near-rings generated by infra-endomorphisms of groups. In *Discrete Mathematics Symposium, 1/30–2/1, 1991*. Nat'l Chiao Tung University, Hsinchu, 1991.

[75] Y. Fong and Y. N. Yeh. Near-rings generated by infra-endomorphisms of groups. *Contr. General Algebra* **8** (1992), 63–69.

[76] H. Freedman. On endomorphisms of primary abelian groups. *J. London Math. Soc.* **43** (1968), 305–307.

[77] A. Fröhlich. Distributively generated near-rings I. Ideal theory. *Proc. London Math. Soc.* **8** (1958), 76–94.

[78] A. Fröhlich. Distributively generated near-rings II. Representation theory. *Proc. London Math. Soc.* **8** (1958), 95–108.

[79] A. Fröhlich. The near-ring generated by the inner automorphisms of a finite simple group. *J. London Math. Soc.* **33** (1958), 95–107.

[80] A. Fröhlich. On groups over a d.g. near-ring I. Sum constructions and free r-groups. *Quart. J. math. Oxford Ser. II* (1960), 193–210.

[81] A. Fröhlich. On groups over a d.g. near-ring II. Categories and functors. *Quart. J. Math. Oxford Ser. II* (1960), 211–228.

[82] A. Fröhlich. Non-abelian homological algebra I. Derived functors and satellites. *Proc. London Math. Soc.* **11** (1961), 239–275.

[83] A. Fröhlich. Non-abelian homological algebra II. Varieties. *Proc. London Math. Soc.* **12** (1962), 1–28.

[84] A. Fröhlich. Non-abelian homological algebra III. The functors ext and tor. *Proc. London Math. Soc.* **12** (1962), 739–768.

[85] L. Fuchs. Recent results and problems on abelian groups. In *Topics in Abelian Groups.* Scott, Foresman and Co., Chicago, Ill, 1963.

[86] P. Fuchs, G. Hofer, and G. Pilz. Codes from planar near rings. *IEEE Trans. Inform. Theory* **36** (1990), 647–651.

[87] P. Fuchs, C. J. Maxson, M. R. Petter, and K. C. Smith. Centralizer near-rings determined by fixed point free automorphism groups. *Proc. Royal Soc. Edinburgh* (1987), 327–337.

[88] N. J. Groenewald. Different prime ideals in near-rings. *Commun. Algebra* (1991).

[89] H. E. Heatherly. Distributive near-rings. *Quart. J. Math. Oxford (2)* **24** (1973), 63–70.

[90] H. E. Heatherly and J. J. Malone. Some near-ring embeddings. *Quart. J. Math. Oxford Ser.* **21** (1970), 445–448.

[91] I. N. Herstein. Semi-homomorphisms of groups. *Canad. J. Math.* **20** (1968), 384–388.

[92] I. N. Herstein. A note on derivations. *Canad. Math. Bull.* **21** (1978), 369–370.

[93] I. N. Herstein. A note on derivations II. *Canad. Math. Bull.* **22** (1979), 509–511.

[94] M. Holcombe. The syntactic near-rings of a linear sequential machine. *Proc. Edin. Math. Soc.* **26** (1983), 12–25.

[95] S. A. Huq. Semi-homomorphisms of rings. *Bull. Austral. Math. Soc.* **36** (1987), 121–125.

[96] C. U. Jemsen and H. Lenzing. *Model theoretic algebra with particular emphasis on fields, rings, modules.* Gordon and Breach Science Publishers, 1984.

[97] Z. Z. Jia. Brown-McCoy radicals and semisimplicity of distributively generated near-rings.

[98] K. Kaarli. On minimal ideals of distributively generated near-rings. To appear.

[99] K. Kaarli. A note on near-rings with identity. *Tartu Riikl. Ül. Toimetised* **336** (1974), 234–242. (In Russian).

[100] K. Kaarli. Minimal ideals in near-rings. *Tartu Riikliku Ulikooli Toimetised* (1975). Russian.

[101] K. Kaarli. On near-rings generated by endomorphisms of certain groups. *Tartu Riikliku Ulikooli Toimetised* (1978). (Russian).

[102] K. Kaarli. On Jacobson type radicals of near-rings. *Acta Math. Hung.* (1987).

[103] W. F. Ke. On nonisomorphic bibd with identical parameters. *Annals of Discrete Mathematics* **52** (1992), 337–346.

[104] W.-F. Ke and H. Kiechle. On the solutions of the equation $x^m + y^m - z^m = 1$ in a finite field. *Proc. of AMS* (1994), to appear.

[105] W.-F. Ke and K.-S. Wang. On the Frobenius groups with kernel of order 64. *Contributions to General Algebra* **7** (1991), 221–233.

[106] W.-F. Ke. *Structures of circular planar nearrings.* Ph. D. dissertation, Univ. Arizona, Tucson, 1992.

[107] W.-F. Ke and H. Kiechle. Combinatorial properties of ring generated circular planar near-rings. submitted.

[108] W.-F. Ke and H. Kiechle. Automorphisms of certain design groups. *J. Algebra* **167** (1994), 488–500.

[109] W.-F. Ke and H. Kiechle. Characterization of some finite Ferrero pairs. *Proceedings of the 1993 Near-rings and Near-fields Conference, New Bruncswick* (1995), to appear.

[110] D. A. Lawver. Concerning nil groups for near-rings. *Acta Math. Acad. Sci. Hungar* **22** (1972), 373–378.

[111] D. A. Lawver. Existence of near-rings in a special cases (near-rings on $Z_{(p^\infty)}$). Oberwolfach, 1972.

[112] R. Laxton and R. Lockhart. The near-rings hosted by a class of groups. *Proc. Endinburgh Math. Soc* **23** (1980), 69–86.

[113] J. M. Hall and J. K. Senior. *Groups of order 2^n, $(n \leq 6)$.* The Macmillan Company, New York, 1964.

[114] J. J. Malone. Near-rings with trivial multiplications. *Amer. Math. Soc. Monthly* **74** (1967), 1111–1112.

[115] J. J. Malone and C. G. Lyons. Finite dihedral groups and d.g. near-rings I. *Compositio Mathematica* **24** (1972), 305–312.

[116] J. J. Malone and C. G. Lyons. Finite dihedral groups and d.g. near rings II. *Compositio Mathematica* **26** (1973), 249–259.

[117] C. J. Maxson. On the construction of finite local near-rings I. On non-cyclic abelian p-groups. *Quart. J. Math., Oxford Ser.* **21** (1970), 449–457.

[118] C. J. Maxson. On the construction of finite local near-rings II. On abelian p-groups. *Quart. J. Math., Oxford Ser.* **22** (1971), 65–72.

[119] C. J. Maxson and K. C. Smith. The centralizer of a group automorphism. *J. Algebra* (1978), 27–41.

[120] S. K. C. Maxson C. J. The centralizer of a set group automorphism. *Commun. Algebra* (1980), 211–230.

[121] J. D. P. Meldrum. *Near-rings and Their Links with Groups.* No. 134 in Research Note Series. Pitman Publ. Co., 1985.

[122] J. D. P. Meldrum and A. Oswald. Near-rings of mapping. *Proc. Royal Soc. Edinburgh* (1979), 213–223.

[123] J. D. P. Meldrum and M. Zeller. The simplicity of near-rings of mapping. *Proc. Royal Soc. Edinburgh* (1981), 185–193.

[124] R. Mlitz and A. Oswald. Hypersolvable and supernilpotent radicals. *Studia Sci. Math. Hungar.* (1989).

[125] M. C. Modisett. *A characterization of the circularity of certain balanced incomplete block designs.* Ph. D. dissertation, University of Arizona, 1988.

[126] H. Neumann. Near-rings connected with free group. *Proc. inter. Conf. Amsterdam, II* (1954), 46–47.

[127] H. Neumann. On varieties of groups and their associated near-rings. *Math. Z.* **65** (1956), 36–69.

[128] R. S. Pierce. Homomorphisms of primary abelian groups. In *Topics in Abelian Groups.* Scott, Foreman and Co., Chicago, Ill, 1963.

[129] G. Pilz. *Near-rings.* North-Holland/American Elsevier, second, revised edition, 1983.

[130] E. Posner. Derivations in prime rings. *Proc. Amer. Math. Soc.* (1957).

[131] S. D. Scott. Minimal ideals of near-rings with minimal condition. *J. London Math. Soc.* (1974).

[132] R. Stringall. Endomorphism rings of abelian groups generated by automorphism groups. *Acta Math. Acad. Sci. Hungar.* **18** (1967), 401–404.

[133] L. P. Su. Homomorphisms of near-rings of continuous functions. *Pacific J. Math* **38** (1971), 261–266.

[134] L. P. Su. Near-rings of continuous functions. *Chinese Univ. of Hong Kong* (1972), 141–150.

[135] O. Veblen and J. H. MacLagan-Wedderburn. Non-desarguesian and non-pascalian geometries. *Trans. Amer. Math. Soc.* (1907).

[136] S. Veldsman. Varieties and radicals of near-rings. Manuscript.

[137] S. Veldsman. On equiprime near-rings. *Commun. Algebra* (1992).

[138] X. K. Wang. Zero-product-associative reduced near-rings. *J. Math. Res. Exposition* **12** (1992), 569–572.

[139] X. K. Wang. Derivations in prime near-rings. *Proc. AMS* (1994), to appear.

[140] Z. X. Wei. An anticommutativity theorem for near-rings. *Hunan Jiaoyu Xueyuan Xuebao (Journal of Hunan Normal College)* **9** (1991), 7–9.

[141] Z. X. Wei. Some conditions for a near-ring to be a near-field. *Proc. First China-Japan Int. Symp. on Ring Theory (Guilin 1991)* (1992), 157–159.

[142] P. S. Wu. Several remarks on radicals of near-rings. *J. Beijing Normal Univ. (Nat. Science Ed.)* (1979).

[143] P. Yang. Derivations on near-rings and rings. *Acta Sci. Natur. Univ. Jilin* (1991), 21–25.

[144] Y. P. Ye. Maximal ideals in the near-ring of polynomials and Jacobson radicals. *Chin. Ann. Math. Ser. A*, 535–542.

[145] C. M. Zhang. The J-radical of type $*$ and a class of J-type radical for near-rings (Chinese/English summary). *Hunan Shifan Daxue Ziran Kexue Xuebao (Natural Science Journal of Hunan Normal University)* 11 (1988), 189–192.

[146] C. M. Zhang. A J-radical of type 5/2 for near-rings (Chinese/English summary). *Hunan Shifan Daxue Ziran Kexue Xuebao* 11 (1988), 14–18.

[147] C. M. Zhang. The class of J-radicals for weakly-symmetric near-rings. *Acta Sci. Natur. Univ. Nor. Hunan* 14 (1991), 174–176.

[148] Y. M. Zheng. The Hamilton-Cayley theorem over a commutative near-ring. *Acta Math. Sinica* 34 (1991), 316–319.

Theory of Formations

Guo Wenbin   *Department of Mathematics, University of Yangzhou, Jiangsu, P. R. of China*, 225002

All the groups considered in this chapter are finite. A class $\mathfrak{F}$ of groups is called a formation provided: it is closed under homomorphic image and sub-direct product. A formation $\mathfrak{F}$ is said to be local (see Šemetkov (1978)) if there exists a formation function f such that $\mathfrak{F} = \{G \mid G$ is a group and $G/C_G(H/K) \in f(p)$ for all principal factors H/K of G and for all primes p dividing $|H/K|\}$. In this case we say that $\mathfrak{F}$ is locally defined by f and f is a local screen of $\mathfrak{F}$. A non-empty formation $\mathfrak{F}$ is called saturated if it satisfies: if $G/\Phi(G) \in \mathfrak{F}$, then $G \in \mathfrak{F}$. It is well known that any local formation is saturated.

We denote by **P** the set of all prime numbers. Let G be a group. We denote by G_p a Sylow p-subgroup of G. $|G|$ denotes the order of G. $\pi(G)$ denotes the set of all prime divisors of $|G|$. $F(G)$ denoes the Fitting subgroup of G. $\Phi(G)$ denotes the Frattini subgroup of G. Let $\mathfrak{X}$ be a set of subgroups of G. We denote by $<\mathfrak{X}>$ the subgroup of G generated by $\mathfrak{X}$. $\mathfrak{E}$ denotes the formation consisting of a unit element group. $\mathfrak{G}$ denotes the formation of all finite groups. $\mathfrak{G}_\pi$ denotes the formation of all π-groups. $\mathfrak{N}$ denotes the formation of all nilpotent groups. $\mathfrak{A}$ denotes the formation of all abelian groups. $\mathfrak{S}$ denotes the formation of all solvalbe groups. $\mathfrak{U}$ denotes the formation of all supersolvable groups.

Let $\mathfrak{F}$ be an formation. A group G is said to be an $\mathfrak{F}$-group if $G \in \mathfrak{F}$. $G^{\mathfrak{F}}$ denotes the $\mathfrak{F}$-coradical of G, i. e. the intersection of all normal subgroups N with $G/N \in \mathfrak{F}$. $\pi(\mathfrak{F}) = \bigcup_{G \in \mathfrak{F}} \pi(G)$. $\mathfrak{F}_\pi = \mathfrak{F} \cap \mathfrak{G}_\pi$.

Let $\mathfrak{X}$ be a set of groups. We denote by form $\mathfrak{X}$ the formation generated by $\mathfrak{X}$. lform $\mathfrak{X}$ denotes the local formation generated by $\mathfrak{X}$.

Let $\mathfrak{X}$, $\mathfrak{F}$ be two non-empty formations. $\mathfrak{X}\mathfrak{F}$ denotes the set of all group G with $G^{\mathfrak{F}} \in \mathfrak{X}$.

A subgroup H of group G is called a primary subgroup if the order of H is a power of a prime number.

Let $\mathfrak{F}$ be a local formation defined by f. A normal subgroup N of group G is said to be an f-hypercentre in G, if $G/C_G(H/K) \in f(p)$ for all G-principal factors H/K of N and for all primes p dividing $|H/K|$.

§ 1. On Finite Groups in which Some Primary
Subgroups with Hypercentral Condition

The famous Frobenius theorem affirm: A finite group G is p-nilpotent, if $N_G(P)/C_G(P)$ is a p-group for all p-subgroups P of G. Obviously, the theorem can be described as follows: G is p-nilpotent if each p-subgroup of G is a hypercentre in its normalizer. Now the Frobenius theorem has been had the characteristic of formation. This impel us to look for a general law. The following theorem 1. 1 solved the problem.

Theorme 1. 1 (Guo Wenbin and Šemetkov(1992)). Let $\mathfrak{F}$ be a local formation defined by f, $\sigma = \{p \mid p \text{ is a prime and } f(p) \neq \mathfrak{G}\}$ and G be a group. If the $\mathfrak{F}$-coradical of G is σ-solvable, then $G \in \mathfrak{F}$ if and only if P is an f-hypercentre in $N_G(P)$ for all primary subgroups P of G.

Corollary 1. 1. Let G be a p-solvable group. Then G is supersolvable if and only if each non-unity p-subgroup P of G has an $N_G(P)$-principal series such that each quotient factor is a group with order p.

Now we discuss a more general problem than the above, i. e. we study the groups G that satisfies the following conditions: each nonnormal primary subgroup P of G is an f-hypercentre in $N_G(P)$, where f is a local screen of a local formation. The research may be regarded as a continuation of the research on minimal non-$\mathfrak{F}$-groups.

Definition 1. 1(Guo Wenbin and Šemetkov(1992), Guo Wenbin(1991)). Let $\mathfrak{F}$ be a local formation defined by f. A group G is called an $L_{\mathfrak{F}}^*$-group, if $G \notin \mathfrak{F}$ and P is an f-hypercentre in $N_G(P)$ for every nonnormal primary subgroup P of G.

A group G is called p-decomposable if $G = G_P \times G_{P'}$.

Theorem 1. 2 (Guo Wenbin and Šemetkov(1992), Guo Wenbin(1991)). Let G be an $L_{\mathfrak{F}}^*$-group and $G^{\mathfrak{F}}$ be solvable. Then $G^{\mathfrak{F}}$ is nilpotent.

Theorem 1. 3 (Guo Wenbin and Šemetkov(1992),Guo Wenbin(1991)). Let G be a solvable $L_{\mathfrak{N}}^*$-group. Then the following hold:

1) G is a metanilpotent and dispersive group.

2) For every $p \in \pi(F(G)/G^{\mathfrak{N}}Z_\infty(G))$, there exists a nonnormal in G p-subgroup P of F(G) such that $N_G(P)$ is not p-decomposable.

3) If $N_G(P) \nsubseteq F(G)$ for a certain nonnormal in G primary subgroup P of F(G), then there is a primary subgroup of $\mathfrak{N}$-injectors of G whose normalizer in G is nonnilpotent.

Definition 1. 2(Guo Wenbin(1991)). A non-nilpotent group G is called an LZ-group if $N_G(P)$ is $\pi(P)$-decomposable for each nonnormal primary subgroup P of G.

Theorem 1. 4 (Guo Wenbin(1991)). Let G be a non-unit solvable group and $Z_\infty(G) = 1$. Then G is an LZ-group if and only if the following hold:

1) $G = G^{\mathfrak{N}} \rtimes H$, where $G^{\mathfrak{N}} = F(G)$, H is a Carter subgroup of G and a system normalizer of G at same time.

2) H is a cyclic Hall subgroup and each non-unit subgroup of H is non-normal in G.

3) If S is a maximal nilpotent subgroup of G, then $S_{\pi(H)}$ is a Hall subgroup of G, and each q-subgroup of S is normal in G for every $q \in \pi(S) \cap \pi(G:S)$.

4) every nonnormal in G primary subgroup Q of $G^{\mathfrak{N}}$ is included in $Z_\infty(H_P G^{\mathfrak{N}})$ for any $p \in \pi(H) \cap \pi(N_G(Q))$.

Theorem 1. 5 (Guo Wenbin(1991)). A solvable group G is an LZ-group if and only if the following hold:

    1) $G/Z_\infty(G)$ is an LZ-group.

    2) If P is a primary subgroup of $F(G)$ and $N_G(P) \neq G$, then $N_G(P)$ is $\pi(P)$-decomposable.

In the above papers we not only unify the Frobenius theorem and Schmidt theorem, but also break a fresh path.

Let $\mathfrak{F}$ be a local formation closed under subgroups. $\Phi_{\mathfrak{F}}(G)$ denotes the intersection of all maximal $\mathfrak{F}$-subgroups of G. $\mathfrak{F}_p$ denotes the class of groups $\{G \mid G/O_{p'p}(G) \in f(p)\}$, where f is a local screen of $\mathfrak{F}$ }.

Theorem 1. 6 (Chen Chongmu(1994)). Let $\mathfrak{F}$ be a local formation and G be a solvable group. If each element of $G^{\mathfrak{F}}$ with order a prime or 4 is included in $\Phi_{\mathfrak{F}}(G)$, then $G \in \mathfrak{F}$.

Theorem 1. 7 (Chen Chongmu(1994)). Let $\mathfrak{F}$ be a local formation. Then $G \in \mathfrak{F}_p$ if and only if $G/O_{p'}(G) \in \mathfrak{F}$.

§ 2. On Finite Groups With Given Normalizers of Sylow Subgroups

In the last few years a series of papers were devoted to investigation of finite groups with a given condition on Sylow subgroups. It was proved by Brianchi M, Mauri, A. G. B. and Hauck, P. (1986) that if the normalizer of any non-unit Sylow subgroup of a finite group G is nilpotent then G is nilpotent. It was noted by Fedri V. and Serens, L(1988) that the formation $\mathfrak{u}$ of all supersolvable groups does not have this property. In other words, if the normalizer of any non-unit Sylow subgroup of a finite group G is supersolvable then G may be not supersolvable. From the above the question has emerged: Which formations have this property? Which formations do not?

A formation is called an Š-formation if each noncyclic minimal non-$\mathfrak{F}$-group is a Schmidt group.

First, we see that the formation $\mathfrak{N}$ of all nilpotent groups is an S-closed (i. e. closed under subgroups) local Š-formation and the formation $\mathfrak{u}$ is not an Š-formation. Therefore the question arises: Which S-closed local Š-formations have the property mentioned above? The problem was set by Professor Šemetkov L. A. In the following theorem 2. 1 this problem is decided completely for the class of solvale groups.

Let $\mathfrak{F}$ be a formation. $N^{\mathfrak{F}}$ denotes the class of all groups G in which the normalizer $N_G(G_p)$ of each non-unit Sylow p-subgroup G_p of G belongs to $\mathfrak{F}$ for all primes p dividing $|G|$.

From theorem 2. 1 to corollary 2. 7 we assume that all groups are solvable.

Theorem 2. 1 (Guo Wenbin (1994a)). Let $\mathfrak{F}$ be an S-closed local Š-formation. Then $N^{\mathfrak{F}} \subseteq \mathfrak{F}$ if and only if $\mathfrak{F}_\sigma \in \{\mathfrak{S}_\sigma, \mathfrak{N}_\sigma\}$ for any two-element subset σ of $\pi(\mathfrak{F})$.

Corollary 2. 1 (Guo Wenbin (1994a)). Let $\mathfrak{F} = \mathfrak{N}_\pi$, where π is a set of primes. Then $N^{\mathfrak{F}} \subseteq \mathfrak{F}$.

Corollary 2. 2 (Guo Wenbin(1994a)). Let $\mathfrak{F} = \underset{i \in I}{\times} \mathfrak{S}_{\pi_i}$, where $\pi_i \bigcap \pi_j = \Phi$ for any diffirent i, j in I. Then $N^{\mathfrak{F}} \subseteq \mathfrak{F}$.

Corollary 2. 3 (Guo Wenbin(1994a)). If $\mathfrak{F}$ is the formation of all p-nilpotent groups, then $N^{\mathfrak{F}} \not\subseteq \mathfrak{F}$.

Let ψ be a linear order on the set of all prime numbers. A group G of order $p_1^{\alpha_1} p_2^{\alpha_2} \cdots p_n^{\alpha_n}$, where $p_1 \psi p_2 \psi \cdots \psi p_n$, is called ψ-dispersive if it has a normal subgroup of order $p_1^{\alpha_1} p_2^{\alpha_2} \cdots p_i^{\alpha_i}$ for any $i = 1. 2, \cdots, n-1$. It is known that the class of all ψ-dispersive groups is an S-closed local Š-formation.

Corollary 2. 4 (Guo Wenbin(1994a)). Let $\mathfrak{F}$ be the class of all ψ-dispersive groups. Then $N^{\mathfrak{F}} \not\subseteq \mathfrak{F}$.

Corollary 2. 5 (Guo Wenbin(1994a)). Let $\mathfrak{F}$ be an S-closed ψ-dispersive local Š-formation. Then $N^{\mathfrak{F}} \subseteq \mathfrak{F}$ if and only if $\mathfrak{F} \subseteq \mathfrak{N}$.

Theorem 2. 2 (Guo Wenbin(1994b)). Let $\mathfrak{F} = C_\pi \mathfrak{N}$ be the class of all groups such that each Hall π-subgroup is nilpotent. Then $\mathfrak{F}$ is an Š-formation and $N^{\mathfrak{F}} \subseteq \mathfrak{F}$.

Besides the above results about Š-formation the following theorems are gotten.

Theorem 2. 3 (Guo Wenbin(1994b)). Let f be a full and integrated local screen of a local formation $\mathfrak{F}$. If $N^{\mathfrak{F}} \subseteq \mathfrak{F}$, then $N^{f(p)} \subseteq f(p)$ for all $p \in \pi(\mathfrak{F})$.

Let $\mathfrak{X}$ be a class of groups. $M(\mathfrak{X})$ denotes the class of all minimal non-$\mathfrak{X}$ groups. Let

$$m(\mathfrak{X}) = \max\{|\pi(G)| \mid G \in M(\mathfrak{X})\}$$

and be called m-rank of $\mathfrak{X}$.

It is easily verified that $\{\check{S}\text{-formations}\} \subsetneqq \{\mathfrak{F} \mid \mathfrak{F}$ is a formation and m$(\mathfrak{F}) \leqslant 2\}$.

Let δ be a relation between elements of the set $\mathbf{P}$ of all prime numbers, and δ be reflexive and symmetric. We define a class $\mathfrak{F}_\delta$ of groups:

$\mathfrak{F}_\delta = \{G \mid G$ is a proup, and $G_p \subseteq N_G(G_q), G_q \subseteq N_G(G_p)$ for all p, q in $\pi(G)$ and $(p, q) \in \delta\}$.

Theorem 2. 4 (Guo Wenbin(1996a)). $\mathfrak{F}_\delta$ is a saturated fromation closed under subgroups and an $\check{S}$-formation.

Theorem 2. 5 (Guo Wenbin(1996a)). Let $\mathfrak{F}$ be a local formation closed under subgroups and m($\mathfrak{F}$) $\leqslant 2$. Then $N^{\mathfrak{F}} \subseteq \mathfrak{F}$ if and only if there exists a reflexive and symmetric relation δ between elements of the set $\mathbf{P}$ of all prime unmbers such that $\mathfrak{F} = \mathfrak{F}_\delta$.

Corollary 2. 6 (Guo Wenbin(1996a)). The set of formations $\mathfrak{F}$ with $N^{\mathfrak{F}} \subseteq \mathfrak{F}$ and m($\mathfrak{F}$) $\leqslant 2$ is equal to the set of $\check{S}$-formations $\mathfrak{X}$ with $N^{\mathfrak{X}} \subseteq \mathfrak{X}$.

Let ψ be a linear order. $\mathfrak{D}_\psi$ denotes the local formation of all finite ψ-dispersive groups. $\mathfrak{N}^2$ denotes the formation of all finite metanilpotent groups.

Theorem 2. 6 (Guo Wenbin, to appear(a)). Let $\mathfrak{D}$ be a local formation, $\mathfrak{D} \subseteq \mathfrak{D}_\psi$ and f be a local screen of $\mathfrak{D}$. And let $G \in N^{\mathfrak{D}}$, $|G| = p^\alpha q^\beta$ and $p \psi q$. If the group C_p whose order is p does not belong to $f(q)$, then $G \in \mathfrak{N}$.

Theoirem 2. 7 (Guo Wenbin, to appear(a)). Let $\mathfrak{D}$ be a local formation as in theorem 2. 6, $p, q \in \pi(\mathfrak{D})$ and $p \psi q$. If $C_p \in f(q)$, then $\mathfrak{S}_{\{p,q\}} \subseteq SN^{\mathfrak{D}}$, where $SN^{\mathfrak{D}}$ is the set of all subgroups of all $N^{\mathfrak{D}}$-groups.

Theorem 2. 8 (Guo Wenbin, to appear (a)). Let $\mathfrak{F}$ be a local formation closed under subgroups, $\mathfrak{F} \subseteq \mathfrak{D}_\psi \cap \mathfrak{N}^2$, and let $G \in N^{\mathfrak{F}}$, $|\pi(G)| \geqslant 3$, let p be the ψ-minimal prime factor of $|G/G^{\mathfrak{F}}|$ and $p^s = |G/G^{\mathfrak{F}}|_p$. Then the nilpotent length of $G \leqslant 2s+2$.

Corollary 2. 7 (Bryce, Fedri and Serena (1991)). Let $G \in N^{\mathfrak{U}}/\mathfrak{U}$, and $|\pi(G)| \geqslant 3$. Then the nilpotent length of $G \leqslant 2t+2$, where p^t is the highest power of the smallest prime p dividing $|G/G^{\mathfrak{U}}|$.

Theorem 2. 9 (Chen Chongmu, to appear). Let G be a finite group, and $p \geqslant 5$ be a prime. If $N_G(Gp)/C_G(Gp)$ is a p-group, and $N_G(Gq)/C_G(Gq)$ is a p'-group for all $q \neq p$, then G is p-decomposable.

A group G is said to be perfectly separable if each subgroup of G has a complement in G.

Theorem 2. 10 (Guo Wenbin (1994b)). Let $\mathfrak{F}$ be a formation of all perfectly separable groups. If the normalizer of each nonnormal Sylow subgroup of a group G belongs to $\mathfrak{F}$, then the $\mathfrak{F}$-coradical $G^{\mathfrak{F}}$ of G is a nilpotent group.

Theorem 2. 11 (Guo Wenbin (1994b)). A group is perfectly separable if and only if the normalizer of each Sylow subgroup of G is perfectly separable.

Theorem 2. 12 (Guo Wenbin (1993b)). Let G be a finite group. If the normalizer of each Sylow subgroup of G has a Hall complement, then G is a dispersive grorp. Moreover, if Fitting subgroups of these complements are cyclic groups, then G is supersolvable.

Theorem 2. 13 (Guo Wenbin (1994c). If the normalizers of all Sylow subgroups of a group G have prime index, then G is solvable.

§ 3. The Complement of a Subformation

A subformation $\mathfrak{F}_1$ of formation $\mathfrak{F}$ is said to have a complement in $\mathfrak{F}$ if there is a subformation $\mathfrak{F}_2$ of $\mathfrak{F}$, such that $\mathfrak{F}_1 \cap \mathfrak{F}_2 = \mathfrak{E}$, and $\mathfrak{F} = \text{form}$ $(\mathfrak{F}_1 \cup \mathfrak{F}_2)$, where $\mathfrak{E}$ is the formation of unit element groups.

Skiba, A. N. was the first to investigate the problem of whether a subformation of a formation has a complement (see Skiba, A. N. (1981)). Later, Eidinov, M. N. (1984) and Vedernikov, V. A. (1990) gave a description of formations in which every proper subformation has a complement. For many algebra systems the work in this area has been tackled by Skiba, A. N. and Šemetkov, L. A. (1991). Vedernikov, V. A. (1990) proposed the problem of describing a local formation in which every local subformation has a complement. But Vedernikov obtained only that "a local formation $\mathfrak{F}$ is generated by some simple groups if evevy local Fitting subformation gererated by a sole group has a complement." Now we have the following theorem.

Theorem 3. 1 (Guo Wenbin, to appear (b)). Let $\mathfrak{F}$ be a non-unit local formation. Then the following conditions are equivalent:

1) Every subformation of type $\mathfrak{N}_p$ (where $p \in \pi(\mathfrak{F})$) has a complement in $\mathfrak{F}$.

2) Formation $\mathfrak{F}$ is nilpotent (i. e. $\mathfrak{F} \subseteq \mathfrak{N}$).

3) The lattice $L_l(\mathfrak{F})$ of all local subformations of $\mathfrak{F}$ is Boolean.

This theorem solves Vedernikov's problem completely. Vedernikov obtained only a necessary condition under rather strong hypotheses. And from our theorem, it may be seen that Vedernikov's description of necessity is not quite sufficient.

Corollary 3. 1. A local formation $\mathfrak{F}$ is nilpotent if and only if every loacl Fitting subformation generated by a sole group of $\mathfrak{F}$ has a complement in $\mathfrak{F}$.

Corollary 3. 2. A group G is nilpotent if and only if every local subformation of lformG has a complement.

§ 4. The Groups Generated by Subnormal Subgroups

It is known that the group generated by nilpotent subnormal subgroups is nilpotent, but the group generated by two supersolvable subnomal subgroups may not be supersolvable. Baer, R. (1957) proved that the product of two supersolvalbe normal subgroups is supersolvable if and only if the derived subgroup of the product is nilpotent. K. Doerk and T. Hawkes (1978) proved that if G is a solvable group and $G = A \times B$ then $G^{\mathfrak{F}} = A^{\mathfrak{F}} \times B^{\mathfrak{F}}$. The following results continue and develop this research.

Theorem 4. 1 (Fen Yun (1986)). Let $\mathfrak{F}$ be a local formation. If there are two normal subgroups of G in $\mathfrak{F}$ with indexes prime to each other, then G $\in \mathfrak{F}$.

Theorem 4. 2 (Guo Wenbin, to appear(c)). Let $\mathfrak{F}$ be a local formation. If $G=<A,B>$, where A. B are subnormal subgroups of G, and $A \in \mathfrak{F} \cap \mathfrak{N}$, $B \in \mathfrak{F}$, then $G \in \mathfrak{F}$.

Corollary 4. 1. Let $G=<A,B>$, where A is a nilpotent subnormal subgroup of group G, and B is a sapersolvable subnormal subgroup of G. Then G is supersolvable.

Corollary 4. 2. Let $G=<A,B>$, where A,B are subnomal subgroups of group G. If A is nilpotent and the derived subgroup of B is nilpotent, then the derived subgroup of G is nilpotent.

Let $\mathfrak{F}$ be a non-empty formation. A subgroup K of G is said to be $\mathfrak{F}$-subnormal in G if there is a chain of subgroups

$$G=K_0 \supset K_1 \supset \cdots \supset K_n = K$$

such that K_i is a maximal subgroup of K_{i-1} and $K_{i-1}^{\mathfrak{F}} \subseteq K_i$ for $i=1,2,\cdots n$.

Theorem 4. 3 (Guo Wenbin, to appear(c)). Let $\mathfrak{X}$ be a local formation closed under subgroups. Then the follouing conditions are equivalent:

1) If $\mathfrak{F}$ is a local subformation of formation $\mathfrak{N} \mathfrak{A}$, then $\mathfrak{F}$ contains all $\mathfrak{X}$-groups $G=<A,B>$, where A,B are $\mathfrak{F}$-subnormal $\mathfrak{F}$-subgroups of G.

2) $\mathfrak{X} \subseteq \mathfrak{N} \mathfrak{A}$

Corollary 4. 3 (Guo Wenbin, to apper(c)). Let $G=<A,B>$, where A, B are $\mathfrak{F}$-subnormal supersolvable subgroups of G. Then G is supersolvable if and only if the derived subgroup G' of G is nilpotent.

Corollary 4. 4 (Baer(1957)). Let $G=N_1 N_2$, where $N_i(i=1,2)$ are supersolvable normal subgroups of G. Then G is supersolvable if and only if $G \in \mathfrak{N} \mathfrak{A}$.

Corollary 4. 5 (Asaad and Shaalan(1989)). Suppose H and K are supersolvable subgroups of G, the derived subgroup G' of G is nilpotent and $G=KH$. Suppose further that H is quasinormal in K and K is quasinormal in H. Then G is supersolvable.

So the results of Baer and the results of Asaad and Shaalan are achieved by theorem 4. 3.

Theorem 4. 4 (Guo Wenbin and Kamornikov(1994)). Let $\mathfrak{F}$ be a formation, G be a group with $\pi(\mathfrak{F})$-solvable $\mathfrak{F}$-coradical. If the $\mathfrak{F}$-coradical of an $\mathfrak{F}$-abnormal subgroup U of G is subnormal in G, then U contains an $\mathfrak{F}$-projector of G.

Theorem 4. 5 (Guo Wenbin and Kamornikov (1994)). A group G is metanilpotent if and only if $N_G(G_p)^{\mathfrak{N}}$ is subnormal in G for all $p \in \pi(G)$.

REFERENCES

Šemetkov, L. A. (1978). Formations of finite groups, Moscow Nauka, Main Editorial Board for Physical and Mathematical Literature, Moscow, p. 272.

Doerk, K. and Hawkes, T. (1978). On the residual of a direct product, Arch. Math. **30**,458−468.

—— (1992). Finite soluble groups, Walter de Gruyter Berlin New York, p. 891.

Guo Wenbin and Šemetkov, L. A. ,(1992). On finite grorps with hypercentral condition, Doklady Akad. Nank BSSR,**36**:485−486.

Guo Wenbin, (1991). On finite gronps in which some primnry subgroups have hypercentral condition, Doctor Thesis, The Archive of Supreme Attestation Commission of USSR.

—— (1993),(a). Finite groups with formation condition for normalizers of primary subgroups, Gomel State University,Republic Belarus, Prep. **11**:1 −33.

—— (1993),(b). On normalizers of Sylow subgroups I, Doklady Akad, Nauk BSSR, **37**:22−24.

—— (1994),(a). On finite groups with given normalizers of Sylow subgroups I, Chinese Science Bulletin, **39**:1952−1955.

—— (1994),(b). Finite groups with given Sylow−normalizers, Chinese Annals of Mathematics, Ser. A, **15**:627−631.

—— (1994),(c). On normalizers of Sylow subgroups Ⅱ , Gomel State University, Republic Belarus,Prep. **12**:1−6.

—— (1996),(a). On finite groups with given normalizers of Sylow subgroups Ⅲ , Acta Math. Sinica, Peking,**39**(2).

—— (to appear),(a). Nilpotent length of finite soluble groups with given Sylow−normalizers.

—— (to appear),(b). On local formations in which every subformation of type $\mathfrak{N}$p has a conplement.

—— (to appear),(c). The groups generated by subnormal subgroups.

Guo Wenbin and Kamornikov, C. F. ,(1994). Abnormal subgroups with subnormal Coradicals, Problem of Mathematics and Informations, **1** Gomel. p. 31.

Chen Chongmu, (1994). Some theorems about local formations, Chinese Annals of Mathematics, Ser. A, **15**:665—670.

—— (to appear). Finite groups with p—decomposable Sylow normalizers.

Bianchi, M. ,Mauri, A. G. B. and Hauck,P. (1986). On finite groups with nilpotent Sylow normalizers, Arch. Math. **47**:193—197.

Fedri, V. and Serens,L. (1988). Finite Soluble groups with supersoluble Sylow normalizers, Arch. Math. **50**:11—18.

Bryce, R. A. , Fedri, L. and Serena, L. (1991). Bounds on the Fitting length of finite soluble groups with supersoluble Sylow normalizers, Bull. Austral. Math. Soc. **44**:19—31.

Skiba, A. N. (1981). On formations with given systems of subformations, Subgroup structure of finite groups, Minsk, Nauka and Technic, 155—180.

Eidinov, M. E. (1984). On formation with subformations which have complements, Ⅸ all Soviet Union symposium on theory of groups, Thesis report, Moscow, 101.

Vedernikov, V. A. (1990). Completely factorizable formations of finite groups, Problems in Algebra, Minsk, University Press,**5**:28—34.

Skiba, A. N. and Šemetkov, L. A. (1991). On formations with systems of subformations which have complements, Ukraine Math. J. **32**:101—118.

Baer, R. (1957). Classes of finite groups and their properties, Illinois J. Math. **1**:115—118.

Fan Yun, (1986). F—stable nature and F—critical nature, Acta Mathematica Sinica, China, **29**:117—126.

Asaad, M. and Shaalan, A. (1989). On the supersolvability of finite groups, Arch. Math. **53**:318—326.

Some Recent Developments in
Filtered Ring and Graded Ring Theory
in China

LI HUISHI   Department of Mathematics, Shaanxi Normal University, 710062 Xian, P.R. China

Abstract

This paper is a rough survey of some recent results concerning filtered
ring and graded ring theory in China.

§1 NONCOMMUTATIVE ZARISKIAN FILTERED RING THEORY

1.1. Zariskian filtrations

Noncommutative Zariskian filtered ring theory has its root in the algebraic analysis theory
and the study of enveloping algebras of Lie algebras and rings of differential operators (e.g.
[Be], [Ka], [Bj1], [Shar], [Gin], etc...). Since the late 1960s, filtered rings with various additional conditions have been widely studied, e.g. the filtered rings with good Noetherian
filtration in [Bj1], the filtered rings with comparison condition and the closure condition in
[Bj2], the filtered rings with Σ-Noetherian condition in [Ve2], the filtered rings with radical
condition in [AVO], etc... . Motivated by a remarkable study of the Rees ring associated to a
given filtered ring in [AVVO] and by some interesting results concerning the positively filtered
Noetherian rings in [Bj1], Li Huishi and coauthors have systematically studied those filtered
rings R that have a filtration $\cdots \subset F_{-1}R \subset F_0R \subset F_1R \subset \cdots$ satisfying the conditions: (1)
the Rees ring $\tilde{R}$ of R is left (right) Noetherian; (2) $F_{-1}R$ is contained in the Jacobson radical

of F_0R (that is the left (right) *Zariskian filtered ring* first named and studied in [LVO1]. For a general theory of Zariskian filtrations we refer to [LVO7]). Comparing with the classical commutative Zariski ring theory the following characterization of a noncommutative Zariskian filtered ring is obtained (see [LVO1], [LVOW], [Li1] and [LVO7]):

Theorem. Let R be a filtered ring with (increasing) filtration $FR = \{F_nR, F_nR \subseteq F_{n+1}R, F_nRF_mR \subseteq F_{n+m}R, \ n, \ m \in \mathbb{Z}\}$ consisting of additive subgroups of R such that $R = \cup_{n \in \mathbb{Z}} F_nR$ (i.e., FR is exhaustive), $G(R) = \oplus_{n \in \mathbb{Z}} F_nR/F_{n-1}R$ the associated graded ring of R and $\tilde{R} = \oplus_{n \in \mathbb{Z}} F_nR$ the Rees ring of R. Let R-filt be the category of filtered R-modules and filtered R-morphisms of degree zero. Then the following statements are equivalent:

(1) $F_{-1}R$ is contained in the Jacobson radical of F_0R and $\tilde{R}$ is left Noetherian;

(2) FR is separated, i.e., $\cap_{n \in \mathbb{Z}} F_nR = 0$, $G(R)$ is left Noetherian, $F_{-1}R$ is contained in the Jacobson radical of F_0R and every good filtration FM on $M \in R$-filt has the left Artin-Rees property (recall that a filtration FM on M is said to be good if there exist $m_1, \cdots, m_s \in M$ and $k_1, \cdots, k_s \in \mathbb{Z}$ such that $F_nM = \sum_{i=1}^s F_{n-k_i}Rm_i$ for every $n \in \mathbb{Z}$, and FM is said to have the Artin-Rees property if for every finitely generated R-submodule $N = \sum_{i=1}^s Rx_i$ of M there exists a $c \in \mathbb{Z}$ such that for every $n \in \mathbb{Z}$ we have $F_nM \cap N \subseteq \sum_{i=1}^s F_{n+c}Rx_i$);

(3) FR is separated, $G(R)$ is left Noetherian, $F_{-1}R$ is contained in the Jacobson radical of F_0R and FR has the left Artin-Rees property;

(4) $G(R)$ is left Noetherian and the completion $\hat{R}$ of R with respect to the FR-topology on R is a faithful flat right R-module;

(5) $G(R)$ is left Noetherian, good filtrations in R-filt induce good filtrations on R-submodules and good filtrations are separated;

(6) $G(R)$ is left Noetherian and for any $M \in R$-filt with good filtration FM, if N is any R-submodule of M with an arbitrary filtration FN (which is not necessarily good!) then each part $F_nN, \ n \in \mathbb{Z}$, of FN is closed in the FM topology of M, or in other words, for each $n \in \mathbb{Z}$ we have $F_nN = \cap_{p \in \mathbb{Z}}(F_nN + F_pM)$;

(7) $G(R)$ is left Noetherian, $F_{-1}R$ is contained in the Jacobson radical of F_0R and for every left ideal L of R with good filtration FL we have $F_nL = \cap_{p \in \mathbb{Z}}(F_nL + F_pR), \ n \in \mathbb{Z}$.

Zariskian filtered rings characterized above include enveloping algebras of finite dimensional Lie algebras over a field, rings of differential operators studied in [Be], [Ka], [Bj1] and [Shar], in particular the ring $\mathcal{E}_p$ of microdifferential operators that has a non-discrete non-complete filtration. All complete filtered rings with Noetherian associated graded rings are Zariskian. More Zariskian filtrations may be obtained by using dehomogenization of gradings to filtrations introduced in [LVO5] and by taking the localizations of a given Noetherian filtered ring at the lifting Ore sets (see [Li8]).

1.2. Lifting structures of a Zariski ring

Let R be a left Zariski ring with filtration FR and $G(R)$ the associated graded ring of R.

Concerning the structure properties of a Zariski ring lifted from its associated graded ring we have the following results (see [Li1], [LVO1], [LVO6]):

(a) If $G(R)$ is gr-Artinian semisimple (simple) then R is Artinian semisimple (simple); if $G(R)$ is a gr-skewfield then R is a skewfield.

(b) If $G(R)$ has finite Krull dimension then R has finite Krull dimension and $K.\dim R \leq K.\dim G(R)$.

(c) If $G(R)$ is left gr-regular (i.e., every finitely generated graded $G(R)$-module has finite projective dimension) then R is left regular (i.e., every finitely generated R-module has finite projective dimention). Moreover, $gl.\dim R \leq gr.gl.\dim G(R) \leq gl.\dim G(R)$.

(d) If $G(R)$ is left gr-hereditary then R is left hereditary; if $G(R)$ is Von Neuman gr-regular then R is Von Neuman regular.

(e) If R is also right Zariskian and $G(R)$ is a gr-Auslander regular ring then R is an Auslander regular ring (recall that R is Auslander regular if R is left and right Noetherian, $gl.\dim R < \infty$ and for every finitely generated R-module M and any submodule N of $\text{Ext}^k_R(M, R)$, $k \geq 0$ in $\mathbb{Z}$, we have $\text{Ext}^j_R(N, R) = 0$ for all $j < k$. The gr-Auslander regularity is defined intrinsically in the category of graded modules and graded morphisms).

A result concerning lifting projective covers for modules over a Zariski ring is obtained in [Wang]:

(f) Let R be a Zariski ring with filtration FR. Suppose that every finitely generated graded projective $G(R)$-module is stably free and $M \in R$-filt has a good filtration FM. Then

(i) if $G(M)$ has a projective cover in $G(R)$-gr, then this projective cover is finitely generated and M has a finitely generated filtered projective cover in R-filt;

(ii) if $G(R)$ is gr-semiperfect, then every $M \in R$-filt with good filtration FM has a minimal finitely generated filtered projective resolution in R-filt.

1.3. K_0 of rings with Zariskian filtrations

Concerning the K_0 group of a Zariski ring, we have the following results (see [Li1], [LVVO], [LVO6]):

Let R be a left Zariski ring with filtration FR and $G(R)$ the associated graded ring of R.

(a) If $G(R)$ is left gr-regular and every finitely generated graded projective $G(R)$-module is gr-stably free, then every finitely generated projective R-module is stably free.

(b) If $G(R)$ is left gr-regular (i.e., every finitely generated graded $G(R)$-module has finite projective dimension) then there is an embedding of K_0-groups: $K_0(R) \hookrightarrow K_0(G(R))$.

1.4. Zariski rings with Auslander regular associated graded rings

Let R be a (left and right) Zariski ring with filtration FR. Suppose that $G(R)$ is an Auslander regular ring; then the following results have been obtained:

(a) ([Li2]) For any finitely generated R-module, $j_R(M) = j_{G(R)}(G(M)) = j_{\widetilde{R}}(\widetilde{M}) = j_{\widehat{R}}(\widehat{M})$, where $j_-(-)$ represents the grade number of the corresponding module.

(b) ([Li2] Generalized Roos theorem) If $G(R)$ is commutative and every maximal ideal of $G(R)$ has the same height, then for every finitely generated R-module M with good filtration FM, $j_R(M) +$ K.dim$(G(M)) =$ gl.dim$G(R)$.

(c) ([Wu]) If $G(R)$ is commutative and every maximal graded ideal of $G(R)$ has the same height ω', say, then for any finitely generated R-module M with good filtration FM, $j_R(M) + d'(M) = \omega'$, where $d'(M) = \sup\{$K.dim$G(M)_P$, P a maximal graded ideal of $G(R)\}$.

Using the result of (c) above, a long standing mistake in [Bj1] is corrected (see [Wu] for details).

Concerning the holonomic and pure module over a Zariski ring R with $G(R)$ being Auslander regular, the following (d)–(f) are due to [Li2]:

(d) A finitely generated R-module M with good filtration FM is pure (i.e., for any submodule N of M, $j_R(M) = j_R(N)$) if and only if $\widetilde{M}$ is pure if and only if $\widehat{M}$ is pure.

(e) If $G(R)$ is commutative, then a finitely generated R-module M with good filtration FM is holonomic (i.e., $j_R(M) =$ gl.dim$R$) if and only if ht$p(=$ the height of $p) =$ gl.dim$R$ for every minimal prime divisor in $\min(J(M))$, where $J(M) = \sqrt{\text{Ann}_{G(R)}(G(M))}$ and $\min(J(M))$ is the set of all minimal prime divisors of $J(M)$.

(f) If $G(R)$ is commutative, then for every finitely generated R-module M with good filtration FM, $j_R(M) = \inf\{$ht$p$, $p \in \min(J(M))\}$.

A combination of the above (c) and [Li2] yields the following:

(g) ([LVO6]) If $G(R)$ is commutative and every maximal graded ideal of $G(R)$ has the same height$(=$ gl.dim$G(R))$, then a finitely generated R-module M with good filtration FM is pure if and only if the characteristic variety $\min(J(M))$ of M is geometrically pure (i.e., $\dim(\min(J(M))) =$ K.dim$_{G(R)}(G(M)) =$ K.dim$(G(R)/p_i)$ for every $p_i \in \min(J(M)))$ and Ass$(G(R)) = \min(J(M))$.

1.5. Zariski rings with Auslander-Gorenstein associated graded rings

Some results we mentioned in 1.4. above have been generalized in [Wang] and [Zhou] to the case where $G(R)$ is an Auslander-Gorenstein ring (i.e., $G(R)$ has finite injective dimension and satisfies the Auslander condition).

(a) Let R be a (left and right) Zariski ring. Suppose that $G(R)$ is a commutative Auslander-Gorenstein ring with pure dimension ω (i.e., every maximal ideal of $G(R)$ has the same height ω). It is proved in [Wang] that for every finitely generated R-module M with good filtration FM, $j_R(M) +$ K.dim$(G(M)) = \omega$.

Replacing the pure dimension by graded pure dimension (i.e., every graded maximal ideal of $G(R)$ has the same height), it is proved in [Zhou] that the same equality holds for every finitely generated R-module M with good filtration FM.

(b) Let R be a (left and right) Zariski ring. Suppose that $G(R)$ is a commutative Auslander-

Gorenstein ring with pure dimension ω; then it is proved in [Wang] that a finitely generated R-module M with good filtration FM is holonomic if and only if $\mathrm{ht}p = \mathrm{inj.dim}R$ for every $p \in \min(J(M))$.

In [Zhou], without the pure dimensional assumption the same result is proved.

(c) ([Wang]) Let R be a (left and right) Zariski ring. Suppose that $G(R)$ is a commutative Auslander-Gorenstein ring but with pure dimension ω. Then for any finitely generated R-module M with good filtration FM, $j_R(M) = \inf\{\mathrm{ht}p, p \in \min J((M))\}$.

1.6. Some results concerning (quantized)Weyl algebras and enveloping algebras of Lie algebras over rings

The following (a)–(c) are from [Wang].

(a) Let k be a field of characteristic 0 and R a commutative Noetherian k-algebra; if $\mathrm{inj.dim}R < \infty$ then $\mathrm{inj.dim}A_n(R) = \mathrm{K.dim}A_n(R) = n + \mathrm{inj.dim}R$.

(b) Let k be a field of characteristic $p(> 0)$ and R a commutative affine k-algebra. Then every simple module over the n-th Weyl algebra $A_n(R)$ has an Artinian injective hull.

(c) Every simple module over $A_n(\mathbb{Z})$ has an Artinian injective hull.

The following (d)–(f) are from [LVO8].

Let D be a commutative Noetherian regular domain of dimension d.

(d) Let $A_n(D)$ be the n-th Weyl algebra over D.

    (i) if there is some maximal ideal ω of D such that $\mathrm{char}(D/\omega) = p \neq 0$ then $\mathrm{gl.dim}A_n(D) = 2n + d$.

    (ii) If for all maximal ideals ω of D $\mathrm{char}(D/\omega) = 0$, then $\mathrm{gl.dim}A_n(D) = n + d$.

(e) Let $A_1(D, q)$ be the quantized Weyl algebra over D (in the sense of Goodearl) with $q \neq 1$ a unit in D.

    (i) $\mathrm{gl.dim}A_1(D, q) = 2 + d$ if $q - 1$ is not contained in the Jacobson radical $J(D)$ of D, or if $q - 1 \in J(D)$ but $\mathrm{char}(D/\omega) = p \neq 0$ for some maximal ideal ω of D.

    (ii) $\mathrm{gl.dim}A_1(D, q) = 1 + d$ if $q - 1 \in J(D)$ and $\mathrm{char}(D/\omega) = 0$ for all maximal ideals ω of D.

(f) let g be an n-dimensional Lie algebra over D (this means that G is a free D module of rank n with a Lie product), and $U_D(g)$ the enveloping algebra of g. Then $\mathrm{gl.dim}U_D(g) = n + d$.

1.7. Lifting Ore sets of Noetherian filtered rings and some local-global results

Let R be a filtered ring with filtration FR such that the Rees ring $\tilde{R}$ of R is left Noetherian. In [Li8] it is observed that if T is a left Ore set of the associated graded ring $G(R)$ consisting of homogeneous elements then the saturated set $S = \{s \in R, \sigma(s) \in T\}$ is a left Ore set of R. The localization of R at this saturated Ore set has some interesting properties which can be used to simplify certain important microlocal properties of the microlocalization $Q_\Sigma^\mu(R)$ of R at a multiplicatively closed subset Σ with $\sigma(\Sigma) = T$ being an Ore set in $G(R)$ in the literature e.g. [AVVO]. In particular, $S^{-1}R$ is a left Zariskian filtered ring (although R is not Zariskian). Moreover, some local-global results are obtained in [Li8].

(a) Holonomic modules with regular singularities revisited.

In a purely algebraic study of holonomic $\mathcal{D}$-modules with regular singularities (or with R.S. for short), the stalk E_p of analytic micro-local differential operators at a point p has been replaced by an algebraic microlocalization $E_p(R)$ of a filtered E-ring R at a prime ideal $p \in \operatorname{Spec}^s(G(R))$, and a microlocal characterization of holonomic modules with regular singularities was given first in [VE2] by using microlocalizations of E-rings, then in [LVO1] by using microlocalizations of strongly filtered rings. In both [VE2] and [LVO1] the localized pseudo-norm has been heavily used. By using only the exactness of the localization functor a local-global description of the holonomic modules with regular singularities is given in [Li8]:

Theorem. Let R be a left Zariskian filterd ring with filtration FR such that $G(R)$ is a commutative $\mathbb{Q}$-algebra. Let M be a holonomic R-module with good filtration FM. Then the following statements are equivalent:

(1) M has R.S. as an R-module;

(2) $S_p^{-1}M$ has R.S. as a $S_p^{-1}R$-module for all $p \in \min(J(M))$;

(3) $S_p^{-1}M$ has R.S. as a $S_p^{-1}R$-module for all $p \in \operatorname{Spec}^s(G(R))$,

where $\operatorname{Spec}^s(G(R))$ is the set of all graded prime ideals of $G(R)$, $J(M) = \sqrt{\operatorname{Ann}_{G(R)}G(M)}$, $\min(J(M))$ denotes the set of all minimal prime divisors of $J(M)$ and S_p denotes the saturated Ore set corresponding to the homogeneous Ore set $h(G(R) - p)$ for $p \in \operatorname{Spec}^s(G(R))$.

(b) Local description of dimension theory and Auslander regularity (Gorenstein property).

In [Sao] some microlocal-global results concerning various dimensions and the Auslander regularity (Gorenstein property) of a Zariskian filtered ring R with commutative associated graded ring $G(R)$ have been obtained. By using the localizations of R at the lifting Ore sets S_p and the faithful flatness of the R-module $\oplus S_p^{-1}R$ it is proved in [Li8] that under the same assumptions as in [Sao] there exist local-global descriptions for various dimensions and the Auslander regularity (Gorenstein property) of R:

Theorem. Let R be a Zariskian filtered ring with filtration FR such that $G(R)$ is commutative.

(1) If inj.dim$R$ denotes the injective dimension of R then inj.dim$R = \sup\{$ inj.dim$S_p^{-1}R$, $p \in \operatorname{Spec}^s(G(R))\} = \sup\{$ inj.dim$S_{\mathcal{M}}^{-1}R$, $\mathcal{M}$ maximal in $\operatorname{Spec}^s(G(R))\}$.

(2) If R has finite global dimension gl.dim$R$ then gl.dim$R = \sup\{$ gl.dim$S_p^{-1}R$, $p \in \operatorname{Spec}^s(G(R))\} = \sup\{$ gl.dim$S_{\mathcal{M}}^{-1}R$, $\mathcal{M}$ maximal in $\operatorname{Spec}^s(G(R))\}$.

(3) If $G(R)$ is a gr-semilocal ring, i.e. it has only a finite number of maximal graded ideals, then K.dim$R = \sup\{$K.dim$S_{\mathcal{M}}^{-1}R$, $\mathcal{M}$ maximal in $\operatorname{Spec}^s(G(R))\} = \sup\{$K.dim$S_p^{-1}R$, $p \in \operatorname{Spec}^s(G(R))\}$ where K.dim$R$ denotes the Krull dimension of R in the sense of Gabriel-Rentschler.

Theorem. Let R be as in the above theorem.

(1) R is Auslander-Gorenstein if and only if $S_p^{-1}R$ is Auslander-Gorenstein for every $p \in$

$\text{Spec}^{\mathfrak{g}}(G(R))$, if and only $S_{\mathcal{M}}^{-1}R$ is Auslander-Gorenstein for all maximal graded $\mathcal{M} \in \text{Spec}^{\mathfrak{g}}(G(R))$.

(2) Suppose that R has finite global dimension. Then R is Auslander regular if and only if $S_p^{-1}R$ is Auslander regular for every $p \in \text{Spec}^{\mathfrak{g}}(G(R))$, if and only if $S_{\mathcal{M}}^{-1}R$ is Auslander regular for all maximal graded $\mathcal{M} \in \text{Spec}^{\mathfrak{g}}(G(R))$.

§2   GRADED RING THEORY

2.1. Homological dimension of graded rings

(a) (see [Li1] or [LVO4]) Let A be a (group)G-graded ring and X a regular non-unit homogeneous normalizing element of A. Suppose that A is left Noetherian and $X \in J^{\mathfrak{g}}(A)$, where $J^{\mathfrak{g}}(A)$ denotes the graded Jacobson radical of A (this is defined to be the largest proper graded ideal of A such that its intersection with A_e is in the Jacobson radical of A_e, where e is the neutral element of G; hence when A is trivially graded $J^{\mathfrak{g}}(A)$ coincides with the real Jacobson radical of A). Then gr.gl.dim$A \le 1+$ gr.gl.dim(A/XA) and the equality holds in case A/XA has finite graded global dimension.

(b) (see [Li1] or [LVO4]) Let A be a $\mathbb{Z}$-graded ring and let X be a regular non-unit homogeneous normalizing element of A. Suppose that A is left Noetherian and $X \in J^{\mathfrak{g}}(A)$. Then gl.dim$A \le 1+$ gl.dim(A/XA) and the euqality holds in case A/XA has finite global dimension.

(c) In [Li7], the "c-mixed gradation" is introduced on $A[t, \sigma]$, where $A = \oplus_{\mathfrak{g} \in G}A_{\mathfrak{g}}$ is any (group)G-graded ring, $A[t, \sigma]$ is the skew polynomial ring over A in variable t, σ is a graded ring automorphism of A and c is any element in the centre $Z(G)$ of G, and the following results are proved:

  (1) with respect to the c-mixed gradation, gr.gl.dim$A[t, \sigma] \le 1+$ gr.gl.dim$A$ and the equality holds in case gr.gl.dim$A < \infty$; if $G = \mathbb{Z}$ then the equality always holds;

  (2) with respect to the c-mixed gradation, if A is left and right Noetherian then gr.inj.dim$A[t, \sigma] = 1+$ gr.inj.dim$A$;

  (3) if X is a regular non-unit normalizing homogeneous element of degree $c \in Z(G)$ which is contained in the graded Jacobson radical $J^{\mathfrak{g}}(A)$ of A and A is a left and right Noetherian ring, then gr.inj.dim$A = 1+$ gr.inj.dimA/XA; if $G = \mathbb{Z}$ and the gradation on A is left limited we also have inj.dim$A = 1+$ inj.dimA/AX.

(d) ([Li9]) Let $A = \oplus_{n \ge 0}A_n$ be a positively graded Noetherian ring. If $A_0/J(A_0)$ is semisimple, where $J(A_0)$ is the Jacobson radical of A_0, then gl.dim$A = $ p.dim$_A(A_0/J(A_0))$; if A_0 is a (not necessarily commutative) local ring with maximal ideal ω, then gl.dim$A = $ p.dim$_A(A_0/\omega) = $ inj.dim$_A(A_0/\omega)$, where p.dim$_A(A_0/J(A_0))$ denotes the projective dimension of $A_0/J(A_0)$ as an A-module (note that $A_0/J(A_0) \cong A/(J(A_0) + \oplus_{n>0}A_n)$), and inj.dim$_A(A_0/\omega)$ denotes the injective dimension of A_0/ω as an A-module.

(e) ([LVO8]) Let $A = \oplus_{n \in \mathbf{z}} A_n$ be a $\mathbb{Z}$-graded D-algebra over the commutative regular local domain D such that $A_0 = D$. Let ω be the maximal ideal of D generated by a regular sequence $(x_1, ..., x_d)$ and put $\overline{A} = A/\omega A$. Suppose that A is left Noetehrian. If A is free as a D-module then gl.dim$A \leq$ gl.dim$\overline{A} + d$ and the equality holds whenever gl.dim$\overline{A} < \infty$.

2.2. Certain graded homological properties are ungraded properties

(a) ([Li4]) Let A be a left Noetherian $\mathbb{Z}$-graded ring. If A is gr-regular (i.e., every finitely generated graded A-module has finite projective dimension) then A is regular (i.e., every finitely generated A-module has finite projective dimenison).

In [LVO4], the "good graded filtration" was introduced for graded modules over a graded ring with compatible double gradations. After developing some homological properties for good graded filtrations, the following two results are obtained:

(b) ([Li6]) Let A be a left Noetherian $\mathbb{Z}$-graded ring. If A is gr-stable (i.e., every finitely generated graded projective A-module is gr-stably free) then A is stable (i.e., every finitely generated projective A-module is stably free).

(c) ([LVO4]) Let A be a $\mathbb{Z}$-graded ring. If A is gr-Auslander regular (this is defined intrinsically in the category of graded A-modules) then A is Auslander regular.

2.3. Applications of the foregoing results

(a) ([Li4]) Let A be a $\mathbb{Z}$-graded ring and X a regular non-unit homogeneous normalizing element of A. Suppose that A is left Noetherian and $X \in J^s(A)$. If A/XA is left gr-regular then A is left regular.

(b) ([Li4]) Let A be a strongly $\mathbb{Z}$-graded ring. If A_0 is left Noetherian then: A_0 is left regular if and only if A is left regular if and only if $A^+ = \oplus_{n \geq 0} A_n$ is left regular if and only if $A^- = \oplus_{n \leq 0} A_n$ is left regular.

(c) [Li7] if $A = \oplus_{n \in \mathbf{z}} A_n$ is strongly $\mathbb{Z}$-graded, then gl.dim$A^- =$ gl.dim$A^+ = 1+$ gl.dimA_0 where $A^- = \oplus_{n \leq 0} A_n$, $A^+ = \oplus_{n \geq 0} A_n$.

(d) [Li7] if $A = \oplus_{n \in \mathbf{z}} A_n$ is strongly $\mathbb{Z}$-graded and A_0 is left and right Noetherian, then inj.dim$A^- =$ inj.dim$A^+ = 1+$ inj.dimA_0.

(e) [LVO7] Let R be a ring and I an ideal of R. Let FR be the I-adic filtration on R and $G(R)$ the associated graded ring of R. Suppose that $G(R)$ is as a graded ring isomorphic to A^+ or to A^- for some strongly $\mathbb{Z}$-graded ring A (e.g. I is an invertible ideal of R). (1) If R/I is left Noetherian and regular then $K_0(R/I) \cong K_0(G(R))$; (2) If $I \subseteq J(R)$, R/I is regular and $\tilde{R}$ is left Noetherian , then there is an embedding of K_0-groups: $K_0(R) \hookrightarrow K_0(R/I) = K_0(G(R))$.

(f) Let R be a left Zariski ring with filtration FR.

 (1) ([Li4]) If $G(R)$ is left gr-regular then the Rees ring $\tilde{R}$ of R is left regular.

 (2) (see [Li1] or [LVO4]) gr.gl.dim$\tilde{R} \leq 1+$ gr.gl.dim$G(R)$; gl.dim$\tilde{R} \leq 1+$ gl.dim$G(R)$; The two equalities hold if $G(R)$ has finite (gr-)global dimension.

(g) ([Li6]) Let A be a strongly $\mathbb{Z}$-graded ring. If A_0 is left Noetherian, left regular and every finitely generated projective A_0-module is stably free then the same is true for A.

(h) ([Li6]) Let A be a $\mathbb{Z}$-graded ring and X a regular non-unit homogeneous normalizing element of A. Suppose that $X \in J^s(A)$ and $\cap_{n \geq 1} X^n A = 0$, or in other words, the XA-adic filtration on A is separated. If every finitely generated graded projective A/XA-module is gr-stably free then every finitely generated graded projective A-module is gr-stably free; If furthermore A is left Noetherian and A/XA is left regular then every finitely generated projective A-module is stably free.

(i) ([Li6]) Let R be a filtered ring with filtration FR. Suppose that FR is separated (i.e., $\cap_{n \in \mathbb{Z}} F_n R = 0$) and $F_{-1} R \subseteq J(F_0 R)$. If every finitely generated graded projective $G(R)$-module is gr-stably free then every finitely generated graded projective $\tilde{R}$-module is gr-stably free. If furthermore $\tilde{R}$ is left Noetherian (i.e., FR is left Zariskian) and $G(R)$ is left regular then every finitely generated projective $\tilde{R}$-module is stably free.

(j) ([LVO6]) Let A be a strongly $\mathbb{Z}$-graded ring. Then: A_0 is Auslander regular if and only if A is Auslander regular if and only if $A^+ = \oplus_{n \geq 0} A_n$ is Auslander regular if and only if $A^- = \oplus_{n \leq 0} A_n$ is Auslander regular.

(k) ([LVO4]) Let R be a left and right Noetherian ring and I an invertible ideal of R with respect to some overring S of R. Suppose that I is contained in the Jacobson radical $J(R)$ of R. Consider the I-adic filtration FR on R, if R/I is an Auslander regular ring then R, $G(R)$, $\tilde{R}$ and the generalized Rees ring $\check{R}(I)(= \oplus_{n \in \mathbb{Z}} I^n t^n \subseteq S[t, t^{-1}])$ are Auslander regular rings.

(l) ([LVO6]) Let R be a left and right Noetherian $\mathbb{Z}$-graded ring and I a graded ideal of R. Suppose that I is invertible with respect to some overring S of R and that I is contained in the graded Jacobson radical $J^s(R)$ of R. Consider the I-adic filtration FR on R; if R has finite global dimension and R/I is an Auslander regular ring then R and $\tilde{R}$ are Auslander regular rings.

(m) ([LVO3]) Let R be a left and right Zariski ring with filtration FR. Suppose that $G(R)$ is Auslander regular; then $\tilde{R}$ is Auslander regular.

(n) ([Li7]) Let A be a ring and I an invertible ideal of A, $\check{A}(I) = \oplus_{n \in \mathbb{Z}} I^n t^n$ the generalized Rees ring of A and $R = \check{A}(I)/(1 - t^{-1})\check{A}(I)$ the dehomogenization of $\check{A}(I)$ with respect to t^{-1} in the sense of [LVO5].

    (1) $\text{gl.dim} G_I(A) = 1 + \text{gl.dim} A/I$ where $G_I(A)$ is the associated graded ring of A with respect to the I-adic filtration on A;

if A is left Noetherian, $\text{gl.dim} A/I < \infty$ and I is contained in the Jacobson radical $J(A)$ of A, then

    (2) $\text{gr.gl.dim} \check{A}(I) = 1 + \text{gr.gl.dim} G(R) = 1 + \text{gl.dim} A/I$;

    (3) $\text{gl.dim} A = 1 + \text{gl.dim} A/I = \text{gl.dim} G_I(A)$;

    (4) $\text{gl.dim} \check{A}(I) = 1 + \text{gl.dim} G(R) \leq 2 + \text{gl.dim} A/I$;

    (5) $\text{gl.dim} \tilde{A} = 1 + \text{gl.dim} G_I(A) = 2 + \text{gl.dim} A/I$ where $\tilde{A}$ resp. $G_I(A)$ is the Rees ring

resp. the associated graded ring of A with respect to the I-adic filtration on A.

(o) ([Li7], [Wang]) Let R be a left and right Zariski ring with filtration FR. Then gr.inj.dim$\tilde{R} = 1+$ gr.inj.dim$G(R)$, inj.dim$\tilde{R} = 1+$ inj.dim$G(R)$.

(p) ([Li7]) Let A be a ring and I an invertible ideal of A. With notation as above, if A is left and right Noetherian then

> (1) inj.dim$G_I(A) = 1+$ inj.dimA/I where $G_I(A)$ is the associated graded ring of A with respect to the I-adic filtration on A;

if furthermore I is contained in the Jacobson radical $J(A)$ of A, then

> (2) gr.inj.dim$\check{A}(I) = 1+$ gr.inj.dim$G(R) = 1+$ inj.dimA/I;
>
> (3) inj.dim$A = 1+$ inj.dim$A/I = $ inj.dim$G_I(A)$;
>
> (4) inj.dim$\check{A}(I) = 1+$ inj.dim$G(R) \leq 2+$ inj.dimA/I;
>
> (5) inj.dim$\tilde{A} = 1+$ inj.dim$G_I(A) = 2+$ inj.dimA/I where $\tilde{A}$ resp. $G_I(A)$ is the Rees ring resp. the associated graded ring of A with respect to the I-adic filtration on A.

(q) The 3-dimensional Sklyanin algebra $S(K)$ over a field K is the graded algebra defined by generators X, Y, Z considered to be homogeneous of degree 1, and the ideal of quadratic relations is generated by the relations

$$aXY + bYX + cZ^2 = 0$$
$$aXZ + bZY + cX^2 = 0$$
$$aZX + bXZ + cY^2 = 0$$

Let F be the prime subfield of K, $k = F(a, b, c)$ and D any regular domain in K and containing k as a subring. By ([LVO8] Theorem 3.5.) it is proved that for a, b, c such that $\mu = $ gl.dim$S(k)$ (the corresponding Sklyanin algebra over k) we have gl.dim$S(D) = d + \mu$ where $d = $ dim$D$.

§3   SMASH PRODUCT

(a) Let G be an arbitrary group. In [Liu1], the total G-graded ring E of the linear transformations of M was introduced, where M is any k-vector space over a division ring k, $M = \oplus_{g \in G} M_g$ is a decomposition of M by additive subgroups and $E = \Sigma_{g \in G} E_g$, $E_g = \{\varphi \in \text{End}_k M, \varphi(M_h) \subseteq M_{hg}, h \in G\}$. A graded subring $A = \oplus_{g \in G} A_g$ (not necessarily containing 1) of E is called G-graded dense over $M = \oplus_{g \in G} M_g$ if for any g, any finitely many k-independent $x_1, \cdots, x_n \in M_g$, any h and any $y_1, \cdots, y_n \in M_{gh}$ there exists $a \in A_h$ such that $x_i a = y_i$, $i = 1, \cdots, n$. Obviously, if A is G-graded dense then it is G-graded primitive. Conversely, it is proved in [Liu1] that any G-graded primitive ring may be obtained in such a way, in other words, a G-graded primitive ring must be a G-graded dense ring.

(b) Let $A = \oplus_{g \in G} A_g$ be a (group)G-graded ring and $A \# G$ be the smash product of A on G. In [LiuV] and [Liu2] the following results are proved: (1) $A \# G$ is semiprimitive if and only if A is semiprimitive; (2) $A \# G$ is primitive if and only if A is graded uniformly primitive with all $A_h \neq 0$; (3) $A \# G$ is simple if and only if A is strongly G-graded and A_e is simple; (3) $A \# G$ is a prime ring if and only if for any nonzero graded left and right ideal L, R, of A, $L_g \neq 0$, $R_g \neq 0$ for all $g \in G$, and A_e is a prime ring if and only if $\sum_{g \in G} R_g L_{g^{-1}} \neq 0$ if and only if $\sum_{g \in G} R_g L_{g^{-1}h} \neq 0$ for all $h \in G$.

(c) If $\mathcal{R}$ is a radical class of associative rings, in [FS1] two associated radical classes of graded rings, denoted $\mathcal{R}^G$, $\mathcal{R}_{\text{ref}}$, are considered. It is shown that if $\mathcal{R}$ is special resp. normal, then both $\mathcal{R}^G$ and $\mathcal{R}_{\text{ref}}$ are graded special resp. graded normal. A graded version of ADS theorem and the termination of the Kurosh lower graded radical construction are also discussed in this paper.

(d) Let $A = \oplus_{g \in G} A_g$ be a (group)G-garded ring and $A \# G$ the smash product of A on G. In [FS2] some equivalent characterizations of graded essential ideals of A, essential ideals of A_e and essential ideals of $A \# G$ are obtained, and moreover, some applications to graded prime essential rings, graded irredundant subdirect sums and graded essentially nilpotent rings are also given.

([FCC]) Let $A = \oplus_{g \in G} A_g$ be a (group)G-graded ring and $A \# G$ the smash product of A on G. Then $A \# G$ is primitive if and only if A_e is primitive and the gradation of A is faithful if and only if A is graded primitive and the gradation of A is faithful.

REFERENCES

[AVO] A. Awami and F. Van Oystaeyen, On filtered rings with Noetherian associated graded rings, "Ring Theory", (J.L. Buesco, P. Jara, B. Torrecillas, ed.), LNM, 1328, 8–27, Springer-Verlag 1988.

[AVVO] M.J. Asensio, M. Van den Bergh and F. Van Oystaeyen, A new algebraic approach to microlocalization of filtered rings, *Trans. Amer. Math. Soc. Vol. 316, No. 2*, 1989, 537–555.

[Be] I.N. Bernstein, Algebraic theory of $\mathcal{D}$-modules, *Preprint*, 1982.

[Bj1] J-E. Björk, "Rings of differential operators", Math. Library 21, North Holland, Amsterdam, 1979.

[Bj2] J-E. Björk, Filtered Noetherian rings, *Math. Surveys and monographs, No. 24*, 1987, 59–97.

[Bor] A. Borel, "Algebraic $\mathcal{D}$-modules", Academic press, London-New York, 1987.

[FCC] Huixiang Chen, Hongjin Fang and Chuanren Cai, G-prime ideals of smash product $A \# G$ and graded prime ideals of A, *J. Yangzhou Teachers College, Vol. 12, No. 3*, 1992, 21–25.

[FS1] Hongjin Fang and Patrick Stewart, Radical theory for graded rings, *J. Austral. Math. Soc. (Series A) No. 52*, 1992, 143–153.

[FS2] Hongjin Fang and Patrick Stewart, Graded rings and essential ideals, *to appear*.

[Gi] V. Ginsburg, Characteristic varieties and vanishing cycles, *Invent. Math. 84*, 1986, 327-402.

[Ka] M. Kashiwara, "Algebraic study of systems of partial differential equations", Master's thesis, University of Tokyo, 1971.

[Li1] Huishi Li, "Noncommutative Zariskian filtered rings", Ph. D. thesis, UIA Antwerp, 1989.

[Li2] Huishi Li, Note on pure module theory over Zariskian filtered ring and the generalized Roos theorem, *Comm. Alg. Vol. 19*, 1991, 843–862.

[Li3] Huishi Li, Note on microlocalization of filtered rings and the embedding of rings in skew fields, *Bull. Math. Soc. Belgique, (serie A), Vol. XLIII*, 1991, 49–57.

[Li4] Huishi Li, Noetherian gr-regular rings are regular, *China Annals of Mathematics, 15B: 4*, 1994, 463—468.

[Li5] Huishi Li, Rees rings of grading filtration and an application to Weyl algebras, *Comm. Alg., Vol.21, No. 8*, 1993, 2967–2972.

[Li6] Huishi Li, On the stability of graded rings, *J. Alg., Vol. 169*, 1994, 274–286.

[Li7] Huishi Li, Note on the homological dimension of graded rings, *Comm. Alg., Vol. 22, No. 15*, 1994, 6225–6237.

[Li8] Huishi Li, Lifting Ore sets of Noetherian filtered rings and applications, *J. Alg.*, 1995, to appear.

[Li9] Huishi li, Global dimension of graded local rings, To appear.

[LVO1] Huishi Li and F. Van Oystaeyen, Strongly filtered rings, applied to Gabber's integrability theorem and modules with regular singularities, "Proc. Sem. Malliavin", LNM. 1404, Springer-Verlag, 1988.

[LVO2] Huishi Li and F. Van Oystaeyen, Filtrations on simple Artinian rings, *J. Alg. Vol. 132*, 1990, 361–376.

[LVO3] Huishi Li and F. Van Oystaeyen, Zariskian filtrations, *Comm. Alg. Vol. 17*, 1989, 2945–2970.

[LVO4] Huishi Li and F. Van Oystaeyen, Global dimension and Auslander regularity of graded rings, *Bull. Math. Soc. Belgique, (serie A), Vol. XLIII*, 1991, 59–87.

[LVO5] Huishi Li and F. Van Oystaeyen, Dehomogenization of gradings to Zariskian filtrations and applications to invertible ideals, *Proc. Amer. Math Soc. vol. 115, No. 1*, 1992, 1–11.

[LVO6] Huishi Li and F. Van Oystaeyen, Sign gradation on group ring extensions of graded rings, *J. of pure and applied algebra, Vol. 85*, 1993, 311–316.

[LVO7] Huishi Li and F. Van Oystaeyen, "Zariskian filtrations", Monograph, Kluwer Academic Publishers, 1995.

[LVO8] Huishi Li and F. Van Oystaeyen, Reductions and global dimension of quantized algebras over a regular commutative domain, *To appear.*

[LVOW] Huishi Li, F. Van Oystaeyen and E. Wexler-Kreindler, Zariski rings and flatness of completion, *J. Alg. Vol. 138*, 1991, 327–339.

[LVV1] Huishi Li, M. Van den Bergh and F. Van Oystaeyen, Note on the K_0 of rings with Zariskian filtrations, *K-Theory, Vol. 3*, 1990, 603–606.

[LVV2] Huishi Li, M. Van den Bergh and F. Van Oystaeyen, The global dimension and regularity of Rees rings for non-Zariskian filtration, *Comm. Alg. Vol. 18*, 1990, 3195–3208.

[Liu1] Shaoxue Liu, the structure of group graded primitive rings, *Kexue Tongbao, No. 22* 1990, 1696–1698.

[Liu2] Shaoxue Liu, Two results on shmash products, *Kexue Tongbao, No. 34*, 1989, 967–968.

[LiuV] Shaoxue Liu and F. Van Oystaeyen, Group graded rings, smash products and additive categories, "Perspective in ring theory", Kluwer academic publishers, 1988, 299–310.

[NVO] C. Năstăsescu and F. Van Oystaeyen, "Graded ring theory", Math. Library, 28, North Holland, Amsterdam, 1982.

[Sao] M. Saorin, Microlocal properties of filtered rings, *J. Alg., Vol. 140* 1991, 141–159.

[Sch] P. Schapira, "Microdifferential systems in the complex domain", Springer Verlag, Berlin, 1985.

[Ve] A. Van den Essen, Modules with regular singularities over filtered rings, *P.R.I. for Mathematical Science, Kyoto Univ., Vol. 22, No. 5*, 1986.

[Wang] Zhixi Wang, "Modules over Auslander-Gorenstein filtered ring and injective dimension", Ph. D. thesis, Beijing Normal University, 1992.

[Wu] Quanshui Wu, On the formula $d(M) + j(M) = 2n$ over the rings $\mathcal{D}_n(\widehat{\mathcal{D}}_n)$ and $\mathcal{E}_p$, *Science in China, Vol. 36, No. 12*, 1993, 1409–1416.

[Zhou] Meng Zhou, "Auslander-Gorenstein filtered ring and microlocalization", Ph. D. thesis, Beijing Normal University, 1993.

Representation Theory of Finite-Dimensional Algebras in China

SHAO-XUE LIU
Department of Mathematics, Beijing Normal University, China

PU ZHANG
Department of Mathematics, University of Science and Technology of China, China

Representation theory of finite dimensional algebras, which originated at the end of the 1960's, has made rapid progress in the last decades. In the mid of the 1980's Professor Shaoxue Liu began a seminar on this theory in Beijing Normal University; after almost ten year's effort, a research group in this field has been formed in China. The aim of this article is mainly to report some of the works of this group up to 1992.

During the period of developing representation theory of algebras in China, many foreign colleagues have kindly provided their help and suggestions. Some of them have given lectures in Beijing: M. Auslander, V. Dlab, D. Happel, I. Reiten, C. M. Ringel K. W. Roggenkamp, A. Skowronski and L. Unger. We take this opportunity to thank all of them.

Contents

1   The structure of Auslander-Reiten components

Let A be an Artin algebra with the Auslander-Reiten quiver $\Gamma(A)$ and the stable Auslander-Reiten quiver $\Gamma_s(A)$, $\tau = D\text{Tr}$ the Auslander-Reiten translation. An indecomposable A-module M is called stable (resp. periodic) provided $\tau^n M \neq 0$ for all $n \in \mathbb{Z}$ (resp. $\tau^m M = M$ for some $m \in \mathbb{Z}$). A component of the Auslander-Reiten quiver Γ_A which consists of stable modules is called a regular component. Riedtmann [Rm] introduced the concepts of (stable) valued translation quivers. The typical examples of those quivers are respectively $\Gamma(A)$ and $\Gamma_s(A)$. Another example of stable valued translation quiver is $\mathbb{Z}\vec{\Delta}$. Note that any connected stable valued translation quiver is of the form $\mathbb{Z}\vec{\Delta}/G$, where $\vec{\Delta}$ is a connected valued oriented tree and G is an admissible automorphism group of $\mathbb{Z}\vec{\Delta}$.

If A is a representation-finite algebra over an algebraically closed field k, then Riedtmann proved in [Rm] that the components of $\Gamma_s(A)$ are of the forms $\mathbb{Z}\vec{\Delta}/G$, where Δ is a Dynkin diagram. In their paper [HPR], Happel-Preiser-Ringel proved: if C is a component of the stable Auslander-Reiten quiver of an Artin algebra containing periodic module, then C is either of the form $\mathbb{Z}\vec{\Delta}/G$, or else $\mathbb{Z}A_\infty/ < \tau^n >$, where Δ is a Dynkin diagram, and A_∞ is the infinite diagram $\bullet - \cdots - \bullet - \bullet \cdots$. In fact, the existence of a periodic vertex in a connected valued stable translation quiver C implies that all vertices are periodic; in this case C is said to be periodic, otherwise non-periodic. The periodic regular components must be the regular tubes, i.e. $\mathbb{Z}A_\infty/ < \tau^n >$ for $n \geq 1$. The structure of periodic stable valued translation quivers with subadditive functions is also given in [HPR]. But what is the case for the non-periodic translation quivers with subadditive functions? This was settled by Y. B. Zhang in her doctoral thesis [55] under the direction of C. M. Ringel:

THEOREM ([55])   Let C be a non-periodic connected valued stable translation quiver with a non-zero subadditive function f with values in N_0. Then either C is smooth and f is additive and bounded, or else $C = \mathbb{Z}\vec{\Delta}$ for some valued quiver $\vec{\Delta}$.

The proof [55] is rather complicated and technical. She considered the first homology group of the orbit graph of C and some additive function on it measuring the difference between the numbers of forward and backward arrows in any walk. In order to write C in the form $\mathbb{Z}\vec{\Delta}$, it is necessary to find a suitable orientation on the orbit graph of C. The above theorem has the following useful corollary

THEOREM ([55])   Let C be a non-periodic component of the stable Auslander-Reiten quiver of an Artin algebra A. Then, $C = \mathbb{Z}\vec{\Delta}$ for some valued quiver $\vec{\Delta}$ without oriented cycles.

It is then natural to ask what kind of valued quivers $\vec{\Delta}$ can actually occur in $C = \mathbb{Z}\vec{\Delta}$, where C is a regular component of $\Gamma(A)$. If $\vec{\Delta}$ is a finite wild quiver with $|\Delta_0| \geq 3$, then it is well known that $H = k\vec{\Delta}$ has a regular tilting module $_H T$, and the connecting component of the tilted algebra $B = \text{End}_H T$ is of the form $\mathbb{Z}\vec{\Delta}$ (see [R2]). Let $\vec{\Delta}$ be a connected symmetrizable valued quiver without oriented cycle and assume that after deletion of finitely many vertices and arrows one obtains a disjoint union of quivers of type A_∞; then there exists an algebra with regular component of the form $\mathbb{Z}\vec{\Delta}$ (see [CR]).

S.P. Liu [17] introduced the concepts of the left and right degrees of irreducible maps; this turned out to be useful in dealing with the possible shapes of Auslander-Reiten components. He used these concepts to give a new (non-combinatorial) proof of the Happel-Preiser-Ringel theorem which states that periodic regular Auslander-Reiten components are regular tubes; the proof does not even use Riedtmann's structure theorem for stable translation quivers.

In [R3] Ringel proposed the following open problem: "Let A be a finite-dimensional algebra over an algebraically closed field. Is it true that all but finitely many components of $\Gamma(A)$ are of the form $\mathbb{Z}A_\infty$, $\mathbb{Z}A_\infty/<\tau^n>$ and $\mathbb{Z}D_\infty$?" Y.B. Zhang ([56]) used the spectral radius of the Coxeter translations of some labeled trees to prove that any non-periodic regular component of the Auslander-Reiten quiver of an Artin algebra, whose growth number is less than $c = \sqrt[3]{\frac{1}{2} + \sqrt{\frac{23}{108}}} + \sqrt[3]{\frac{1}{2} - \sqrt{\frac{23}{108}}}$, is of the form $\mathbb{Z}A_\infty$, $\mathbb{Z}A_\infty^\infty$, $\mathbb{Z}B_\infty$, $\mathbb{Z}C_\infty$ or $\mathbb{Z}D_\infty$. This gives a partial solution to the problem above.

C.C. Xi ([29]) determined the structures of a finite-dimensional connected wild hereditary algebra A with the growth number $\rho(A) < c$.

Ringel [R3] has conjectured that in an Auslander-Reiten component P only finitely many indecomposable modules can have the same composition factors. Y. B. Zhang proved in [57] this is true for hereditary algebras, i.e. the indecomposable modules in any AR-component of a wild hereditary artin algebra are uniquely determined by their dimension vectors.

S.P. Liu showed in [17] that a maximal connected left (or right) stable subquiver of $\Gamma(A)$ containing some special τ-orbits is either a regular tube or can be embedded into $\mathbb{Z}A_\infty$. Liu's result is as follows:

THEOREM ([17]) Let Γ be a maximal connected left stable subquiver of $\Gamma(A)$. Assume that there is a module X in Γ such that $\ell(\tau^n X)$ does not tend to infinity as n tends to infinity . Then either Γ is a regular tube or there is an infinite path $\cdots \to X_{i+1} \to X_i \to \cdots \to X_1 \to X_0$ in Γ with the properties that

1) The path meets each τ-orbit in Γ exactly once;

2) For any integer $i \geq 0$, the arrow $X_{i+1} \to X_i$ has trivial valuation in $\Gamma(A)$;

3) X_0 has exactly one immediate predecessor X_1 in Γ, and for each integer $i > 0$, X_i has exactly two immediate predecessors X_{i+1} and τX_{i-1} in Γ.

As a consequences Liu showed that Ringel's above conjecture is also true for the components with only finitely many τ-orbits. And then he got that: a finite-dimensional algebra A over an algebraically closed field is representation-finite if and only if $\Gamma(A)$ admits only finitely many τ-orbits.

Recall that an indecomposable module M is said to be directing, provided that M does not belong to any cycle $M = M_0 \xrightarrow{f_1} M_1 \longrightarrow \cdots \xrightarrow{f_r} M_r = M$ $(r \geq 1)$, where M_i are all indecomposable and f_i are non-zero non-isomorphisms. A directing module has many pleasant properties. For example, an algebra A which has a sincere directing module must be a tilted algebra.(Recently this concept was generalized to arbitrary modules (not necessarily indecomposable, [HR2])).

An Auslander-Reiten component C which consists entirely of directing modules have been studied by Skowronski-Smalϕ [SS]: such a component C can have only finite many τ-orbits, and if in addition C is regular, then C is the connecting component of some convex subalgebra B of A with B a tilted algebra. It follows that an algebra can have only finitely many components which consists of directing modules. Peng-Xiao ([21])

and Skowronski ([Sk]) independently proved the following result

THEOREM ([21], [Sk])   The Auslander-Reiten quiver of an Artin algebra admits at most finitely many DTr-orbits containing directing modules.

In order to give a criterion for an Auslander-Reiten component being a directing component (i.e. the component in which every indecomposable module is directing), P. Zhang [52] introduced the notion of quasi-slice, then proved that an AR-component C with a quasi-slice Σ is directing if and only if the modules in Σ are directing. For an A-module M, we may form the one-point extension $B = A[M]$, and we call the component of $\Gamma(B)$ which contains the new-added indecomposable projective B-module the extension component. The following theorem provides a method of constructing the new directing component:

THEOREM ([52])   Let C_i be a directing component of $\Gamma(A)$ for $1 \leq i \leq m$, M a directing module (in the sense of [HR2]) with all indecomposable direct summands lying in the union $\bigcup_{i=1}^{m} C_i$. If M is not the proper predecessor in $A - mod$ of any projective A-module, then the extension component of $\Gamma(B)$ is directing.

In [24] Peng considered the algebras over which the indecomposable projective modules are all directing: If A is such an algebra with an idempotent e, so is A/AeA; and A is representation-finite if and only if its Tits form is weakly positive definite. The following result was also proved in [24]: let A be a finite-dimensional algebra with directing projective and directing injective modules, C a component of $\Gamma(A)$ which contains at least one non-stable module; then C is a directing component.

Recall that an A-module M is said to be the middle of a short chain if there is some indecomposable A-module X such that $\mathrm{Hom}_A(X, M) \neq 0$ and $\mathrm{Hom}_A(M, \tau X) \neq 0$. By definition, an indecomposable A-module M is on a short cycle provided there is an indecomposable A-module N and nonzero nonisomorphisms $f : M \to N$ and $g : N \to M$. Short chains and short cycles were extensively studied by Reiten-Skowronski-Smaløf in [RSS1, RSS2] and Happel-Liu [HL]. In particular, these two concepts are the same for indecomposables, i.e. let M be indecomposable; then M is the middle of a short chain if and only if M is on a short cycle, see [HL].

2   Some invariants of modules

One of the basic problems in representation theory of algebras is to look for suitable invariants to determine indecomposable modules. The typical invariant of a module M is the dimension vector $\underline{\dim} M$ of M. Some criteria for indecomposable modules to be determined by their composition factors (i.e. if M, N are indecomposable with $\underline{\dim} M = \underline{\dim} N$, then $M \cong N$) can be found, for example, in [DR1, BS, AR1].

Now let the base field be algebraically closed. J.Y. Guo [2] considered two classes of algebras: algebras stably equivalent to representation-finite hereditary algebras and representation-finite self-injective algebras. Some sufficient and necessary conditions for the indecomposable modules over these algebras to be determined by their composition factors were given. Two algebras $\wedge$ and $\wedge'$ are said to be stably equivalent, if the categories $\wedge-\underline{\text{mod}}$ and $\wedge'-\underline{\text{mod}}$ are equivalent, where $\wedge-\underline{\text{mod}}$ denotes the category which has the same objects as $\wedge-\text{mod}$, and the morphism set is given by $\underline{\text{Hom}}(X, Y) = \text{Hom}(X, Y)/P(X, Y)$, where $P(X, Y)$ is the additive subgroup of $\text{Hom}(X, Y)$ consisting of the maps which factors through some projective module. Let A be an algebra with the Gabriel quiver Q_A; the vertex i in Q_A is called a node provided $P(i)$ is not simple and for any pair of indecomposable $P(j)$, $P(h)$ and nonzero maps $f : P(j) \to P(i)$, $g : P(i) \to P(h)$, we have $fg = 0$. Let i be a node of Q_A; then a path in Q_A of the form $j \to i \to h$ is called an hindrance. A subquiver of Q_A consisting of arrows starting from a vertex j is called a joint of Q_A with the source of j.

THEOREM ([2]) Let A be an algebra stably equivalent to a representation-finite hereditary algebra. A sufficient and necessary condition for any indecomposable modules to be determined by their composition factors is that the Gabriel quiver Q_A satisfies the following conditions:

1) Two end points of a hindrance–free walk of Q_A with more than one arrow do not coincide.

2) The vertex set of two joint connected subquivers without a common arrow are not the same.

THEOREM ([3]) Let A be a connected representation-finite self-injective algebra with $\Gamma_s(A) \cong \mathbb{Z}\vec{\Delta}/\Pi$, where Δ is a Dynkin diagram with n vertices. Then every indecomposable $A-$module is determined by its composition factors if and only if A is either a local algebra or else the number of non-isomorphic indecomposable projective $A-$modules is larger than n.

From the above result we know that not all representation-finite self-injective algebras have the property that indecomposable modules are determined by their composition factors, so J. Xiao considered some other more precise invariants to determine the indecomposable modules. For definitions of the Loewy factors $L\underline{\dim}\, M$ and the socle factors $S\underline{\dim}\, M$ of a module M we refer to, e.g. [37]. The stable AR quiver $\Gamma_s(A)$ of a representation-finite self-injective algebra A has the form $\mathbb{Z}\vec{\Delta}/\pi$, where Δ is a Dynkin diagram, and is called the Cartan class of A.

THEOREM ([37]) Let A be a self-injective algebra with the Cartan class A_n; then every indecomposable module M is determined by $L\underline{\dim}\, M$ and also by $S\underline{\dim}\, M$. Moreover, $L\underline{\dim}\, M$ is a $(0,1)$-matrix.

A typical example of this kind algebra is a Brauer block of a group algebra $k[G]$ with defect group a cyclic p–group, where G is a finite group and k is an algebraically closed field of characteristic p.

The self-injective algebras of Cartan class D_n are divided into two types : two-cornered algebra and three-cornered algebra, according to the type of the configuration of $\mathbb{Z}D_n$; for notations we refer to the original paper [BLR]. For a two-cornered algebra A, Xiao [41] proved that the indecomposable $A-$module M is determined by $L\underline{\dim}\, M$ and also by $S\underline{\dim}\, M$.

THEOREM ([41]) Let A be a two-cornered algebra, X and Y two indecomposable non-simple modules; then $X \cong Y$ if and only if $\text{top}X \cong \text{top}Y$, $\text{soc}\, X \cong \text{soc}\, Y$.

For a three-cornered algebra, Xiao [42] also got similar results. The formulas for computing the Loewy factors and the socle factors of indecomposable modules over the representation-finite self-injective algebras with Cartan class A_n and D_n were also given in [37, 41, 42].

Riedtmann's theorem about the structure of stable Auslander-Reitan quiver of representation-finite self-injective algebra A over an algebraically closed field can be extended to an arbitrary field; in this case $\Gamma_s(A)$ is of form of $\mathbb{Z}\vec{\Delta}/\pi$, where Δ is a diagram of the classes A_n, D_n, E_n ($n = 6, 7, 8$), B_n, C_n, F_4, G_2,. Yang-Xiao [46] determined the structure of indecomposable modules over a representation-finite self-injective algebra with Cartan class B_n and C_n over a perfect field k, and the k-species of representation-finite self-injective algebras with Cartan class B_n or C_n. If k is an arbitrary field, Xiao-Yang also considered this question for a self-injective algebra with Cartan class B_n or C_n; they determined the configurations of $\mathbb{Z}B_n$ and $\mathbb{Z}C_n$, and the admissible automorphism group π.

THEOREM ([45]) Let k be an arbitrary field and A a representation-finite self-injective k-algebra with Cartan class B_n or C_n, and $\Gamma(A) = (\mathbb{Z}B_n)_C/\pi$ or $\Gamma(A) = (\mathbb{Z}C_n)_C/\pi$. If $\pi = (\tau^r)^{\mathbb{Z}} \cdot (2n-1)/r$, then A is a standard algebra.

Let A be a k-algebra; the trivial extension algebra $T(A) = A \ltimes D(A)$ is defined as: the addtive group of $T(A)$ is $A \oplus D(A) = A \oplus \mathrm{Hom}_k(A, k)$, and the multiplication is $(a, \varphi)(b, \psi) = (ab, a\psi + \varphi b)$, $\forall a, b \in A$, $\varphi, \psi \in D(A)$. One basic result about representation-finite trivial extension algebras is: if A is a basic connected finite-dimensional k-algebra, then $T(A)$ is representation-finite of Cartan class Δ if and only if A is an iterated tilted algebra of Dynkin type Δ. Xiao [38] gave a sufficient and necessary conditions for a representation-finite selfinjective algebra to be a trivial extension algebra. Xiao-Zhang [44] studied the properties of indecomposable modules over the trivial extension algebra of an iterated tilted algebra.

Let A be an iterated tilted algebra of type $\vec{\Delta}$; the repetitive algebra $\hat{A}$ has the additive structure $(\underset{i \in \mathbb{Z}}{\oplus} A_i) \oplus (\underset{i \in \mathbb{Z}}{\oplus} Q_i)$ with $A_i = A$, $Q_i = D(A)$ for $i \in \mathbb{Z}$, whose multiplication is defined as follows.

$$(a_i, \varphi_i)_i \cdot (b_i, \psi_i)_i = (a_i b_i, a_{i+1}\psi_i + \varphi_i b_i)_i.$$

for $(a_i, \varphi_i)_i$, $(b_i, \psi_i)_i \in \hat{A}$. Let v be the Nakayama automorphism $\hat{A} \to \hat{A}$; then $T(A) \cong \hat{A}/v$ and v induces the Galois covering functor $\pi : \hat{A} \to T(A)$ and an automorphism of $\hat{A}$-mod. Then Happel's theorem in [H2] says that $\hat{A} - \underline{\mathrm{mod}} \cong D^b(A)$, and $\Gamma_s(T(A)) \cong \Gamma(D^b(k\vec{\Delta}))/ < T^2\tau >$, where $\hat{A} - \underline{\mathrm{mod}}$ is the stable category of $\hat{A}$-mod; $B^b(A)$ is the derived category of A, and $T^2\tau$ is just the automorphism of $\hat{A}$ induced by the Nahayama functor v. We denote by π the covering functor from $\hat{A}$-mod to $T(A)$-mod induced by $\pi : \hat{A} \to T(A)$. The indecomposable $T(A)$-module M is said to be on platform, if there is $X \in \hat{A}$-mod such that $\pi(X) = M$ and X as an object of $\hat{A} - \underline{\mathrm{mod}}$ belongs to a component of form $\mathbb{Z}\vec{\Delta}$ of $\Gamma_s(\hat{A} - \underline{\mathrm{mod}}) \cong \Gamma(D^b(k\vec{\Delta}))$.

THEOREM ([44]) Let $T(A)$ be the trivial extension of an iterated tilted algebra of type $\vec{\Delta}$, X a $T(A)$-module on platform; then the number of isoclasses of the $T(A)$-modules on platform which have the same dimension vector as X is at most n, where n is the number of vertices of $\vec{\Delta}$.

THEOREM ([44]) Let $T(A)$ be as above, X and Y the $T(A)$-modules on platform;

then the followings are equivalent.
1) $X \cong Y$;
2) $\text{top} X \cong \text{top} Y$, and $\text{soc} X \cong \text{soc} Y$;
3) $L\underline{\dim} X = L\underline{\dim} Y$;
4) $S\underline{\dim} X = S\underline{\dim} Y$.

3 Tilting modules and tilted algebras

We recall the definition of tilting module (see e.g. Happel-Ringel [HR1]). Assume that A is a finite-dimensional algebra over a field k; a module $_A T$ is called a tilting module provided the following conditions are satisfied : (i) the projective dimension of T is at most 1; (ii) $\text{Ext}_A^1(T, T) = 0$ and (iii) there exists an exact sequence $0 \rightarrow$ $_A A \rightarrow T_0 \rightarrow T_1 \rightarrow 0$, where $T_0, T_1 \in \text{add} T$, the module class which consists of all possible direct sums of direct summands of $_A T$. Note that under the conditions (i) and (ii), (iii) is equivalent to (iii'): the number of the non-isomorphic indecomposable direct summands of T is just the number of simple A-modules. Note that any algebra A has at least one tilting module, i.e. $_A A$. And a selfinjective algebra has a unique tilting module. Assem constructed such an non-selfinjective algebra which has a unique tilting module; and Happel gave a sufficient and necessary condition for such an algebra, see [H2] III. 2.14. Lin-Peng [8] proved that A has a unique tilting module if and only if every simple A-module $S(a)$ is a direct summand of $\text{top} I(b)$, where $I(b)$ is an indecomposable injective module.

For a tilting triple (A, T, B) (i.e., T is a tilting module over A, and $B = \text{End}(T)$), we may form the module classes: $\mathcal{J}(T)$, $\mathcal{F}(T)$, $\mathcal{X}(T)$, and $\mathcal{Y}(T)$ (see [R1]); then $(\mathcal{J}(T), \mathcal{F}(T))$ is a torsion pair in A-mod and $(\mathcal{X}(T), \mathcal{Y}(T))$ is a torsion pair in B-mod. By definition, a tilting module T is separating provided that $(\mathcal{J}(T), \mathcal{F}(T))$ is splitting in A-mod , i.e. any indecomposable A-module either belongs to $\mathcal{J}(T)$, or to $\mathcal{F}(T)$; T is a splitting tilting module provided that $(\mathcal{X}(T), \mathcal{Y}(T))$ is splitting in B-mod. Note that if A is hereditary, then any tilting A-module is splitting. P. Zhang ([50]) proved that a tilting module T over a hereditary algebra A is separating if and only if T is a slice (in the sense of Ringel [R1]) in A-mod.

Slices provide an inner charcterization to the tilted algebras. Recall that an algebra B is called a tilted algebra if there exists a tilting triple (A, T, B) with A hereditary. This class of algebras is extensively studied by many people for their importance in the representation theory of algebras. For a tilting triple (A, T, B) with $A = k\vec{\Delta}$, if Δ is a Dynkin diagram, it is clear that B is representation-finite. If Δ is an Enclidean diagram, then B is represantation-finite if and only if T has a non-zero preprojective direct summand and a non-zero preinjective direct summand; and in other cases, B is of tame type, see [R1]. If Δ is wild, the representation type of B is not so easy to see. Kerner [K] has given a method to determine the representation type of B; in [25] Peng gave another algorithm:

Let A be a hereditary algebra, $_A T$ a tilting module; then there is an integer $s \geq 0$, a sequence of hereditary algebras $(_0 A, _1 A, \cdots, _s A)$, a sequence of module $(_0 T, _1 T, \cdots, _s T)$ and a sequence of non-negative integers $(m_0, m_1, \cdots, m_{s-1})$, such that

(a) $_0 A = A$, $_0 T = T$, $_i T$ is a tilting $_i A$-module , and $_s T$ has no preinjective direct summands in mod $_s A$,

   (b) If $s > 0$, then $_i T =_i T' \oplus_i T''$ for $0 \le i < s$, such that for any indecomposable direct summand X of $_i T'$, $\tau_i^{m_i} X$ is a non-zero projective $_i A$-module ; and any indecomposable direct summand Y of $_i T''$, $\tau_i^{m_i} Y$ is not a project $_i A$-module (where τ_i is the relative Auslander-Reiten translation). Denote by e_i the idempotent of $_i A$ corresponding to projective $_i A$-module $_i T'$; then $_{i+1} A =_i A/ < e_i >$, $_{i+1} T = \tau_i^{m_i+1} {}_i T''$.

THEOREM ([25])   With the above notations one has
   (i) B is representation-finite if and only if $_s A = 0$ and $A_t = 0$
   (ii) B is of tame type if and only if $_s A$ or A_t is tame, and no one is wild.
   (iii) B is of wild type if and only if $_s A$ or A_t is wild.
   The shape of Auslander-Reiten component of a hereditary algebra A is well known, see [R1]; it has one preprojective component and one preinjective component, and all other components are quasi-serial, i.e., the regular tubes or the components of form $\mathbb{Z}A_\infty$. It is natural to ask what the shapes of components of a tilted algebra look like. It is well known that any tilted algebra has a connecting component, which is a directing component. The structure of a regular component of a tilted algebra which is not a connecting component is also known: it is also quasi-serial. Ringel [R3] asked the following question : what are the possible structures of non-regular components of tilted algebras? Strauss [S] proved that any tilted algebra has exactly one preprojective component and one preinjective component. S.P. Liu [18] gave a description of the other components of a tilted algebra by using the relative Auslander-Reiten sequence in $\mathcal{J}[T]$ and $\mathcal{F}[T]$, and ray insertions and coray insertions, which are introduced in [R1]. Let us recall some notions.
   Let Γ be a translation quiver. A vertex x of Γ is called a ray vertex if there is an infinite sectional path

$$x = x[1] \to x[2] \to \cdots \to x[n] \to \cdots$$

in Γ, such that for each integer $i > 0$, the path $x \to \cdots \to x[i + 1]$ is the only sectional path of length i in Γ which starts at x. For a ray vertex in Γ as above and integer $n > 0$, we may define a new translation quiver $\Gamma[x, n]$, and then define a translation quiver $\Gamma[x_0, n_1][x_1, n_1] \cdots [x_s, n_s]$ by induction, which is called a translation quiver obtained from Γ by ray insertions, see [18]. There is a dual concept of a coray vertex and coray insertions.

THEOREM ([18])   Let A be a hereditary connected Artin algebra. Let T be a tilting A-module and $B = \text{End}_A T$, C a component of $\Gamma(B)$ other than the connecting component of B. If C is neither a preprojective component nor a preinjective component, then either C is quasi-serial or C is obtained from a quasi-serial translation quiver by ray insertions or by coray insertions.
   Liu also determined the shapes of components of $\Gamma(B)$ which lie completely in $\mathcal{Y}(T)$, where (A, T, B) is a tilting triple with T a splitting tilting module.
   A partial tilting module is a direct summand of a tilting module. If A is a hereditary algebra and $_A T$ a partial tilting module, then $B = \text{End}_A T$ is also a tilted algebra. By definition, a concealed algebra B is an endomorphism algebra of a preprojective tilting module over an hereditary algebra. This is a tilted algebra which has two components containing a slice. P. Zhang [51] proved that: If M is a preprojective partial tilting

module over a concealed algebra, then $B = \text{End } M$ is either a tilted algebra of Dynkin type, or a concealed algebra. So one can check the representation type of B by its quadratic form. Using this one can easily prove : Let A be a tame concealed algebra, $T = T_0 \oplus T_1$ a tilting module with T_0 nonzero preprojective and T_1 regular; then $B_0 = \text{End }_A T_0$ is also a tame concealed algebra. But the situation in the case of A being wild concealed is quite different: We have examples to show that B_0 can be of Dynkin type, tame and wild respectively.

P. Zhang [47,48] has studied the structure and representation of one-point extension of a tilted algebra by so-called foremodules (the modules cogenerated by a slice module), especially by indecomposable foremodules. This class of algebras contains the class of tilted algebras as a proper subclass and also has global dimension at most 2. Let (A, T, B) be a tilting triple with A being a hereditary algebra, $_A M \in \mathcal{J}(T)$, $_B N = \text{Hom }_A(T, M)$; then $(A[M], T \oplus (M, k, e), B[N])$ is also a tilting triple; and $\tilde{T} = T \oplus (M, k, e)$ is a splitting tilting $A[M]$-module if and only if $\text{Hom }_A(F(T), \tau M) = 0$; and when $\tilde{T}$ is a splitting module, $B[N]$ has some similar properties as a tilted algebra. Let $I(a)$ be an indecomposable injective module over a hereditary algebra A; then $A[I(a)]$ has the same representation type as A; by this result some classes of representation-finite tilted algebras with slices of wild type were constructed in [48].

4   Stable equivalence and triangle equivalence

Let A be a finite-dimensional k-algebra, $C(A)$ the category of complexes over A-mod. and $K(A)$ the homotopy category of $C(A)$. Note that $K(A)$ is a triangulated category with the triangles of mapping cones. A morphism in $K(A)$: $X^\bullet \xrightarrow{f} y^\bullet$ is called a quasi-isomorphism if the induced morphism $f^n : H^n(X^\bullet) \to H^n(Y^\bullet)$ is a group isomorphism for $n \in \mathbb{Z}$, where $H^n(X^\bullet)$ denotes the n-th homology group of $X^\bullet$. We denote by S the set of all quasi-isomorphisms in $K(A)$; then S is a saturated multiplicative system in $K(A)$ (see [I]). Thus we can construct a fractional category $S^{-1}K(A)$; this is called the derived category of A and denoted by $D(A)$. In most cases we are interested in $D^b(A) = (S \cap K^b(A))^{-1} K^b(A)$, the derived category of bounded complexes over A-mod. If $gl.d.A < \infty$, then $D^b(A)$ is rather simple: $D^b(A) \cong K^b(A-\text{proj.})$, the homotopy category of bounded complexes over A-projective modules. If $\mathcal{A}$ is a locally bounded Frobenius k-algebra (i.e.,there exists a complete set of orthogonal primitive idempotents $\{e_x \mid x \in I\}$ such that $\mathcal{A}e_x$ and $e_x\mathcal{A}$ are finite dimensional over k for $x \in I$, and the projective-modules coincide with the injective modules), then the stable category $\mathcal{A} - \underline{\text{mod}}$ has a structure of triangulated category. In this way the stable category $\widehat{A} - \underline{\text{mod}}$ of the repetitive algebra $\widehat{A}$ of a finite-dimensional k-algebra A is a triangulated category.

Happel [H1,H3] proved that for a finite dimensional algebra A, $D^b(A)$ and $\widehat{A} - \underline{\text{mod}}$ are triangle-equivalent if and only of A has finite global demension, and that two algebras which are tilt-equivalent are derived equivalent. Thus if A is an iterated tilted algebra of type $\vec{\Delta}$, then $D^b(A) \cong D^b(k\vec{\Delta})$; moreover Happel [H2] determined the stable Auslander-Reiten quiver of $T(A)$. Peng-Xiao in [22, 23] considered the converse of the Happel's above results. In [22], they obtained the following result.

THEOREM ([22])   If $\mathcal{A}$ is a locally bounded Frobenius algebra such that there is a triangle-equivalence $\mathcal{A} - \underline{\text{mod}} \cong D^b(k\vec{\Delta})$, then there exists some tilted algebra A of type $\vec{\Delta}$ such that $\mathcal{A} \cong \hat{A}$.

This result can be essentially considered as a generalization of Bretscher-Läser-Riedtmann's main result in [BLR] in the Dynkin case. Associated with the works of Hughes-Waschbüsch in [HW] and Tachickawa-Wakamatsu in [TW], there are some interesting corollaries as follows. For any finite dimensional k-algebra B, if B is an iterated titled algebra of type $\vec{\Delta}$, then $\hat{B} \cong \hat{A}$ for some tilted algebra A of type $\vec{\Delta}$; they also got that if B is an iterated tilted algebra and e an idempotent, then eBe is also an iterated tilted algebra.

In [23], the following result is proved

THEOREM [23]   Assume that $\mathcal{A}$ is a finite-dimension symmetric k-algebra such that $\mathcal{A} - mod \cong T(k\vec{\Delta}) - \underline{\text{mod}}$; then $\mathcal{A} \cong T(A)$ for some tilted algebra A of type $\vec{\Delta}$.

As a direct consequence, if B is an iterated tilted algebra of type $\vec{\Delta}$, then $T(B) \cong T(A)$ for some tilted algebra A of type $\vec{\Delta}$. In [52], P. Zhang determined the structure of $T(B)$ with a $\mathbb{Z}\vec{\Delta}$-components: Let B be an iterated tilted algebra; then $T(B)$ has a $\mathbb{Z}\vec{\Delta}$-component if and only if $T(B) \cong T(A)$, where A is a tilted algebra with the connecting component $\mathbb{Z}\vec{\Delta}$.

Alperin and Auslander-Reiten have conjectured that algebras which are stably equivalent should have the same number of non-projective simple modules. This has been settled for representation-finite algebra (see Martinez [M1]) and for some other cases, but in general remains open. It is sufficient to verify the conjecture for self-injecture algebras ([M2]). A.P.Tang considered this conjecture for 3-nilpotent (i.e. $\text{rad}^2 A \neq 0$, $\text{rad}^3 A = 0$) self-injective algebras.

Let $\vec{Q}$ be a quiver, $\vec{Q}^s$ its separating quiver. Tang [28] associated a basic connected 3-nilpolent self-injective algebra A with a class $(\vec{\Delta}, \varphi_0, \cdots, \varphi_{m-1}, I))$ in the following way: Let $A = k\vec{Q}/I$; then $\text{soc}\, A = R^2$, where $R = \text{rad}\, A$, and $V = A/\text{soc}\, A$ is a basic connected 2-nilpotent algebra with the same Gabriel quiver as A, thus V is stably equivalent to the hereditary algebra $k\vec{Q}^s$. Let $\vec{\Delta}_0, \cdots, \vec{\Delta}_{m-1}$ be all the connected components of the separating quiver $\vec{Q}^s$; then $\vec{\Delta}_i \cong \vec{\Delta}$ or $\vec{\Delta}_i \cong \vec{\Delta}^{op}$ for $0 \leq i \leq m-1$ and some quiver $\vec{\Delta}$. Using anti-antomorphisims $\varphi_i : \vec{\Delta}_i \rightarrow \vec{\Delta}_{i+1}$, for $0 \leq i \leq m-1$, Tang recovered $\vec{Q}$ from $\vec{\Delta}$; in this sense the 3-nilpotent self-injective algebra A is determined by a class $(\vec{\Delta}, \varphi_0, \cdots, \varphi_{m-1}, I)$. If in addition $m \geq 3$, she verified the conjecture:

THEOREM ([28])   Assume that $\wedge$ is stably equivalent to the 3-nilpotent basic connected self-injective algebra A, and A is given by $(\vec{\Delta}, \varphi_0, \cdots, \varphi_{m-1}, I)$ in the above way. If $m \geq 3$, then $\wedge$ and A have the same number of non-projective simple modules.

5   Quasi-hereditary algebras

As pointed out by Cline-Parshall-Scott (e.g. see [CPS]), the quasi-hereditary algebra provides a bridge between representation theory of finite-dimensional algebras and that of Lie theory. Also, Dlab and Ringel have made an extensive studies on this class of algebras.

There are several equivalent definitions of quasi-hereditary algebras. The usual one is in terms of hereditary ideals, see [PS,DR2]. A quasi-hereditary algebra may have many ideals such that the corresponding factor algebras are not quasi-hereditary. C. C. Xi [36] showed that an algebra A with $A/\text{rad}^n A$ quasi-hereditary for some $n \geq 2$, is quasi-hereditary itself. Recall that a module M is called semilocal if M is a direct sum of local modules (i.e., the module has a unique maximal submodule). Lin-Xi proved in [9] that for a semilocal module with Loewy length m, then $\text{End}_A\left(\bigoplus_{i=1}^m M/N^i M\right)$, where N is radical of A, is quasi-hereditary. This generalizes a result in [DR3], which is the special case of $M =_A A$.

Dlab-Ringel [DR1] gave an example showing that the quasi-hereditary class is not closed under tilting. Let (A, T, B) be a tilting triple. In [53] P.Zhang considered the following questions: under what conditions, is $B = \text{End}_A T$ quasi-hereditary? And if A is quasi-hereditary, what properties does B possess? Let $T = \bigoplus_{i=1}^n T(i)$; one can construct the module $K(i)$ for $1 \leq i \leq n$ from $T(i)$. In the case of $K(i) \in G(_A T)$, the answers to the above questions were given in [53].

Schur algebras are important quasi-hereditary algebras. Let Σ_r be the symmetric group; then $V^{\otimes r}$ is a Σ_r-module under the permutation action, and the Schur algebra $S_k(n, r)$ can be defined as the endormorphism algebra of $k\Sigma_r$-module $V^{\otimes r}$. Note that if char $k = 0$, or char $k = p > r$, then $S_k(n, r)$ is semisimple.

The structure of the Schur algebras $S_k(p, p)$, for p a prime was given by C.C. Xi in [33]. Note that $S_k(n, r)$ with $n \geq r$ is Morita equivalent to $S_k(r, r)$ ([Gr]). So Xi's result can be stated in the following

THEOREM ([33]) Let k be an algebraically closed field with char $k = p > 0$; then each block of the Schur algebra $S_k(n, p)$ with $n \geq p$ is either simple or Morita equivalent to the path algebra of

$$
\circ \xrightarrow[\beta_1]{\alpha_1} \circ \xrightarrow[\beta_2]{\alpha_2} \circ \cdots \circ \xrightarrow[\beta_m]{\alpha_m} \circ \qquad m \geq 1
$$

module the ideal generated by

$$
\alpha_i \alpha_{i+1}, \ \beta_i \beta_{i+1}, \ \alpha_{i+1}\beta_{i+1} - \beta_i \alpha_i, \ 1 \leq i \leq m-1, \ \alpha_1 \beta_2
$$

where m depends only on p. Moreover, there is only one non-simple block.

The proof of this theorem uses the fact that $s(S(p, p)) = s(k\Sigma_p) + 1$, where $s(A)$ denotes the number of simple A−modules and the structure of the connected basic symmetric algrbra A with A/M quasi-hereditary where M is the socle of an indecomposable projective left ideal of A. For a non-simple connected basic algebra A, Xi [33] also gave a sufficient and necessary condition for A being symmetric with an indecomposable module M such that $\text{End}_A(_A A \oplus M)$ is quasi-hereditary.

In a recent paper [35] Xi has studied the q-Schur algebras.

Fix an ordering of simple A-modules. We denote by $\Delta(i)$ the maximal quotient of the indecomposable projective module $P(i)$ with composition factors of the form $E(j)$ with $j \leq i$, and $\Delta = \{\Delta(1), \cdots, \Delta(n)\}$; dually, we have $\nabla = \{\nabla(1), \cdots, \nabla(n)\}$. If $\mathcal{X}$ is a set of modules, let $\mathcal{F}(\mathcal{X})$ be the set of modules which have a filtration with factors in $\mathcal{X}$. Then A is said to be quasi-hereditary with respect to the fixed ordering provided: a) $\text{End}\,\Delta(i)$

is a division ring for $1 \leq i \leq n$; b) $_A A \in \mathcal{F}(\Delta)$, see [DR4, R4]. Note that under the condition a), b) is equivalen to c): $\operatorname{Ext}_A^2(\Delta(i), \nabla(j)) = 0$ for $1 \leq i, j \leq n$. Let A be quasi-hereditary with a fixed ordering; then the intersection $\mathcal{F}(\Delta) \cap \mathcal{F}(\nabla) = add_A T$, where $_A T$ is a generalized tilting and generalized cotilting module, which is called the characteristic module of A ([R4]). This module T is of particular interest; its endormorphism algebra is called the Ringel dual of A.

In [R4] Ringel proved that given an ordering of simple A-modules, A is quasi-hereditary if and only if there exists a module $T = \bigoplus_{i=1}^{n} T(i)$ with $T(i)$ indecomposable such that (i) $(\underline{\dim} T(i))_i = 1$; (ii) $\bigoplus_{j=1}^{n} T(j)$ is a generalized tilting A/J_{i+1}-module for $1 \leq i \leq n$. Note that the characteristic module of a quasi-hereditary algebra satisfies these two conditions. P.Zhang [53] proved the following

THEOREM ([53])   Let A be an algebra with a fixed ordering of simple $A-$modules. If T is an $A-$module satisfing the two conditions above, then T is exactly the characteristic module of A.

6   Hall algebras

The Hall algebra of a finitary ring has been introduced by C. M. Ringel; there is a strong relation between Hall algebras and the quantum groups, see [R5-6]. Let R be a finitary ring, $N_1, \cdots, N_t$, and M finite R-modules. Denote by $F_{N_1 \cdots N_t}^M$ the number of filtrations

$$M = U_0 \supseteq U_1 \supseteq \cdots \supseteq U_t = 0$$

of M such that $U_{i-1}/U_i \cong N_i$ for $1 \leq i \leq t$. Let $H(R)$ be the $\mathbb{Q}$-vector space with basis $(u_{[M]})_{[M]}$ (indexed by the set of isoclasses $[M]$ of all finite R-modules M), with a multiplication

$$u_{[N_1]} u_{[N_2]} = \sum_{[M]} F_{N_1 N_2}^M u_{[M]}$$

then $H(R)$ is an associative $\mathbb{Q}$-algebra with $1 = u_{[0]}$. In general, $H(R)$ is not commutative.

Let $\vec{\Delta}$ be a cyclic quiver with n vertices, and k a field of p elements; denote by $\mathcal{T}$ the category of locally nilpotent finite dimensional representations of $\vec{\Delta}$ over k. Then J.Y.Guo [7] determined the center of $H(\mathcal{T})$

THEOREM ([7])   The centre of $H(\mathcal{T})$ is exactly the subalgebra of $H(\mathcal{T})$ generated by $c = \sum_{\underline{\dim} M = (1, \cdots, 1)} (1 - p)^{n(M)} u_{[M]}$, where $n(M)$ is the number of indecomposable summands of the module M.

In [4] Guo considered the relation between isomorphism of the species and that of the corresponding Hall algebras.

In [5, 6] Guo considered the structure of the Hall algebra $H(A)$ of a cyclic serial algebra A. The Hall polynomials and the base of the Hall algebra $H_Q(A) = H(A) \otimes_Z Q$ were given. Guo also proved that the Hall algebra $H(A)$ coincides with the Loewy algebra $L(A)$ which is the subalgebra of $H(A)$ generated by the semisimple modules. $H_Q(A)$ is a filtered ring and the associated graded ring is an iterated, skew polynomial ring; thus both are Noetherian without zero divisor.

7   Uniserial modules over commutative Artin ring

It is well known in commutative algebra that a commutative Artin ring is a direct product of a finite number of commutative Artin local rings. In the following R denotes a commutative Artin local ring; then $R \cong S/\underline{a}$ for some commutative noetherian regular local ring (S, m) and some m-primary ideal $\underline{a} \subseteq m^2$. Denote by e-dim R the embedding dimension of R, i.e. the minimal number of generators of the maximal ideal of R, and assume that e-dim $R > 1$.

In his doctorial thesis [62] under the direction of M. Auslander, J. G. Luo studied uniserial R-modules and other related modules, which often provide important information about the ring itself. An uniserial R-module U is said to be maximal provided that U is not a proper submodule of any uniserial module.

THEOREM ([62])   Suppose R contains an infinite field. Let $d = e$-$\underline{\dim}$ R and $n =$ Loewy length of R. Then all maximal uniserial R-modules have length n if and only if $R \simeq K[[X_1, \cdots, X_d]]/(X_1, \cdots, X_d)^n$, where K is the residue field of R. In this case, R can be embedded in a direct sum of a finite number of maximal uniserial R-modules.

In case $R = K[[X_1, \cdots, X_d]]/(X_1, \cdots, X_d)^n$ for some field K, the automorphism group of R provides an effective tool; this is because that every automorphism of $K[[X_1, \cdots, X_d]]$ induces an automorphism of R. Using this technique the following was proved in [62]:

If R contains an infinite field, let U, V be nonisomorphic uniserial R-modules; then U, V belong to different components in the AR-quiver of R. And if $\ell_R(U) \geq 2$, then the component containing U is stable.

In case $e - \underline{\dim}$ $R = 2$ and let U, V be maximal uniserial R-modules, one has the following

THEOREM ([62])   For any nonsplit exact sequence

$$\mathcal{E} :\ 0 \to V @>f>> E @>g>> U \to 0$$

we have

(1)   up to isomorphism, $\mathcal{E}$ is uniquely determined by the $\text{End}(V)^{(op)}$-submodule $\text{Im} \text{Hom}_R(V, g)$ of $\text{Hom}_R(V, U)$.

(2) E is indecomposable.

(3) $DTr_R E = E$.

Note that the modules fixed by DTr are of particular interest because of a result of Hoshino, which states that if Λ is an Artin algebra and M is an indecomposable Λ-module with $DTrM \simeq M$, then either Λ is local Nakayama or the AR-component containing M is a homogeneous 1-tube. Luo determined the uniserial R-modules U satisfing $D\text{Tr}\, U \simeq U$.

Denote by m the maximal ideal of S, where $R = S/a$, a criterion when E_g decomposes was given in [62], where

$$0 \to S/m \to E_g \to Tr_R(S/m) \to 0$$

is an almost split sequence. As a consequence one gets: Suppose R contains an algebraically closed field; then R is tame if and only if $e\text{-dim}R = 2$ and E_g decomposes.

8   Some related topics

Gabriel's theorem ([G]), which says that any basic connected finite dimensional algebra over an algebraically closed field k is of the form $A = k\vec{\Delta}/I$ for some finite quiver $\vec{\Delta}$ and some admissible ideal I of the path algebra $k\vec{\Delta}$, also provides a new point of view to some classical problems in finite-dimensional algebras, or more generally, in ring theory, in terms of quivers and relations.

In [12] Liu-Luo-Xiao considered the isomorphism problem for path algebra of quivers; in [11] Liu studied the relations between the algebraic properties of path algebras and the combinatorial properties of the corresponding quivers. These results were extended by Liu in [13] to the tensor rings over valued graphs. An (oriented, in general, infinite) valued graph is a triple (Σ, d, g), where Σ is a set of vertices and both d and g refer to functions defined on $\Sigma \times \Sigma$ with values which are cardinal numbers (possibly zero and ∞). A collection $(D_i, {}_iM_j, i, j \in \Sigma)$, where D_i is a division ring and ${}_iM_j$ a $D_i - D_j$−bimodule satisfying the condition

$$[{}_iM_j : D_i] = d_{ij} \quad \text{and} \quad [{}_iM_j : D_j] = d_{ij}$$

is said to be a modulation of the valued graph (Σ, d, g). The tensor ring $T = T(\Sigma, D, M)$ is defined as

$$T = D \oplus M \oplus M^{(2)} \oplus ... \oplus M^{(n)} \oplus ...$$

where $M^{(n+1)} = M^{(n)} \underset{D}{\otimes} M$, $D = \underset{i}{\oplus} D_i$, $M = \underset{i,j}{\oplus} {}_iM_j$, with the multiplication induced by tensor products. The following theorem describes the isomorphism of two tensor rings in terms of the one of the valued graphs.

THEOREM ([13])   Let $\Sigma = (\Sigma, d, g)$ and $\Sigma' = (\Sigma', d', g')$ be two valued graphs, and suppose that $(D, M) = (D_i, {}_iM_j, i, j \in \Sigma)$ and $(D', M') = (D'_\alpha, {}_\alpha M'_\beta, \alpha, \beta \in \Sigma')$ are their modulations respectively. Write $T = T(\Sigma, D, M)$ and $T' = T(\Sigma', D', M')$. If $f : T \to T'$ is a ring isomorphism, then there is a bijective map $\theta : \Sigma \to \Sigma'$ such that for all $i, j \in \Sigma$,

$$D_i \simeq D'_{\theta(i)} \quad \text{and} \quad {}_iM_j \simeq_{\theta(i)} M_{\theta(j)}$$

Sufficient and necessary conditions for a tensor ring $T = T(\Sigma, D, M)$ to be left Artinian, prime, primary, semiprimitive were also given in [13]. The Jacobson radical $J(T)$ (which coincides with the Baer radical $B(T)$) is also calculated. We quote only the following result.

THEOREM ([13])   The tensor ring $T = T(\Sigma, D, M)$ is left noetheian if and only if Σ satisfies the following conditions

(1) Σ is finite;

(2) d_{ij} are finite for $i, j \in \Sigma$;

(3) $d_{i_1 i_2} \cdot d_{i_2 i_3} \cdots d_{i_n i_1} \neq 0$ implies that $d_{i_t i_{t+1}} = 1$ and $d_{j i_{t+1}} = 0$ for $j \neq i_t$, where $1 \leq t \leq n$.

For a finite quiver $\vec{\Delta}$ (which may have oriented cycles), Xiao ([43]) proved that $k\vec{\Delta}$ is hereditary, and any (left) projective $k\vec{\Delta}$-module is a direct sum of the modules of form $(k\vec{\Delta}) \cdot e$, where e is a vertice in $\vec{\Delta}$.

We denote by $H_i(A)$ (resp. $H^i(A)$) the Hochschild homology (resp. cohomology) groups of A. Let $B = A[M]$ (resp. $B = [M]A$) be the one-point extension (resp. coextension) of A. The relation of $H^i(B)$ and $H^i(A)$ has been given by Happel ([H]). Liu-Zhang proved in [15] that $H_i(B) \cong H_i(A)$ for $i > 0$ and $H_0(B) \cong H_0(A) \oplus k$. (From this one can also easily get that $H_i(A) = 0$ for $i > 0$ and $H_0(A) \doteq k^{n(A)}$ for a triangular algebra A (i.e., an algebra whose Gabriel quiver has no oriented cycle), which was first given in [C]). We call a cycle α of length m a basic cycle if α has exactly m distinct vertices. Let $A = k\vec{\Delta}/I$ be a monomial algebra (i.e. I is generated by some paths in $\vec{\Delta}$); then $H_0(A) \cong k^{n(\vec{\Delta})}$ if and only if for any cycle $\alpha = a_1 a_2 \cdots a_m$, $\{\alpha, t\alpha, \cdots, t^{m-1}\alpha\} \cap I \neq \emptyset$, where a_i is an arrow for $1 \leq i \leq m$, and t is the operator defined as $t\alpha = a_2 \cdots a_m a_1$ (see [15]). We call an algebra $k\vec{\Delta}/J^t$ $(t \geq 2)$ a t-truncated algebra, where J is the ideal generated by all arrows in $\vec{\Delta}$. For integers t and n we denote by $a(n,t)$ (resp. $b(n,t)$) the number $t[\frac{n}{2}] + u(n+1)$ (resp. $t[\frac{n}{2}] + 1 - u(n)$), where $u(n) = \begin{cases} 1 & n \text{ is even} \\ 0 & n \text{ is odd} \end{cases}$; then we have

THEOREM ([16])   Let A be a t-truncated algebra; then

1) If $\vec{\Delta}$ does not contain any cycle of length ℓ for $a(n,t) \leq \ell \leq b(n,t)$, then $H_n(A) = 0$.

2) If $H_n(A) = 0$, then $\vec{\Delta}$ does not contain any basic cycle of length ℓ for $a(n,t) \leq \ell \leq b(n,t)$.

In deformation theory of algebras, $H^2(A) = 0$ is of particular interest, since this implies that A has only trivial deformation. Zhang ([54]) has determined the structures of triangular t-truncated algebras A with $H^2(A) = 0$.

In the case of $t = 2$, we can do some more: Let $A = k\vec{\Delta}/J^2$ be a triangular algebra, $n \geq 2$ an integer, then $H^n(A) = 0$ if and only if $\vec{\Delta}$ does not contain the subquiver $\tilde{\mathbb{A}}_{1,n}$.

References

1. Deng, Bang-Ming and Xiao, Jie, Dimension vectors, Loewy factors and Socle factors are invariants of AR quiver, Kexue Tongbao, 22(1991), 1764–1767 (In Chinese).

2. Guo, Jin-Yun, On the indecomposable modules of algebras stably equivalent to hereditary algebra, Comm. Algebra, 16(9)(1988), 1749–1757.

3. Guo, Jin-Yun, Sufficient and necessary conditions for indecomposable modules

over self-injective algebra determined by their composition factors, Acta Mathmatica Sinica (1989) (In Chinese).

4. Guo, Jin-Yun, The isomorphism of Hall algebras, J. Algebra, 116(1994), 1–33.

5. Guo, Jin-Yun, Hall polynomials and bases of Hall algebras of cyclic serial algebra, to appear in Science in China (In Chinese).

6. Guo, Jin-Yun, Hall algebra of cyclic serial algebra, Science in China (Series A), 24(20)(1994), 139–149.

7. Guo, Jin-Yun, The centre of the Hall algebra of a cyclic quiver, to appear.

8. Lin, Ya-Nan and Peng, Lian-Gang, An effective criterion for uniqueness of the tilting modules, Kexue Tongbao, 37(4)(1992), 489–291 (In Chinese).

9. Lin, Ya-Nan and Xi Chang-Chang, Semilocal modules and quasi-hereditary algebras, Arch. der Math. 60(1993), 512–516.

10. Liu, Shao-Xue, On my visit to Europe, Inshan Xuekan 1(1988), 5–10 (In Chinese).

11. Liu, Shao-Xue, Geometrical properties of directed graphs and algebraic properties of their path algebras, Acta Math. Sinica, 31(1988), 483–487 (In Chinese)

12. Liu, Shao-Xue, Luo, Yun-Lun and Xiao, Jie, Isomorphism of path algebras, Bull. of Beijing Normal University, 3(1986), 13–20 (In Chinese).

13. Liu, Shao-Xue, Isomorphism problem for tensor algebras over valued graphs, Science in China (Series A), 34(3)(1991), 267–272.

14. Liu, Shao-Xue and Xi, Chang-Chang, Finite $\triangle$-good module categories of quasi-hereditary algebras, preprint.

15. Liu, Shao-Xue and Zhang, Pu, On the Hochschild homology of some finite dimensional algebras, In "Representations of Algebras", CMS Conference Proceedings, Vol.14 (1992), 353–360.

16. Liu, Shao-Xue and Zhang, Pu, Hochschild homology of truncated algebras, Bull. London Math. Soc., 26(1994), 427–430.

17. Liu, Shiping, Degrees of irredicible maps and the shapes of Auslander-Reiten quivers, J. London Math. Soc., 45(2)(1992), 32-54.

18. Liu Shiping, The connected components of the Auslander-Reiten quiver of a tilted algebra, J. Algebra 161(1993), 505–523.

19. Liu, Shiping, Infinite radicals in standard Auslander-Reiten component, 166(1994), 245–254.

20. Peng, Lian-Gang, On a class of tilted algebras, Kexue Tongbao, 36(4) (1991), 247–249. (In Chinese)

21. Peng, Lian-Gang and Xiao, Jie, On the number of DTr-orbits containing directing modules, Proc. Amer. Math. Soc., 118(3)(1993), 753–756.

22. Peng, Lian-Gang and Xiao, Jie, Invariability of repetitive algebras of tilted algebras under stable equivalences, J. algebra, 170(1994), 54–68.

23. Peng, Lian-Gang and Xiao, Jie, Invariability of trivial extensions of tilted algebras under stable equivalences, to appear in J. Algebra.

24. Peng, Lian-Gang, Projective representation directed algebras, Chinese J. of Contemporary Math. 14(3)(1993), 231–238.

25. Peng, Lian-Gang, A criterion for the representation type of tilted algebras, Science in China (Series A) 36(7)(1993), 812–815.

26. Peng, Lian-Gang, Repetitive algebras of iterated tilted algebras, Acta Math. Sinica, New Series, 9(2)(1993), 148–151.

27. Peng, Lian-Gang, Wild algebras with preprojective components, Chinese Science Bull. 38(8)(1993), 676–678.

28. Tang, Ai-Ping, Radical cubic zero self-injective algebras, to appear.

29. Xi, Chang-Chang, On wild hereditary algebras with small growth numbers, Comm. Algebra, 18(10)(1990), 3413–3422.

30. Xi, Chang-Chang, Die vektorraumkategorie zu einem unzerlegbaren projektiven Model einer fubulären Algebra, Manuscript Math., 69 (3)(1990), 223–235.

31. Xi, Chang-Chang, Die Vektorraumkategorie zu einem unzerlegbaren projektiven Modul einer Algebra, J. Alg., 139(1991), 355–363.

32. Xi, Chang-Chang, Die Vektorraumkategorie zu einem Punkt einer zahmen verkleideten Algebra, J. Alg. 145(1992), 279–305.

33. Xi, Chang-Chang, The structureof Schur algebras $S_k(n,p)$ for $n \geq p$, Can. J. Math. 4(3)(1992), 665–672.

34. Xi, Chang-Chang, Minimal elements of the poset of a hammock. J. London Math. Soc. 46(2)(1992), 228–238.

35. Xi, Chang-Chang, On representation types of q-Schur algebras, J. Pure Appl. Algebra, 84(1993), 83–94.

36. Xi, Chang-Chang, Quasi-hereditary of algebras and their factor algebras, Proc. Amer. Math. Soc., 119(3)(1993), 727–729.

37. Xiao, Jie; Guo, Jin-Yun and Zhang, Ying-Bo, Loewy factors of indecomposable modules over self-injective algebras of class An, Science in China (Series)A, 33(8)(1990), 897–908.

38. Xiao, Jie, On indecomposable modules over a representation-finite trivial extension algebra, Science in China (Series)A, 34(2)(1991), 129–137.

39. Xiao, Jie, Artin rings relating to Schur Lemma, Kexue Tongbao, 33(2)(1988), 165–167 (In Chinese).

40. Xiao, Jie, A characterization of representation-finite Artin algebra, Adance in Mathematics (China) (2)(1988), 169–172 (In Chinese).

41. Xiao, Jie, Indecomposable modules over representation-finite self-injective algebras of type D_n (I), Acta Math. Sinica, 32(5)(1989), 659–677.

42. Xiao, Jie, Indecomposable modules over representation-finite selfinjective algebras of type D_n (II), Acta Math. Sinica, 33(2)(1990), 214–232.

43. Xiao, Jie, Projective module over path algebras and their localizations, 12A(1991), 144–148 (In Chinese).

44. Xiao, Jie and Zhang, Pu, One Class of representations over trivial extensions of iterated tilted algebras, Tsukuba J. Math., 17(1) (1993), 131–141.

45. Xiao, Jie and Yang, Ri-Xin, Construction of representation-finite self-injective Artin algebras of class B_n and C_n, Acta Math. Sinica, (New Series) 9(3)(1993), 290–306.

46. Yang, Ri-Xin, and Xiao, Jie, Indecomposable representations over representation-finite algebras of type B_n and C_n, Chinese Ann. of Math., 13A(1992), 76–90. (In Chinese)

47. Zhang, Pu, One class of one-point extensions of tilted algebras (I), Acta Mathematica Sinica, 34(4)(1991). (In Chinese)

48. Zhang, Pu, One class of one-point extensions of tilted algebras (II), Acta Mathematica Sinica, 34(5)(1991). (In Chinese)

49. Zhang, Pu, The structure of Auslander-Reiten sequence of a tilted algebra, Chinese Quarterly Jornal of Mathematics, 3(1992). (In Chinese)

50. Zhang, Pu, Separating tilting modules, Chinese Science Bulletin (Kexue Tongbao), 36(20)(1991), 1524–1526.

51. Zhang, Pu, Endomorphism algebras of preprojective partial tilting modules, Chinese Science Bulletin (Kexue Tongbao), 37(1992), 865–867.

52. Zhang, Pu, Directing components of Auslander-Reiten Quiver,, Science in China (Series A), 9(1992), 897–904.

53. Zhang, Pu, Quasi-hereditary algebra and tilting module, Comm. Algebra, to appear.

54. Zhang, Pu, Triangular truncated rigid algebras, Science in China (A) 24(11)(1994), 1121–1125.

55. Zhang, Ying-Bo, The structure of stable components, Can. J. Math., 43(3)(1991), 652–672.

56. Zhang, Ying-Bo, Eigen values of coxeter transformations and the structure of regular components of an Auslander-Reiten quiver, Comm. Algebra, 17(10)(1989), 2347–2362.

57. Zhang, Ying-Bo, The modules in any component of the AR-quiver of a wild hereditary algebra are unique determined by their composition factors, Archiv Math. 53(1989), 250–251.

58. Zhang, Ying-Bo, Xiao, Jie, Representation theory of algebras: an introduction and survey, Advance of Mathematics, 22(6)(1993), 481–501 (In Chinese).

59. Zhang, Ying-Bo and Lin, Ya-Nan, The bocses corresponding to tame hereditary algebras, preprint.

60. Lin, Ya-Nan and Zhang, Ying-Bo, Some examples of bocses corresponding to tame algebras with one parameter, Bull. of Beijing Normal University, 29(3)(1993), 285–290 (In Chinese).

61. Zhang, Ying-Bo, The structure of the ring with only one unitary indecomposable module. Bull. of Beijing Normal University, 29(1)(1993), 35–37 (In Chinese).

62. Luo, Jiangang, Uniserial modules over commutative artin ring, Doctoral Thesis, Brandis University, 1993.

AR1. M. Auslander and I. Rerten, Modules determined by their composition factors, Illinois J. Math. 29(2)(1985), 280–301.

AR2. M. Auslander and I. Reiten, Applications of contravariantly finite subcategories, Adv. Math. 86(1991), 111–152.

AHR. I. Assem, D. Happel and O. Roldan, Representation-finite trivial extension algebras, J. Pure Appl. Algebra 33(1984), 235–242.

BS. ϕ. Bakke and S.O. Smalϕ, Modules with the same socles and tops as a directing modules are isomorphic, Comm. Algebra 15(1987), 1–9.

BLR. O. Bretscher, C. Läser and C. Riedtmann, Selfinjective and simply connected algebras, Manus. Math. 36(1981), 253–307.

C. C. Cibils, Hochschild homology of an algebra whose quiver has no oriented cycles, Representation Theory I. Ottawa 1985, Springer LNM 1177, 55–59.

CR. W. Crawley-Boevey and C.M. Ringel, Algebras whose Auslander-Reiten components have large regular components, J. Algebra, 153(1992), 494–516.

CPS. E. Cline, B. Parshall and L. Scott, Finite dimensional algebras and highest weight categories, J. reineangew Math. 391(1988), 85–99.

DR1. V. Dlad and C.M. Ringel, Indecomposable representations of graphs and algebras, Mem. Amer. Math. Soc. 173(1976).

DR2. V. Dlab and C.M. Ringel, Quasi-hereditaty algebras, Illinois J. Math. 33(1989), 280–291.

DR3. V. Dlab and C.M.Ringel, Every semiprimary ring is the endomorphism ring of a projective module over a quasi-hereditary ring, Proc. AMS. 107(1989), 1–5.

DR4. V. Dlab and C.M. Ringel, The module theoretic approach to quasi-hereditary algebras, London Math. Soc. 168 "Representations of algebras and related topics', Cambridge Univ. Press (1992), 200–224.

G. P. Gabriel, Unzerlegbare Darstellungen I, Manus. Math. 6(1972), 71–103.

Gr. J.A. Green, Polynomial Representations of GL_n, Springer LNM 830, 1980.

H1. D. Happel, On the derived category of a finite-dimensional algebra, Comment. Math. Helv. 62(1987), 339–389.

H2. D. Happel, Triangulated Categories in the representation theory of finite-dimensional algebras, Cambridge University Press 119, 1988.

H3. D. Happel, Auslander-Reiten triangles in derived categories of finite-dimensional algebras, Proc. AMS 112(1991), 641–648.

H4. D. Happel, Hochschild cohomology of finite-dimensional algebras, Springer LNM 1404(1989), 108–126.

HL. Happel, Dieter and Liu, Shi-Ping, Module categories without short cycles are of finite type, Proc. AMS 120(2)(1994), 371–375.

HR1. D. Happel and C.M. Ringel, Tilted algebras, Trans. AMS 274(1982), 399-443.

HR2. D. Happel and C.M. Ringel Directing modules, preprint.

HPR. D. Happel, U. Preiser and C.M. Ringel, Vinberg's characterization of Dynkin diagrams using subadditive functions with applications to DTr-periodic modules, Springer LNM 832(1980), 280–294.

HW. D. Hughes and J. Waschbüsch, Trivial extension of tilted algebras, Proc. London Math. Soc. 46(3)(1983), 344–364.

I. B. Iversen, Cohomology of Sheaves, Springer, 1986.

K. O. Kerner, Tilting wild algebras, J. London Math. Soc. 39(1989) 29–47.

M1. R. Martinez-Villa, The stable equivalence for algebras of finete representation type, Comm. Algebra 13(1985), 991–1018.

M2. R. Martinez-Villa, Properties that are left inariant under stable equialence, Comm. Algebra 18(1990), 4141–4169.

PS. B. Parshall and L.L. Scott, Derived Categories, Quasi-hereditary algebras and Algebraic groups. Proc. Ottawa-Moosonee Workshop Algebra. Carleton-Ottawa Math. LNS 3(1988), 1–105.

Rm. C. Riedtmann, Darstellungsköcher, Überlagerungen und zurück, Comment. Math. Helv. 55(1980), 199-224.

R1. C.M. Ringel, Tame Algebras and Integral Quadratic Forms, Springer LNM 1099 (1984).

R2. C.M. Ringel, The regular components of the Auslander-Reiten quiver of a tilted algebra, Chinese Ann. Math. B9(1988), 1–18.

R3. C.M. Ringel, Representation theory of finite dimensional algebras, Durham Lectures 1985, London Math. Soc. LNS 116(1986), 7–79.

R4. C.M. Ringel, The category of modules with good filtrations over a quasi-hereditary algebras has almost split sequences, Math. Z. 208(1991), 209–223.

R5. C.M. Ringel, Hall algebras and quantum groups, Invent. Math. 101(1990), 583–592.

R6. C.M. Ringel, Hall algebras, In: Topics in Algebra, Banach Centre Publ. 26(1990), 583–592, Warszawa.

RSS1. I. Reiten, A. Skowroński and S.O. Smalф, short chains and short cycles of modules, Proc. AMS, 117(2)(1993), 343–354.

RSS2. I. Reiten, A. Skowroński and S.O. Smalф, Short chains and regular components, Proc. AMS, 117(3)(1993), 601–612.

Sk. A. Skowronski, Regular Auslander-Reiten components containing directing modules, Proc. AMS, 120(1994), 1–26.

SS. A. Skowronski and S.O. Smalф, Directing modules, J. Algebra, 147(1)(1992), 137–146.

S. H. Strauss, On the perpendicular category of a partial tilting modules, J. Algebra 144(1991), 43–66.

Ta H. Tachikawa, Representations of trivial extensions of hereditary algebras, Springer LNM 832 (1980), 579–599.

TW H. Tachikawa and T. Wakamatsu, Tilting functors and stable equivalences for self-injective algebras, to appear.

Regular Semigroups and Their Generalizations

K.P. SHUM   Department of Mathematics, The Chinese University of
Hong Kong, Shatin, N.T., Hong Kong.

GUO YUQI   Institute of Mathematics, Yunnan University, Kunming,
Yunnan, China.

1   Introduction

An element a of a semigroup S is called regular if $a \in aSa$, that is, if $axa = a$ for
some x in S. A semigroup S is called regular if every element of S is regular. The aim
of this article is to give a survey of recent developments in regular semigroups and its
generalizations. Some research on this topic done in China during the last decade will be
particularly mentioned.

Although vigorous research in semigroups began around 1950, the study of semigroups
theory started rather late in China. Before 1970, semigroup theory was almost a blank
area in China; there were only a few articles published in local mathematical journals. The
systematic study of semigroup theory in China began only around 1971 while K.P. Shum,
H.J. Shyr and Y.Q. Guo almost simultaneously published a series of semigroup articles in
local and internatinal journals. Since then, the research on semigroups bloomed; over 250
articles on combinatorial semigroups; ordered semigroups; topological semigroups; algebraic
semigroups; words; automata and languages etc. have been contributed by a number of
Chinese colleagues during the last two decades. "Semigroup" is no longer a strange name
for the Chinese mathematicians. Among all this research, the theory of regular semigroups
and its generalizations is perhaps most worth mentioning as some interesting results have
been achieved in this topic.

In order to keep this expository article within reasonable bounds, we give only a
brief account of the structure of some important subclasses of regular semigroups; rpp
semigroups and quasi-regular semigroups. Some construction techniques of semigroups will
be introduced in detail. For definitions and terminology not mentioned in this paper, the
reader is referred to the text of A.H. Clifford and G.B. Preston [5] or J.M. Howie [18].

---
*This research is partially supported by a National NSF grant, China; STC & EC grant
of Yunnan Province, China, and a UGC (HK) grant # 221. 600. 370

2   Some Classes of Regular Semigorups

In this section, several important subclasses of the class of regular semigroups are introduced. Relationships between these subclasses of regular semigroups are particularly described.

In history, the notion of "regularity" was introduced by John Von Neumann for rings (See "On regular rings", Proc. Nat. Acad. Sci U.S.A. 22 (1936), 296-300); however, "regularity" nowadays played a much more central role in semigroup theory rather than in ring theory. Profound results on regular semigroups have been obtained since 1965 by many authors. Up to the present moment, generalizations of regular semigroups are still attractive topics for further exploration and investigation.

We first note that if a is a regular element of a semigroup S, then from $axa = a$ for some $x \in S$, we immediately know that ax and xa are idempotent elements of S. Thus, it is trivial to see that a regular semigroup containing a unique idempotent must be a group. Hence, groups are regular semigroups, but the class of regular semigroups is much vaster and more extensive than the class of groups. We cite the following examples to show that there are regular semigroups which are not groups.

Example 2.1   Let $S = \{a, b, c, d\}$ be a semigroup with Cayley table

$\cdot$	a	b	c	d
a	a	b	a	a
b	a	b	a	a
c	a	b	c	c
d	a	b	d	d

Then S is a regular semigroup and is a union of its maximal subgroups. But S itself is clearly not a group.

Example 2.2   Let $S = \{a, b, c, d\}$ with Cayley table

$\cdot$	a	b	c	d
a	a	b	a	b
b	a	b	b	a
c	a	b	c	d
d	a	b	d	c

Then , it is easy to see that S is a regular semigroup but clearly it is not a union of its maximal subgroups.

In view of the above two examples, we introduce the notion of completely regular semigroup.

Definition 2.3 [33] An element a of a semigroup S is called <u>completely regular</u> if there exists x in S such that $a = axa$ and $ax = xa$. A semigroup S is completely regular if all its elements are completely regular.

It is well known that a completely regular semigroup is just the union of its maximal subgroups. Thus the semigroup given in example 2.1 is a completely regular semigroup. However, the semigroup S in example 2.2 is a regular semigroup but not completely regular. This illustrates that the class of completely regular semigroups is a proper subclass of the class of regular semigroups.

Obviously, semigroups containing only idempotents are completely regular. This fact leads to the following definition of bands.

Definition 2.4 [2] By a <u>band</u>, we mean a semigroup S in which every element of S is an idempotent.

Each band is certainly a union of groups since every element consitutes a group in its own right. Thus, the class of bands is a special subclass of completely regular semigroups. Properties of bands have been investigated by a number of authors. In particular, a Rees matrix semigroup $M[G; I, \lambda; P]$ is a band if and only if G is a trivial group [33]. As a consequence, a band is a semilattice of rectangular bands. This result was due to Clifford and Mclean [2] and [28]. A general structure theorem for bands was obtained by M. Petrich in 1967. In his paper [33], Petrich gave a construction for arbitrary band. The techniques of Petrich for the construction of bands has inspired fruitful research on quasi-regular semigroups. We will show its implementation and impact in the later part of this paper.

Definition 2.5 A semigroup S with zero is said to be <u>0-simple</u> if $S^2 \neq 0$ and the only ideals of S are $\{0\}$ and S itself. A semigroup S without zero is said to be simple if S° is 0-simple.

Definition 2.6 An 0-simple (simple) semigroup S is called <u>completely 0-simple (simple)</u> if and only if it contains a <u>primitive</u> idempotent.

The concept of primitive idempotent is taken from ring theory; it is a non-zero idempotent which is minimal in the set of non-zero idempotents with respect to the natural partial ordering assigned on E_S where E_S is the set of all idempotents of a semigroup S. In other words, we define a partial ordering in E_S as follows: for $e, f \in E_S, e \leq f$ if and only if $ef = fe = e$. If S is a semigroup without zero, then the minimal elements of E_S

are the primitive idempotents. It was noted in [42] that prime ideals of semigroups can be determined by regular pairs of idempotents.

The following facts are well known (Ref [33]):

(i)   If S is completely simple (0-simple), then every idempotent of S is primitive.

(ii) If S is a completely 0-simple semigroup then S is regular.

Thus, the class of completely 0-simple (simple) semigroups is a particular subclass of regular semigroups.

The commutativity of idempotents with other elements is always an important property of semigroups. Semigroups with this property were firstly studied by A.H. Clifford [2].

Definition 2.7   A regular semigroup S in which the idempotents are central, i.e., in which $ex = xe$ for every $e^2 = e \in S$ and every $x \in S$ is called a Clifford semigroup.

It has been shown that S is a Clifford semigroup if and only if it is regular and every one-sided ideal of S is a two-sided ideal. Evidently, the idempotents of a commutative regular semigroup are necessarily central.

The properties of idempotents of a semigroup are closely related with the structure of the semigroup. In fact, the structure of a semigroup S is greatly influenced by certain identities defined on E_S.

Definition 2.8   By an orthodox semigroup, we mean a regular semigroup S in which E_S forms a subsemigroup of S, i.e., $(ef)^2 = ef$ for any $e, f \in E_S$.

It was noticed by Reilly and Scheiblich [35] that a regular semigroup S is orthodox if and only if each inverse (in Von Neumann sense) of the idempotent e is still in E_S. Also, it is known that Clifford semigroups are special orthodox semigroups and bands are in particular orthodox [3].

One of the most important subclass of orthodox semigroups is the class of inverse semigroups. Inverse semigroups have been widely studied from the very beginning of semigroup theory. M. Petrich has written a 674 pages monography just on this topic [34]. In fact, inverse semigroups provide rich sources for further research in semigroup theory.

Definition 2.9   A semigroup S is called an inverse semigroup if every $a \in S$ possess a unique inverse, i.e., if there exists a unique element a^{-1} in S such that $aa^{-1}a = a, a^{-1}aa^{-1} = a^{-1}$.

Let $V(a)$ be the set of all inverse elements of a regular semigroup S. Then it is known that S is orthodox if and only if $(\forall a.b \in S)[V(a) \cap V(b) \neq \phi] \Longrightarrow V(a) = V(b)$. (see [32]).

The following result shows the properties of Clifford semigroups.

Theorem 2.10 [34] If S is a regular semigroup, then the following statements are equivalent:

(i) S is a Clifford semigroup;

(ii)   In S, $a = axa$ implies $ax = xa$;

(iii) For any $a \in S, aS = Sa$;

(iv) $\mathcal{L} = \mathcal{R}$.

Also, it is known that S is an inverse semigroup if and only if S is regular and the idempotents of S commute. Hence, an inverse semigroup can be viewed as a special regular semigroup.

Although the theory of inverse semigroups has many features in common with the theory of groups, there are still some major differences. One of the important distinctions is that the non-trivial natural partial ordering exists in every inverse semigroup. The Reilly semigroups, for instance, are bisimple inverse semigroups whose idempotents form a chain isomorphic to the set of negative integers. The most interesting subclass of the class of Reilly semigroups are formed by bicyclic semigroups. Also, in the literature, we usually call a completely 0-simple inverse semigroup a Brandt semigroup (Ref [33]).

The following theorem states the relationship between inverse semigroups and groups.

Theorem 2.11 [34] The following conditons on a semigroup S are equivalent:

(i) S is a group;

(ii)   For every $a \in S$, there is a unique a' such that $a = aa'a$;

(iii) For every $a \in S$, there is a unique a'' such that aa'' is an idempotent;

(iv) S is an inverse semigroup and satisfies the implication $a = axa \Longrightarrow x = xax$.

The following theorem describes inverse semigroups.

Theorem 2.12 [34] The following conditions on a regular semigroup S are equivalent:

(I) S is an inverse semigroup;

(II)   For any $a \in S$ and its inverse a', aa' and $a'a$ commute;

(III)   Every $e \in E_S$ has a unique inverse;

(IV)   Every $e \in E_S$ commutes with all its inverses;

(V)   For any $e \in E_S$ and its inverse e', ee' commutes with $e'e$;

(VI) S has an involution which fixes all idempotents of S;

(VII) S satisfies the implication $a = axa = aya \implies xax = yay$.

The class of left Clifford semigroups has been recently investigated by Zhu, Guo and Shum in 1991 [51]. This class of semigroups is a generalized class of Clifford semigroups. Denote the left Clifford semigroups, for brevity, by left c-semigroups.

Definition 2.13 [51]   A semigroup S is called a left c- semigroup if and only if $aS \subseteq Sa$ for all $a \in S$.

The following is an example of left c-semigroup.

Example 2.14 Let $S = \{e, a, b, c\}$ with Cayley table

$\cdot$	e	a	b	c
e	e	a	b	c
a	a	e	c	b
b	b	b	b	b
c	c	c	c	c

Then it can be seen that S is a left c-semigroup but not a Clifford semigroup, since $bS \subsetneq Sb$.

One can easily observe, by definition, that a left c-semigroup is not necessarily a left regular band. Moreover, a band is not necessarily a left c-semigroup.

Since Clifford semigroups and left regular bands are not compatible under set inclusions, the class of left c-semigroups does admit both the classes of Clifford semigroups and left regular bands as its particular subclasses.

The following theorem is a characterization for left c-semigroups.

Theorem 2.15 [51]   Let S be a left c-semigroup. Then the following conditions are equivalent.

(i) S is a left c-semigroup.

(ii) S is regular and $e \in C(eS)$ (the center of eS) for any $e \in E_S$;

(iii) S is regular and $\mathcal{L} = \mathcal{J}$ is a semilattice congruence for the relations $\mathcal{L}$ and $\mathcal{J}$.

(iv) S is a semilattice of left groups.

In order to compare left c-semigroups and Clifford semigroups, we first list herewith some characterizations of Clifford semigroups for reference.

Theorem 2.16 [18]   The following conditions on a semigroup S are equivalent:

(i) S is a Clifford semigroup;

(ii) S is a semilattice of groups;

(iii) S is regular and $e \in C(S)$ (the center of S) for any $e \in E_S$;

(iv) S is regular and $\mathcal{L} = \mathcal{R}$;

(v) S is regular, and $\mathcal{D}^S \cap (E \cap E) = 1_E$.

Theorem 2.17 [51]   Let S be a semigroup. Then the following statements hold:

(i) S is a Clifford semigroup if and only if S is a left c-semigroup as well as an inverse semigroup.

(ii) S is a left regular band if and only if S is a left c-semigroup as well as a band.

(iii) S is a left c-semigroup if and only if every principal factor of S is a left c-semigroup.

Let E_S be the set of all idempotents of a left C-semigroup S. Then, one can show that S satisfies the following conditions for any $x, y \in S, e \in E_S$.

(α) $E^{(1)}$ − reflexivity : $xy \in E_S$ if and only if $yx \in E_S$.

(β) $E^{(2)}$ − reflexivity : $xey \in E_S$ if and only if $yex \in E_S$.

In the literature, $E^{(2)}$-reflexive inverse semigroups were first tackled by 0' Carroll. He called these inverse semigroups <u>strongly E-reflexive semigroups</u> [30]. It is natural to ask whether we can give a description of the semigroups satisfying both $E^{(1)}$-reflexivity and $E^{(2)}$-reflexivity at the same time. This question is still not yet fully answered.

For the description of the relationships between various subclasses of regular semigroups mentioned in this section and section 4, below a Venn diagram for the purpose visualization.

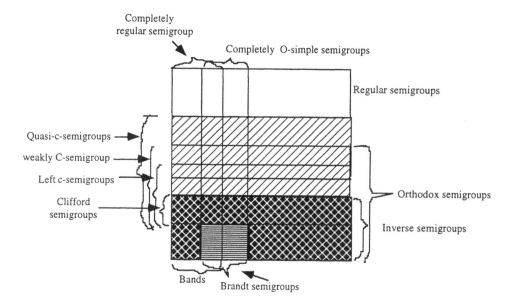

Figure 1

3   Techniques of Constructions

In this section, some methods of construction for certain classes of regular semigroups will be introduced. As is known in theorem 2.16, a Clifford semigroup is a semilattice of groups. It is natural to ask whether we can similarly present some other types of regular semigroups on a semilattice Y. In this connection, we first introduce the concept of <u>strong semilattice</u> considered by M. Petrich [32].

A congruence ρ on a semigroup S is called a <u>semilattice congruence</u> if and only if for all $x, y \in S, xy\rho yx, x^2 \rho x$. In other words, ρ is a semilattice congruence if S/ρ is a semilattice. We then call a semigroup S a semilattice of semigroups belonging to a class $\mathcal{C}$, if there exists a semilattice congruence ρ on S all of whose classes belong to $\mathcal{C}$. We call S a semilattice of semigroups S_α if there exists a homomorphism ϕ from S onto Y such that $S_\alpha = \alpha \phi^{-1}$ for all $\alpha \in Y$. In such case, S is a semilattice $Y = S/\rho$ of semigroups S_α, where $S_{\alpha' s}$ are the ρ-classes $\alpha \in Y$. Thus, for $\alpha, \beta, \in Y$, we have $S_\alpha S_\beta \subseteq S_{\alpha\beta}$ whenever $\alpha\beta \in Y$. Clearly, $S_{\alpha\beta} = S_{\beta\alpha}$.

The concept of strong semliattice of semigroups introduced by Petrich gives a finer picture than semilattice of semigroups because the multiplication of $\cup_{\alpha \in Y} S_\alpha$ can be precisely located. We now illustrate the concept of strong semilattice of semigroups in details.

Suppose that we have a semilattice Y and a family of semigroups S_α indexed by Y, and suppose that, for all $\alpha \geq \beta$ in Y there exists a morphism $\varphi_{\alpha,\beta} : S_\alpha \to S_\beta$ such that

(S1) $(\forall \alpha \in Y), \varphi_{\alpha,\alpha} = 1_{S_\alpha}$;

(S2) $(\forall \alpha, \beta, \gamma \in Y, \ \alpha \geq \beta \geq \gamma), \ \varphi_{\alpha,\beta}\varphi_{\beta,\gamma} = \varphi_{\alpha,\gamma}$.

From the set $S = \cup_{\alpha \in Y} S_\alpha$ and define a multiplication "$*$" in S by $x * y = (x\varphi_{\alpha,\alpha\beta})(y\varphi_{\beta,\alpha\beta})$ for $x \in S_\alpha, y \in S_\beta$. Then by the morphism properties and by the transitivity of (S2), we have for $x \in S_\alpha, y \in S_\beta$ and $z \in S_\gamma$,

$$(x * y) * z = (x\varphi_{\alpha,\alpha\beta}(y\varphi)_{\beta,\alpha\beta}) * z$$

$$\text{and}$$

$$= [(x\varphi_{\alpha,\alpha\beta})(y\varphi_{\beta,\alpha\beta})\varphi_{\alpha\beta,\alpha\beta\gamma}](z\varphi_{\gamma,\alpha\beta\gamma})$$

$$= (x\varphi_{\alpha,\alpha\beta\gamma})(y\varphi_{\beta,\alpha\beta\gamma})(z\varphi_{\gamma,\alpha\beta\gamma}),$$

Similarly, $x * (y * z) = (x\varphi_{\alpha,\alpha\beta\gamma})(y\varphi_{\beta,\alpha\beta\gamma})(z\varphi_{\gamma,\alpha\beta\gamma})$. Thus $(S, *)$ becomes a semigroup, call it a strong semliattice of semigroups S_α and denote it by $S[Y; S_\alpha, \varphi_{\alpha,\beta}]$.

By using the above concept, theorem 2.16 in the previous section can now be strengthened.

Theorem 3.1 [18] A semigroup S is a Clifford semigroup if and only if S is a strong semilattice of groups, that is, $S = [Y; G_\alpha, \varphi_{\alpha,\beta}]$.

We sketch here the proof that a Clifford semigroup S in the expressed in the form $[Y; G_\alpha, \varphi_{\alpha,\beta}]$.

Clearly, $\varphi_{\alpha,\beta}$ is a group homomrphism from $G_\alpha \to G_\beta$. Also, for $\alpha \geq \beta \geq \gamma$, we can check that $\varphi_{\alpha,\beta}\varphi_{\beta,\gamma} = \varphi_{\alpha,\gamma}$. Thus, condition (S1) and (S2) for strong semilattices are satisfied. This shows that $S = [Y; G_\alpha, \varphi_{\alpha,\beta}]$.

In order to build up an analogous result for left c-semigroups, we introduce the concept of $\underline{\xi\text{-product}}$.

First, let $I = \cup_{\alpha \in Y} I_\alpha$ be a semilattice decomposition of a left regular bands I to left zero band I_α. Then, form the product $S_\alpha = T_\alpha \times I_\alpha$, where T_α is a semigroup taken from the strong semilattice $T = [Y; T_\alpha; \varphi_{\alpha,\beta}]$. Consider $S = \cup_{\alpha \in Y} S_\alpha$. Naturally, we aim to make S a semigroup. For this purpose, let $\mathcal{T}_\ell(I)$ be the left transformation semigroup on I. We define a mapping $\xi : S \to \mathcal{T}_\ell(I)$ satisfying the following conditions:

(P1) If $\alpha, \beta \in Y, (a, i) \in S_\alpha, j \in I_\beta$, then $\xi(a, i)j \in I_{\alpha\beta}$;

(P2) If $\alpha, \beta \in Y, \alpha \leq \beta, (a, i) \in S_\alpha, j \in I_\beta$, then $\xi(a, i)j = i$;

(P3) $\xi(a, i)\xi(b, j) = \xi(ab, \xi(a, i)j)$.

By using the above properties of the mapping ξ, we can show, after tedious checking, that $\cup_{\alpha \in Y} S_\alpha$ together with the operation " $*$ ", defined by

$$(a, i) * (b, j) = (ab, \xi(a, i)j)$$

where ab is the semigroup product in T, that is, $ab = a\phi_{\alpha,\alpha\beta}b\phi_{\alpha,\alpha\beta}$, is a semigroup. Denote this semigroup $(S, *)$ by $S = T \times_\xi I$, call it the ξ-product of T and I. It was then proved by Zhu, Guo and Shum [51] that when T is a Clifford semigroup, $S = T \times_\xi I$ is a left c-semigroup.

Hereafter, call $T_S = T = \cup_{\alpha \in Y} G_\alpha$ the $\underline{\text{Clifford component}}$ of S and $I_S = I = \cup_{\alpha \in Y} I_\alpha$ the $\underline{\text{left regular}}$ $\underline{\text{band component}}$ of S.

The following theorem states how to construct a left c-semigroup:

Theorem 3.2 [51]   Any ξ-product $T \times_\xi I$ of a Clifford semigroup $T = [Y; G_\alpha; \phi_{\alpha,\beta}]$ and a left regular band $I = \cup_{\alpha \in Y} I_\alpha$ (the semilattice decomposition to left zero bands I_α) is a left C-semigroup. Conversely, every left c-semigroup S can be constructed by a ξ-product of a Clifford semigroup and a left regular band.

The concept of ξ-product is in fact closely related with the concept of $\underline{\text{spined product}}$. In the literature, spined product was first introduced by N. Kimura [21] in 1958, it has been

The concept of ξ-product is in fact closely related with the concept of spined product. In the literature, spined product was first introduced by N. Kimura [21] in 1958, it has been used to deal with direct product of semigroups having a common homomorphic image. We will point out that for a certain type of left c-semigroups, ξ-product and spined product coincide.

Definition 3.3 Let $\{S_\alpha\}_{\alpha \in \Lambda}$ be a family of semigroups and H a semigroup. Assume that for each $\alpha \in \Lambda$, there exists an homomorphism φ_α of S_α onto H. Let $S = \{(a_\alpha) \in \prod_{\alpha \in \Lambda} S_\alpha | a_\alpha \varphi_\alpha = a_\beta \varphi_\beta\}$ for all $\alpha, \beta \in \Lambda$. Then S is a subsemigroup of the direct product $\prod_{\alpha \in \Lambda} S_\alpha$. We call S the spined product of the semigroups S_α with respect to H and φ_α.

Definition 3.4 [51] A left c-semigroup S is called a strong left c-semigroup if the Green's relation $\mathcal{H}$ on the left c-semigroup S is a congruence.

Remark. A left c-semigroup is not necessarily strong, for instance, in the semigroup given in example 2.14, we have $(e, a) \in \mathcal{H}$ but $(eb, ab) \notin \mathcal{H}$.

The theorem below states that the ξ-product and the spined product of semigroups coincide on strong left c-semigroups.

Theorem 3.5 [51] Let S be a strong left c-semigroup having T_S and I_S as its Clifford component and left regular band component respectively. Then the ξ-product of T_S and I_S is a spined product. On the other hand, given a Clifford semigroup $T = [Y; G_\alpha, \varphi_{\alpha,\beta}]$ and a left regular band I which is composed to the left zero band I_α, that is, $I = \cup_{\alpha \in Y} I_\alpha$. Then, we can form the spined product of T and I with respect to φ and ψ, namely $S = [Y; T, I; \varphi, \psi]$. This semigroup S is a strong left c-semigroup, where φ and ψ are the decomposition homomorphisms of T and I onto the semilattice Y respectively. Moreover, $T = T_S$ and $I = I_S$.

Another useful tool used in the construction of a left c-semigroup is the $\triangle$-product.

To introduce the concept of $\triangle$-product of semigroups, we first consider a semilattice decompositon of a semigroup T to semigroups T_α's on a semlattice Y, that is, $T = \cup_{\alpha \in Y} T_\alpha$. Let $I = \cup_{\alpha \in Y} I_\alpha$ be a partition of the set I (not semigroup!) indexed by Y. Form $S_\alpha = T_\alpha \times I_\alpha$ for each $\alpha \in Y$. If $\alpha \geq \beta$ on Y, define a mapping $\psi_{\alpha,\beta} : S_\alpha \longrightarrow T(I_\beta)$ by $a \longmapsto \psi_{\alpha,\beta}^a$, where $T(I_\beta)$ is the left transformation semigroup on I_β. In addition, we require that $\psi_{\alpha,\beta}$ satisfies the following conditions:

(q_1) if $(\mu, i) \in S_\alpha, i' \in I_\alpha$ then $\psi_{\alpha,\alpha}^{(u,i)} i' = i$;

(q_2) if $(u, i) \in S_\alpha$ and $(v, j) \in S_\beta$, then $\psi_{\alpha,\alpha\beta}^{(u,i)}\psi_{\beta,\alpha\beta}^{(v,j)} =< \psi_{\alpha,\alpha\beta}^{(u,i)}\psi_{\beta,\alpha\beta}^{(v,j)} >$ which is a constant on $I_{\alpha\beta}$;

(q_3) if $(\alpha\beta) \geq \delta$ and $< \psi_{\alpha,\alpha\beta}^{(u,i)}\psi_{\beta,\alpha\beta}^{(v,j)} >= k$, then $\psi_{\alpha\beta,\delta}^{(uv,k)} = \psi_{\alpha\delta}^{(u,i)}\psi_{\beta,\delta}^{(v,j)}$.

Form the set $S = \cup_{\alpha \in Y}S_\alpha$ with an operation "$*$" defined on the set S by $(u, i) * (v, j) = (uv, < \psi_{\alpha,\alpha\beta}^{(u,i)}\psi_{\beta,\alpha\beta}^{(v,j)} >$ for all $(\mu, i) \in S_\alpha, (\nu, j) \in S_\beta$.

Then, we can check that $(S, *)$ is a semigroup. We call $(S, *)$ the $\triangle$-product of T and I, denote this $\triangle$-product of T and I by $T_{\triangle_{Y,\Phi}}I$, where $\Phi = \{\psi_{\alpha,\beta}|\alpha, \beta \in Y, \alpha \geq \beta\}$. This set of mappings Φ is subsequently called the set of structure mappings of $(S, *)$.

Theorem 3.6 [13] Let T be a strong semilattice $[Y; G_\alpha; \varphi_{\alpha,\beta}]$ of groups. Let $I = \cup_{\alpha \in Y}I_\alpha$ be a left regular band which is composed to left zero bands indexed by the semilattice Y. Then the $\triangle$-product of T and I with a set of structure mappings Φ on Y is a left c-semigroup. Conversely, every left c-semigroup can always be constructed by the above $\triangle$- product structure.

Finally, we demonstrate how to construct a left c-semigroup by $\triangle$-product of semigroups.

Step I Let $Y = \{\alpha, \beta, \alpha\beta\}$ be a basic semilattice.

Step II Construct a Clifford semigroup G which is presentable by groups indexed by the semilattice Y. Let $G_\alpha = \{g_0\}, G_\beta = \{b_0, e_0\}$ and $G_{\alpha\beta} = \{c_0, d_0, f_0\}$ be three different groups such that g_0, e_0 and f_0 are the identity elements of the groups G_α, G_β and $G_{\alpha\beta}$ respectively. Also, we require that $b_0^2 = e_0; c_0^2 = d_0, d_0^2 = c_0, c_d d_0 = d_0 c_0 = f_0$. Then by Clifford's theorem, $G = G_\alpha \cup G_\beta \cup G_{\alpha\beta}$ is a Clifford semigroup mounted on Y.

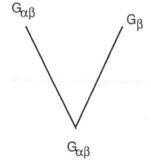

Obviously, the Cayley table of G is of the following form:

	g_0	b_0	e_0	c_0	d_0	f_0
g_0	g_0	f_0	f_0	c_0	d_0	f_0
b_0	f_0	e_0	b_0	c_0	d_0	f_0
e_0	f_0	b_0	e_0	c_0	d_0	f_0
c_0	c_0	c_0	c_0	d_0	f_0	c_0
d_0	d_0	d_0	d_0	f_0	c_0	d_0
f_0	f_0	f_0	f_0	c_0	d_0	f_0

Step III   Let $I_\alpha = \{i, j\}, I_\beta = \{k, l\}, I_{\alpha\beta} = \{m, n\}$ be left zero bands. Form the semigroup $I = \cup_{\alpha \in Y} I_\alpha$.

Step IV   Form the following Cartesian products $S_\alpha = G_\alpha \times I_\alpha, S_\beta = G_\beta \times I_\beta$, and $S_{\alpha\beta} = G_{\alpha\beta} \times I_{\alpha\beta}$. Denote the elements in S_β by $a = (b_0, k), b = (b_0, l), e = (e_0, k)$, and $f = (e_0, l)$; the elements in $S_{\alpha\beta}$ by $c = (c_0, m), d = (c_0, n), x = (d_0, m), y = (d_0, n), u = (f_0, m)$ and $v = f_0, n)$ respectively. Thus, $S_\alpha = \{g, h\}, S_\beta = \{a, b, e, f\}$ and $S_{\alpha\beta} = \{c, d, x, y, u, v\}$. Manacle these semigroups on the semilattice Y as shown below:

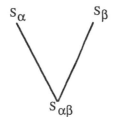

Step V   For any $\gamma, \delta \in Y$ with $\gamma \geq \delta$ and for any $t \in S_\gamma$, define a mapping $\Phi_{\gamma,\delta} : S_\gamma \to \mathcal{J}(I_\delta)$ by $t \to \psi_{\gamma,\delta}^t$. The mapping $\psi_{\gamma,\delta}^t$ is defined by the following:

(i)   For $\alpha \in Y, \psi_{\alpha,\alpha}^t : i' \to i$ for all $t \in S_\alpha$ and $i' \in I_\alpha$.

(ii)   For $\beta \in Y, \psi_{\beta,\beta}^w : i' \to l$ for all $w \in S_\beta$ and $i' \in I_\beta$.

(iii)   For any $i' \in I_{\alpha\beta}, \psi_{\alpha,\alpha\beta}^{(g_0,i)} : i' \to m$, when $(g_0, i) \in S_\alpha$; and $\psi_{\alpha,\alpha\beta}^{(h_0,j)} : i' \to n$ when $(h_0, j) \in S_\alpha$.

(iv)   For any $i' \in I_{\alpha\beta}, \psi_{\beta,\alpha\beta}^{(b_0,k)} = \psi_{\beta,\alpha\beta}^{(e_0,k)} : i' \to m$, when (e_0, k) ;and $(b_0, k) \in S_\beta$; $\psi_{\beta,\alpha\beta}^{(b_0,l)} = \psi_{\beta,\alpha\beta}^{(e_0,l)} : i' \to n$, when (b_0, l) and $(e_0, l) \in S_\beta$.

Then, we can verify that $\psi_{\alpha,\beta}$ satisfies conditions (q1) - (q3).

Step VI Let $S = S_\alpha \cup S_\beta \cup S_\beta$. Then, by the structure mapping $\Phi = \{\psi_{\alpha,\beta} | \alpha, \beta \in Y, \alpha \geq \beta\}$, we can form $S = G_{\triangle Y \cdot \Phi} I$. The Cayley table of $S = G_{\triangle Y \cdot \Phi} I$ can be obtained by Φ:

	g	h	a	b	e	f	c	d	x	y	u	v
g	g	g	u	u	u	u	c	c	x	x	u	u
h	h	h	v	v	v	v	d	d	y	y	v	v
a	u	u	e	e	a	a	c	c	x	x	u	u
b	v	v	f	f	b	b	d	d	y	y	v	v
e	u	u	a	a	e	e	c	c	x	x	u	u
f	v	v	b	b	f	f	d	d	y	y	v	v
c	c	c	c	c	c	c	x	x	u	u	c	c
d	d	d	d	d	d	d	y	y	v	v	d	d
x	x	x	x	x	x	x	u	u	c	c	x	x
y	y	y	y	y	y	y	v	v	d	d	y	y
u	u	u	u	u	u	u	c	c	x	x	u	u
v	v	v	v	v	v	v	d	d	y	y	v	v

The structure map Φ can be expressed by the following figure:

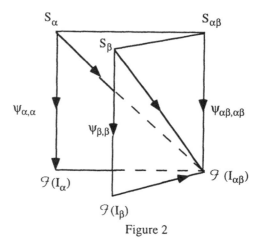

Figure 2

It can be verified that S with the above Cayley table is a left c-semigroup.

Step VII It should be noticed here that $I = \cup_{\alpha \in Y} I_\alpha$, under the structure mapping

Φ, becomes a left regular band. The Cayley table of I is given below:

·	i	j	k	l	m	n
i	i	i	m	m	m	m
j	j	j	n	n	n	n
k	m	m	k	k	m	m
l	n	n	l	l	n	n
m	m	m	m	m	m	m
n	n	n	n	n	n	n

Note (i) In the construction of a semigroup, the most tedieous and laborious work perhaps is the checking of the associative law of the semigroup. However, by using theorm 3.6, we can always construct a left c-semigroup by using the structure mappings. By doing so, there is no need to check the law of associativity.

(ii) It is noticed that left c-semigroups have been generalized further to weakly left c-semigroups in [9] and quasi-c-semigroups in [16]. For the details of these two classes of semigroups, the redear is referred to [9] and [16] for details.

4 Rpp-Semigroups and Semi-spined Products

The study of rpp semigroups is due to Kilp [22] and Fountain [7]. A semigroup S is called rpp if all whose principal right ideals $aS^1(a \in S)$, regarded as right S^1-systems, are projective. It is known that a semigroup S is rpp if and only if for any $a \in S$, the subset $\mathcal{M}_a = \{e \in E | S^1 a \subseteq Se \text{ and for all } x, y \in S^1, ax = ay \implies ex = ey\} \neq \phi$.

An rpp semigroup having all of its idempotents lying in its center is called a c-rpp (that is, Clifford-rpp) semigroup [7]. Obviously, c-rpp semigroups are natural generalizations of Clifford semigroups. Also, it can be easily observed that the class of c-rpp semigroups contains the class of commutative pp semigroups described by Kilp [22] as its subclass. Fountain [7] further characterizes a c-rpp semigroup S by giving a necessary and sufficient condition that S is a strong semilattice of a family of left cancellative monoids.

In this section, we describe the left c-rpp semigroups and state that this class of semigroups includes both the class of c-rpp semigroups and the class of left c-semigroups as its special but independent subclasses.

We start with a set of new Green's relations defined on a given semigroup S, namely, the $\widetilde{Green's}$ relations defined on S (Guo and Trotter call them the (ℓ)-Green's relations in [9]):

Let ρ_a be the <u>inner left translation</u> on S^1 determined by $a \in S$.

Define the following relations on S:

$a\tilde{\mathcal{R}}b$ if $\mathcal{I}m\ \rho_a = \mathcal{I}m\ \rho_b$;

$a\tilde{\mathcal{L}}b$ if $\text{Ker}\ \rho_a = \text{Ker}\ \rho_b$;

$\tilde{\mathcal{H}} = \tilde{\mathcal{R}} \wedge \tilde{\mathcal{L}}, \qquad$ (i.e., $\tilde{\mathcal{R}} \cap \tilde{\mathcal{L}}$)

$\tilde{\mathcal{D}} = \tilde{\mathcal{R}} \vee \tilde{\mathcal{L}}$;

$a\tilde{\mathcal{J}}b$ if $\tilde{\mathcal{J}}(a) = \tilde{\mathcal{J}}(b)$, where $\tilde{\mathcal{J}}(a)$ is the smallest ideal of S containing a and saturating $\mathcal{L}$ (i.e. $\tilde{\mathcal{J}}(a)$ is a union of some $\mathcal{L}$-classes).

It can be observed that $\tilde{\mathcal{R}}$ is just the usual <u>Green's relation</u> $\mathcal{R}$ and $\tilde{\mathcal{L}}$ is just the <u>Green's star</u> <u>relation</u> $\mathcal{L}^*$ described in [46].

We remark here that the $\widetilde{Green's}$ relations on the semigroup S are somewhat analogous to the usual Green's relation on S, for instance, $\tilde{\mathcal{L}}\tilde{\mathcal{R}} = \tilde{\mathcal{R}}\tilde{\mathcal{L}}$. Hence there is an analogy of the "egg box" structure of the usual Green's relations for the $\widetilde{Green's}$ relations. Also, $\tilde{\mathcal{D}} \subseteq \tilde{\mathcal{J}}$.

For handy reference, we give here the following figures which would help us to visualize the relationships between the $\widetilde{Green's}$ relations, the usual Green's relations and the Green$^{*'}s$ relations.

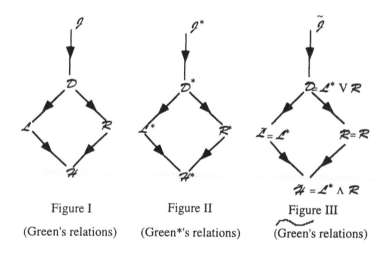

Figure I   Figure II   Figure III

(Green's relations)   (Green*'s relations)   (Green's relations)

We now extend the concept of c-rpp semigroups to left c-rpp semigroups.

First, denote the set of all regular elements of a semigroup S by Reg S. For any $a \in S$, let $M_a = \{e \in E_S | S^1 a \subseteq Se$ and for all $x, y \in S^1, ax = ay \Longrightarrow ex = ey\}$ be a subset of E_S. Call a rpp-semigroup S <u>strongly rpp</u> if there exists a unique idempotent $e \in M_a$ such that $ea = a$. [12]

Definition 4.1 [14] A strong rpp semigroup S satisfying $eS \subseteq Se$ for any $e \in E_S$ is called a left c-rpp semigroup if the following condition (*) is satisfied:

(*) Let $f : a \longmapsto b$ be a <u>1-1 partial left translation</u> acting on $a \in S^1$. Then there exists a $1-1$ partial left translation $g : sa \longmapsto sb$ acting on $sa \in S^1$ which makes the following diagram commute:

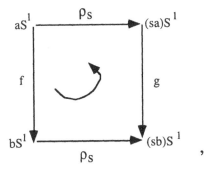

where ρ_s in the diagram is the inner left translation of S^1 determined by $s \in S$.

Note: If the rpp semigroup S in the above definition is replaced by a regular semigroup then the rpp semigroup becomes a left c-rpp semigroup. Since S will turn out to be a completely regular semigroup, S itself, of course, must be a strongly rpp semigroup. Moreover, since $\tilde{\mathcal{L}} = \mathcal{L}$ holds for regular semigroups, every 1-1 partial left translation acting on S^1 must be an inner left translation on S^1 [18]. This explains why the class of left c-rpp semigroup is a generalization of the class of left c-semigroups.

Definition 4.2 [14]   A semigroup S, regarded as a right S-system, is called an <u>0-projective</u> semigroup if $0 \notin S$ or $S = \{0\}$, and S^0 itself is projective.

Theorem 4.3 [14]   The following statements are equivalent for an rpp semigroup S:

(a) S is a left c-rpp semigroup;

(b) S is a strongly rpp semigroup with $\tilde{\mathcal{L}}$ a semilattice congruence on S;

(c) S is a semilattice of $\tilde{L}$-simple, 0-projective and strongly rpp semigroups;

(d) S is a semilattice of direct products of some left cancellative monoids M_α and left zero bands I_α.

From (d), it can be easily observed that a left c-rpp semigroup is not always a c-rpp semigroup. Aslo, since M_α is an arbitrary left cancellative monoid, S is not required to be a left c-semigroup. Thus, the class of left c-semigroups and the class of c-rpp semigroups are two particular subclasses of the class of left c-rpp semigroups. Furthermore, a left c-rpp semigroup also has the following properties:

Theorem 4.4[14] If S is a left c-rpp semigroup, then the following statements hold:

(i) Reg S is a left c-semigroup.

(ii) $\tilde{L} = \tilde{J}$.

The following table sketchs the relationships between the classes $C_c, C_{\ell.c}, C_{c.rpp}, C_{\ell.crpp}$, C_r and C_{rpp} of Clifford semigroups; left c-semigroups; c-rpp semigroups; left c-rpp semigroups; regular semigroups and rpp semigroups respectively.

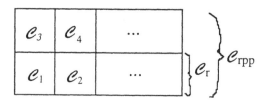

where

$$C_1 = C_c; \ C_1 \cup C_2 = C_{\ell.c}; \ C_1 \cup C_3 = C_{c.rpp}; \ \cup_{i=1}^{4} C_i = C_{\ell.c.rpp}.$$

We now construct an example of class C_4 which can be used to show that there are left c-rpp semigroups that are not regular and not c-rpp.

Example 4.5 [14]   Let $M = \cup_{\alpha \in Y} M_\alpha$ be a strong semilattice of left cancellative monoids M_α (a c-rpp semigroup) with at least one of them not a group and $I = \cup_{\alpha \in Y} I_\alpha$ be a semilattice of left zero bands I_α with at least one of them non-trivial. Form the spined product of M and I with respect to the natural epimorphisms $M \longrightarrow Y$ and $I \longrightarrow Y$. Then the spined product of M and I is a semigroup belonging to the class C_4. In particular, if M is a left cancellative monoid but not a group and I a left zero band with $|I| \geq 2$, then the direct product $M \times I$ is a semigroup in C_4.

We now tackle a generalized concept of spined product of semigroups, namely, the so-called semi-spined product of semigroups.

Let $T = \cup_{\alpha \in Y} T_\alpha$ and $I = \cup_{\alpha \in Y} I_\alpha$ be two semilattice decompositions of the semigroups T and I into its subsemigroups T_α and I_α respectively on a semilattice Y. Construct $S_\alpha = T_\alpha \times I_\alpha$ for each $\alpha \in Y$. Denote $\cup_{\alpha \in Y} S_\alpha$ by S. Define a mapping $\eta : S \to T_\ell(I)$ by $\eta(a,i)j = (a,i)^\sharp j$, where $T_\ell(I)$ is the left transformation semigroup on I. In addition, we require the mapping η to satisfy the following conditions:

(i)   If $(a,i) \in S_\alpha, j \in I_\beta$ then $(a,i)^\sharp j \in I_{\alpha\beta}$;

(ii)   If $(a,i) \in S_\alpha, j \in I_\beta, \alpha \le \beta$, then $(a,i)^\sharp j = ij$, where ij is the semigroup product of i,j in I;

(iii)   If $(a,i) \in S_\alpha, (b,j) \in S_\beta$, then $\eta(a,i) \cdot \eta(b,j) = \eta(ab, (a,i)^\sharp j)$, where ab is the semigroup product of a,b in T.

Define a binary operation "o" on the set $S = \cup_{\alpha \in Y}(T_\alpha \times T_\alpha)$ by $(a,i) \circ (b,j) = (ab, (a,i)^\sharp j)$. It is a matter of routine checking to verify that (S, o) is a semigroup.

Definition 4.6 [14]   The constructed semigroup (S, o) above is called the semi-spined product of the semigroup T and I with respect to η, denote it by $T \times_\eta I$. In fact, η is the semigroup homomorphism from $S = T \times_\eta I$ into $T_\ell(I)$. We call η the semi-spined homomorphism of S.

Note 4.7   When $\eta(a,i) = i_\ell$ for all $(a,i) \in S$, where i_ℓ is the left multiplicative transformation on the semigroup I determined by i, then the semi-spined product of the semigroups T and I, say $T \times_\eta I$, is precisely the spined product of T and I with respect to the natural epimorphisms $T \longrightarrow Y$ and $I \longrightarrow Y$. This explains why the semi-spined product generalizes the spined product.

The following theorem describes how to use semi-spined product to construct left c-rpp semigroups.

Theorem 4.8 [14]   Let $M = [Y; M_\alpha, \varphi_{\alpha,\beta}]$ be a strong semilattice of the c-rpp semigroup M which is decomposed into left cancellative monoids M_α on a semilattice Y. Let $I = \cup_{\alpha \in Y} I_\alpha$ be a semilattice decomposition of a left regular band I into left zero bands I_α on Y as well. Then for any semi-spined homomorphism η, the semi-spined product $S = M \times_\eta I$ is a left c-rpp semigroup. Conversely, any left c-rpp semigroup can be constructed in the above fashion.

We now discuss the relationship between semi-spined products and spined products.

Let $S = \cup_{\alpha \in Y}(M_\alpha \times I_\alpha)$ be a semilattice decomposition of a left c-rpp semigroup S. (Ref. to theorem 4.3(d)). Define a relation $\mathcal{K}$ on S by:

$(a, i)\mathcal{K}(b, j)$ if and only if there exists $\alpha \in Y$ such that $a, b \in M_\alpha$ and $i = j \in I_\alpha$.

Then it is trivial to see that $\mathcal{K}$ is an equivalence relation on S and $\mathcal{H} \subseteq \mathcal{K} \subseteq \tilde{\mathcal{L}}$. Form the semi-spined product $S = M \times_\eta I$ of the C-rpp component $M = \cup_{\alpha \in Y} M_\alpha$ and the left regular band component $I = \cup_{\alpha \in Y} I_\alpha$ together with a semi-spined homomorphism η described in theroem 4.8 such that for some $(a, i) \in M_\alpha \times I_\alpha, (b, j) \in M_\beta \times I_\beta$, $\eta(a, i) : j \to (a, i)^\natural j$. Then, S is a spined product of M and I if and only if $(a, i)^\natural j = ij$ for any $(a, i) \in M_\alpha \times I_\alpha, j \in I_\beta$ and $\alpha, \beta \in Y$, where ij is the semigroup product of i, j in I. Thus, we eventually obtain the following theorem:

Theorem 4.9 [14]   Let $S = \cup_{\alpha \in Y}(M_\alpha \times I_\alpha)$ be a semilattice decomposition of the left c-rpp semigroup described in theorem 4.3(d). Let M_S and I_S be the c-rpp component and the left regular band component of S respectively. Then, the semi-spined product $S = M_S \times_\eta I_S$ is a spined product of M_S and I_S if and only if $\mathcal{K}$ on S is a congruence.

We have also the following special case of left c-rpp semigroups

Theorem 4.10 [12]   Let E be the left regular band which consists of the idempotents of the left c-rpp semigroup S. Then $S = \cup_{\alpha \in Y}(M_\alpha \times I_\alpha)$ is a strong semilattice of the semigroup $M_\alpha \times I_\alpha$ on Y if and only if E is a left <u>normal</u> band, i.e., $efg = egf$ for any $e, f, g \in E$.

We remark that unlike the dual relationship between left c-semigroups and right c-semigroups, the concept of the so-called right c-rpp semigroups defined and investigated by Guo in [10] is not symmetric to the left c-rpp semigroups. The reader is referred to [10] for details.

5   Quasi-regular Semigroups and θ-product

A semigroup S is called <u>quasi-regular</u> if for any $a \in S$ there exists $x \in S$ and a natural number n such that $a^n = a^n x a^n$. Clearly, the class of quasi-regular semigroups contains the class of regular semigroups as it subclass. We describe here how such quasi-regular semigroups can be constructed.

To start with, let the set of regular elements of quasi-regular semigroup S be Reg S. For any $a \in S$, let $r(a) = \min\{n \in N | a^n \in \text{Reg } S\}$. Similar to Clifford semigroup (left c-semigroup), we call a quasi-regular semigroup S a Clifford (left-c)quasi-regular semigroup, if Reg S is an ideal of S and for any $a \in S$, $a^{r(a)}S = Sa^{r(a)}(a^{r(a)}S \subseteq Sa^{r(a)})$. We use $\mathcal{L}^*, \mathcal{R}^*, \mathcal{H}^*$ and $\mathcal{J}^*$ to denote the following equivalence relations on S:

$$a\mathcal{J}^*b \Longrightarrow Sa^{r(a)}S = Sb^{r(b)}S :$$
$$a\mathcal{L}^*b \Longleftrightarrow Sa^{r(a)} = Sb^{r(b)};$$
$$a\mathcal{R}^*b \Longleftrightarrow a^{r(a)}S = b^{r(b)}S;$$
$$\mathcal{H}^* \Longleftrightarrow \mathcal{L}^* \cap \mathcal{R}^*$$

We first deal with the special case when E_S is only an ideal of S.

Definition 5.1 [37] A semigroup S is called E- ideal quasi-regular if S is quasi-regular and E_S is an ideal of S.

Definition 5.2 [37] A semigroup S is called an E-semilattice quasi-regular semigroup if S is quasi-regular as well as E_S is a semilattice and an ideal of S.

The following alike characterizations for E-semilattice quasi-regular semigroups and E-ideal quasi-regular semigroups were obtained by Ren and Guo in [37].

Theorem 5.3 [37] The following statements are equivalent on a semigroup S:

(i) S is an E-semilattice of quasi-regular semigroups;

(ii) S is a semilattice of nil-semigroups such that $ea = ae = a^2e$ for all $a \in S, e \in E_S$;

(iii) S is a torsion semigroup and $ea = ae = a^2e$ for all $a \in S, e \in E_S$;

(iv) $(\forall a \in S)\{\exists m \in N[\forall x \in S \& \forall m' \geq m(a^{m'}x = xa^{m'} = a^{m'}x^2)]\}$;

(v) $(\forall a \in S)\{\exists m \in N[\forall x \in S(a^m x = xa^m = a^m x^2)]\}$;

(vi) S is a torsion semigroup and E_S is a semilattice as well as an ideal of S;

(vii) S is a nil-extension of a semilattice.

Theorem 5.4 [37] The following statements are equivalent on a semigroup S:

(i) S is an E-ideal quasi-reguar semigroup;

(ii) S is a semilattice of nil extensions of rectangular bands and $SeS \supseteq E_S$ for all $e \in E_S$;

(iii) S is a torsion semigroup, and $SaS = \subseteq E_S$ for all $a \in \mathrm{Reg}S$;

(iv) $(\forall\, a \in S)\, \{\exists\, m \in N[\forall\, m' \leq m\ (Sa^{m'}S \subseteq E_S)]\}$;

(v) $(\forall\, a \in S)\, \{\exists\, m \in N\ (Sa^m S \subseteq E_S)\}$;

(vi) S is a torison semigroup and E_S is an ideal of S;

(vii) S is a nil-extension of a band.

Call a semigroup S a <u>left group</u> if for any $a, b \in S$ there exists $x \in S$ such that $xa = b$. According to Petrich [33], a semigroup S is a left group if and only if S is regular and E_S is a left zero band. Because of this result, we call S a <u>quasi left group</u> if and only if S is quasi regular and E_S is a left zero band. Naturally, a quasi regular semigroup S is called a <u>quasi group</u> if $|E_S| = 1$.

Petrich [33] has associated the left groups with bands. Recently, Shum, Ren and Guo have obtained the following characterization for quasi left groups [41].

Theorem 5.5 [41]   Let S be a quasi left group. Then, for some quasi group $\overline{G}$ and left zero band I, $S \cong \overline{G} \times I$ if and only if the following conditions are satisfied:

(i) $\mathcal{H}^*$ is a left zero band congruence on S

(ii)   For any $e \in E_S$, there exists an isomorphism φ_e from H_e^* onto the quasi group $\bar{G}$ such that $x\varphi_e y\varphi_f = (xy)\varphi_e$ whenever $x \in H_e^*$ and $y \in H_f^*$.

In studying the <u>S-system</u> of semigroups, S. Bogdanović [1] introduced the concept of <u>power breaking partial semigroup</u> which turns out to be an important ingredient in the construction of quasi-regular semigroups.

Definition 5.6 [1]   A set Q is called a <u>power breaking partial semigroup</u> if there is a <u>partial binary operation</u> defined on Q such that for any $a, b, c \in Q$, $(ab)c = a(bc)$ whenever $a(bc)$ and $(ab)c$ is in Q; moreover, for any $a \in Q$, there exists an integer $n \in N$ such that $a^n \notin Q$.

If the set E_S is a semilattice, then a mapping φ can be defined from Q into E_S satisfying the following properties:

For every $a, b \in Q$, if $ab \in Q$ then $\varphi(ab) = \varphi(a)\varphi(b)$. Also, define a binary operation "$*$" on $\Sigma = Q \cup^{\bullet} E_S$ (disjoint union) as follows:

(i)   if $a, b \in Q, ab \in Q$ then $a * b = ab$;

(ii) if $a, b \in Q, ab \notin Q$ then $a * b = \varphi(a)\varphi(b)$;

(iii) if $e \in E_S, a \in Q$ then $a * e = e * a = e\varphi(a)$;

(iv) if $e, f \in E_S$ then $e * f = ef$

Denote the above system consisting of Q, E_S, and φ under the operation "$*$" by Σ, that is, $\Sigma = \Sigma(Q, E_S; \varphi, *)$. Then it can be verified that Σ is a semigroup.

The presentation of an E-ideal quasi-regular semigroup is rather complicated as we have to consider the rectangular bands $\{E_\alpha = I_\alpha \times J_\alpha | \alpha \in Y\}$ on the semilattice Y and the left (right) transformation semigroups on the set I, denote it by $\mathcal{T}^*(I)(\mathcal{T}(I))$. The presentation will then be determined by their structure mappings, namely, $\varphi : Q \to \cup_{\alpha \in Y} E_\alpha; \psi_{\alpha,\beta} = \varphi^{-1}(E_\alpha) \to \mathcal{T}^*(I_\beta) \times \mathcal{T}(J_\beta)$ and $\Phi_{\alpha,\beta} : E_\alpha \to \mathcal{T}^*(I_\beta) \times \mathcal{T}(J_\beta)$ for $\alpha \geq \beta$ on Y.

Theorem 5.7 [37]   A semigroup S is an E-ideal quasi-regular semigroup if and only if S is isomorphic to some semigroup of the form $\Sigma = \Sigma(Q, \cup_{\alpha \in Y} E_\alpha; \varphi, \psi, \Phi)$.

We now return to describe the structure of quasi-regular semigroups.

Definition 5.8   A semigroup S is called a quasi-completely regular semigroup if S is a quasi-regular semigroup and Reg S is a subgroup of S.

Definition 5.9   A semigroup S is said to be completely quasi-regular if S itself is quasi-regular and each of its elements, after raising to a certain power, becomes a completely regular element of S. In other words, if $a \in S$, then there exists some $k \in N$ and $x \in S$ such that $a^k = a^k x a^k, a^k x = x a^k$.

It was noticed by Bogdanović [1] that every non-zero idempotent e of a quasi-completely regular semigroup S is $\mathcal{I}^*$-primitive, i.e., e is 0-minimal under the natural partial ordering on E_S the set of all idempotent of S, within the range of a $\mathcal{I}^*$-class. This means that e is the atom under the natural partial ordering $ef = fe = e$ if and only if $e \leq f$ for the idempotents ef on the J^*-class of S. Also, if S is a quasi-regular semigroup with all its idempotents primitive, then S is completely quasi-regular and for any $e \in E_S, G_e = eSe$, where G_e is the maximal subgroup of S generated by e.

The following theorem of quasi-regular semigroups is established.

Theorem 5.10 The following statements are equivalent for a quasi-regular semigroup S:

(i) $E \subseteq C(\text{Reg } S)$, where $C(\text{Reg } S)$ is the center of Reg S;

(ii) S is an orthodox quasi-completely regular semigroup and $\mathcal{I}^*$ is an idempotent-separating congruence on S.

(iii) S is orthodox and $\mathcal{H}^*$ is the least semilattice congruence on S.

(iv) S is orthodox and is a semilattice of quasi-groups.

Definition 5.11[12]   A quasi-regular semigroup satisfying any one of the above statements is called a <u>weakly Clifford quasi-regular</u> semigroup. A weakly Clifford quasi-regular semigroup S is called a <u>Clifford quasi-regular semigroup</u> if Reg S is an ideal of S.

Theorem 5.12 [10]   The following statements are equivalent for a semigroup S:

(i) S is a Clifford quasi-regular semigroup;

(ii) S is a semilattice of quasi-groups and Reg S is an ideal of S;

(iii) S is quasi-regular and $E \subseteq C(S)$ and Reg S is an ideal of S;

(iv) S is quasi-regular, Reg S is an ideal of S and $(\forall a \in S)(a^{r(a)}S = Sa^{r(a)})$;

(v) S is a nil extension of a Clifford semigroup.

It is evident, from (v) that both nil semigroups and Clifford semigroups are two special subclasses of the class of Clifford quasi-regular semigroups; they are respectively non-regular and regular except for the common part of a trivial group.

If S is a semilattice of groups, then it is known that S is orthodox. However, an analogous statement does not hold for semilattice of quasi-groups.

We discribe here the concept of θ-product which is another useful tool in the construction of Clifford quasi-regular semigroups.

Let $T = [Y; T_\alpha, \varphi_{\alpha,\beta}]$ be a strong semilattice of semigroups T_α on a semilattice Y with structure mapping $\varphi_{\alpha,\beta}, \alpha \geq \beta$. Let Q be a power breaking partial semigroup such that

$Q \cap T = \phi$. Define a partial homomorphism $h : Q \to Y$ and denote $h^{-1}(\alpha)$ by Q_α for each $\alpha \in h(Q); Q_\alpha = \phi$ for $\alpha \in Y \backslash h(Q)$. Then form $S_\alpha = Q_\alpha \cup T_\alpha$ for each $\alpha \in Y$.

If $\alpha \geq \beta$, then construct a mapping $\theta_{\alpha,\beta} : S_\alpha \to T_\beta$ satisfying the following conditions:

(T1) $\theta_{\alpha,\beta}|_{T_\alpha} = \varphi_{\alpha,\beta}$;

(T2) if $\alpha \geq \beta \geq \gamma, a \in Q_\alpha$ then $a\theta_{\alpha,\beta}\theta_{\beta,\gamma} = a\theta_{\alpha,\gamma}$;

(T3) if $a \in Q_\alpha, b \in Q_\beta, ab \in Q_{\alpha\beta}$ then $(ab)\theta_{\alpha\beta,\gamma} = a\theta_{\alpha,\gamma}b\theta_{\beta,\gamma}$ when $\alpha\beta \geq \gamma$.

Now, we can check that $S = \cup_{\alpha \in Y}S_\alpha$ forms a semigroup if S is endowed with the following binary opertion "$*$" defined below:

For any $a \in S_\alpha, b \in S_\beta$, we consider the following cases separately:

(F1)   if $a \in Q, b \in Q, ab \in Q$ then $a * b = ab$;

(F2) $a * b = a\theta_{\alpha,\alpha\beta}b\theta_{\beta,\alpha\beta}$, otherwise.

Then, the semigroup $(S, *)$ is called the θ-product of T and Q, denote it by $T \cup_\theta Q$.

Theorem 5.13 [12]   Let $T = [Y; T_\alpha, \varphi_{\alpha,\beta}]$ be a strong semilattice of groups G_α and Q a power breaking partial semigroup. Then the θ-product of T and Q, namely $S = T \cup_\theta Q$ is a Clifford quasi-regular semigroup. Conversely, every Clifford quasi-regular semigroup is constructible by using the θ-product of the semigroups T and Q.

We now construct a Clifford quasi-regular semigroup by θ-product. After studying the following example, we will have a better understanding of Clifford quasi-regular semigroup described in theorem 5.13.

Example 5.14

Step I   Let $G_\alpha = \{e\}, G_\beta = \{f, c | fc = cf = c, c^2 = f\}, G_{\alpha\beta} = \{g, x, y, z | yx = xy = g, x^2 = g, y^2 = z, z^2 = g, g^2 = g, xz = zx = yz = y, zy = x\}$ be groups mounted on a semilattice $Y = \{\alpha, \beta, \alpha\beta\}$. Write $T = \cup_{\lambda \in Y}G_\lambda$.

Step II Let Q be a power breaking partial semigroup given below:

	a	a^2	b	b^2	d	d^2	w
a	a^2	·	a^2	·	·	·	·
a^2	·	·	·	·	·	·	·
b	b^2	·	d^2	·	·	·	·
b^2	·	·	·	·	·	·	·
d	·	·	·	·	d^2	·	·
d^2	·	·	·	·	·	·	·
w	·	··	·	·	·	·	·

Step III Let h be a partial homomorphism that maps from the power breaking partial semigroup Q to the semilattice Y. Denote $Q_\alpha = h^{-1}(\alpha) = \{a, a^2, b, b^2\}$; $Q_\beta = h^{-1}(\beta) = \{d, d^2\}$ and $Q_{\alpha\beta} = h^{-1}(\alpha\beta) = \{w\}$ respectively. Mount Q_α, Q_β and $Q_{\alpha\beta}$ on Y. Then we have the following diagram:

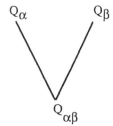

Step IV Let $S_\delta = G_\delta \cup Q_\delta$ for each $\delta \in Y$. Then, we obtain the following semigroups: $S_\alpha = G_\alpha \cup Q_\alpha = \{a, a^2, b, b^2, e\}$; $S_\beta = G_\beta \cup Q_\beta = \{d, d^2, c, f\}$ and $S_{\alpha\beta} = G_{\alpha\beta} \cup Q_{\alpha\beta} = \{w, g, x, y, z\}$.

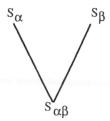

Step VI For any $\alpha, \beta \in Y$ with $\alpha \geq \beta$, construct the following mapping $\theta_{\alpha,\beta} : S_\alpha \to G_\beta$. Thus, the following cases arise:

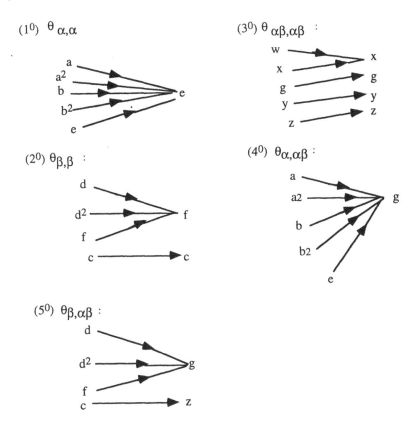

The picture of the mappings from S_α onto G_α, for all $\alpha \in Y$, is displayed below:

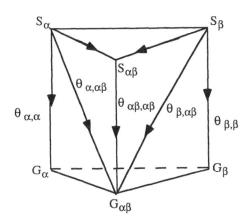

It can be verified that the above mappings all satisfy the required conditions (T1)-(T3) as stated above.

Step VII   Form the θ-product $S = T \cup_\theta Q$, under the mapping θ, the Cayley table of S is now obtained by using the conditions (S1) and (S2). The Cayley table of S is shown below:

	a	a^2	b	b^2	e	d	d^2	c	f	w	g	x	y	z
a	a^2	e	a^2	e	e	g	g	z	g	x	g	x	y	z
a^2	e	e	e	e	e	g	g	z	g	x	g	x	y	z
b	b^2	e	b^2	e	e	y	g	z	g	x	g	x	y	z
b^2	e	e	e	e	e	g	g	z	g	x	g	x	y	z
e	e	e	e	e	e	g	g	z	g	x	g	x	y	z
d	g	g	g	g	g	d^2	f	c	f	x	g	x	y	z
d^2	g	g	g	g	g	f	f	c	f	x	g	x	y	z
c	z	z	z	z	z	c	c	f	c	y	z	y	x	g
f	g	g	g	g	g	f	f	c	f	x	g	x	y	z
w	x	x	x	x	x	x	x	y	x	g	x	g	z	y
g	g	g	g	g	g	g	g	z	g	x	g	x	y	z
x	x	x	x	x	x	x	x	y	x	g	x	g	z	y
y	y	y	y	y	y	y	y	x	g	x	g	g	z	y
z	z	z	z	z	z	z	z	g	z	y	z	y	x	g

It can be easily checked that, in the above table, S is a Clifford quasi-regular semigroup. Again, the most striking thing in the above example is that we have no need to spend time to verify the associative law of multiplication of S! This illustrates that the structure mapping $\theta_{\alpha,\beta}$ is indeed an useful tool in the construction of the Clifford quasiregular semigroups.

6   Generalized Left $\triangle$- Products

The class of left c-quasiregular semigroups contains both the classes of left c-semigroups and c-quasiregular semigroups as its proper subclasses. This type of semigroups has been studied in [38]. In this section, we describe the construction of such semigroups.

Definition 6.1   By a left c-quasiregular semigroup, we mean a quasi-regular semigroup S in which Reg S is both a left c-semigroup and is an ideal of S. In other words, we have $a^{r(a)}S \subseteq Sa^{r(a)}$ for all elements a in S and Reg S is an ideal of S.

The following is an example of left c-quasiregular semigroup.

Example 6.2 Let $S = \{a, b, c, d\}$ with Cayley table

	a	b	c	d
a	a	a	a	a
b	a	a	a	a
c	a	a	a	a
d	a	a	b	b

Then S is a left c-quasiregular semigroup because Reg $S = \{a\}$ is an ideal of $S, aS = Sa; b^2S = Sb^2; c^2S = Sc^2; d^3S = Sd^3$; moreover $bS = Sb; dS = Sd$, but $cS \overset{c}{\neq} Sc$.

Examples of c-quasiregular semigroups which are not necessarily left c-semigroups and examples of left c-semigroups which are not necessarily c-quasiregular semigroups do exist. These examples can be found in [38]. Thus, left c-semigroups and c-quasiregular semigroups are different classes of left c-quasiregular semigroups. It is trivial to see that the class of left c-quasiregular semigroups is in fact a common generalization of the class of left c-semigroups and the class of c-quasiregular semigroups.

To facilitate further study of the left c-quasiregular semigroups, the following two results are rather useful.

Lemma 6.3 [39] The following statements on a semigroup S are equivalent:

(i) S is a quasiregular semigroup in which E_S is a left zero band;

(ii) S is a nil-extension of a left group;

(iii) S is a left Archimedean semigroup with $E_S \neq \phi$.

Lemma 6.4 [51] Let S be an orthodox semigroup. Then the following conditions are equivalent:

(i) S is a left c-semigroup;

(ii) $eS \subseteq Se$ for all $e \in E_S$;

(iii) S is a semilattice of left groups;

(iv) $\mathcal{L}(= \mathcal{T})$ is a semilattice congruence on S;

(v) S is a left inverse semigroup which is also completely regular.

By using the modified versions of the above results, Ren, Guo and Shum gave the following characterization for left c-quasiregular semigroups [38].

Theorem 6.5 [38]   The following statements are equivalent for a semigroup S.

(i) S is a left c-quasiregular semigroup;

(ii) S is quasiregular such that Reg S is both an ideal and a left c-subsemigroup of S;

(iii) S is completely quasiregular such that Reg S is an ideal as well as an orthodox subsemigroup of S. Moreover, $\mathcal{L}^*$ is a semilattice congruence on S;

(iv) S is a semilattice of quasi-left groups. Moreover, Reg S is an ideal of S as well as an orthodox subsemigroup of S;

(v) S is a quasiregular semigroup such that $eS \subseteq Se \subseteq$ Reg S for all $e \in E$;

(vi) S is a quasiregular and for all $a \in S$, there exists some integers $m, n \in N$ such that $a^m S \subseteq Sa^n \subseteq$ Reg S;

(vii) S is a nil-extension of a left c-semigroup.

We now introduce the concept of left generalized $\triangle$-product which is weaker than the $\triangle$-product described in our last section. We will demonstrate here how a left c-quasiregular semigroup can be constructed by using the left generalized $\triangle$-product of a left a c-semigroup T and a power breaking partial semigroup Q. Although the definition of left generalized product of semigroups looks rather complicated, the ideas are more or less parallel to the $\triangle$-product described in §5, only the new ingredient Q is added into it.

Recall that a mapping θ from a power breaking partial semigroup Q into another semigroup is a partial homomorphism [1] if $(ab)\theta = a\theta b\theta$ whenever a, b and $ab \in Q$. Let θ be a partial homomorphism from the power breaking partial semigroup Q into a semilattice Y.

We now proceed our construction via the following steps:

(A) For every $\alpha \in Y$, let $Q_\alpha = \alpha\theta^{-1}$ if $\alpha \in Q\theta$ and write $Q_\alpha = \phi$ if $\alpha \in Y \backslash Q\theta$.

(B) Let $T = [Y; T_\alpha; \varphi_{\alpha,\beta}]$ be a strong semilattice of semigroups T_α. Also, let $I = \cup_{\alpha \in Y} I_\alpha$ be a semilattice partition of the set I on the semilattice Y.

(C) For every $\alpha \in Y$, form the following three subsets of the semigroup S_α, namely, $S_\alpha^{(1)} = T_\alpha \times I_\alpha$; $S_\alpha^{(2)} = Q_\alpha \cup T_\alpha$ and $S_\alpha^{(3)} = Q_\alpha \cup S_\alpha^{(1)}$, where $S = S_\alpha^{(1)} \cup S_\alpha^{(2)} \cup S_\alpha^{(3)}$.

(D) Denote the left transformation semigroup on the set A by $\mathcal{T}(A)$. Then for any $\alpha, \beta \in Y$ with $\alpha \geq \beta$, define the following mappings:

   (a) $\Phi_{\alpha,\beta}^{(1)} : S_\alpha^{(1)} \longmapsto \mathcal{T}(I_\beta)$; by $(u,i) \longmapsto \psi_{\alpha,\beta}^{(u,i)}$;

   (b) $\Phi_{\alpha,\beta}^{(2)} : S_\alpha^{(2)} \longmapsto \mathcal{T}(I_\beta)$ by $a \longmapsto a\theta_{\alpha,\beta}$;

   (c) $\Phi_{\alpha,\beta}^{(3)} : Q_\alpha \longmapsto \mathcal{T}(I_\beta)$ by $a \longmapsto \psi_{\alpha,\beta}^a$.

(E) Impose the following requirements on the above mappings:

   (c1) If $(u,i) \in S_\alpha^{(1)} = T_\alpha \times I_\alpha, i^* \in I_\alpha$, then $\psi_{\alpha,\alpha}^{(u,i)} i^* = i$;

   (c2) If $\alpha, \beta \in Y, a \in S_\alpha, b \in S_\beta$ and if the product ab is not in Q, then $\psi_{\alpha,\alpha\beta}^a \psi_{\beta,\alpha\beta}^b$ is a constant map acting on $I_{\alpha\beta}$. Denote the constant value of this mapping by $\langle (\psi_{\alpha,\alpha\beta}^a \psi_{\beta,\alpha\beta}^b) \rangle$;

   (c3)

      (i) $\Phi_{\alpha,\beta}^{(2)}|_{T_\alpha} = \varphi_{\alpha,\beta}$;

      (ii) Let $\alpha, \beta, \gamma \in Y$ with $\alpha \geq \beta \geq \gamma$. Then $a\theta_{\alpha,\beta} \cdot \theta_{\beta,\gamma} = a\theta_{\alpha,\gamma}$ for $a \in Q_\alpha$.

   (c4) Let $\alpha, \beta, \delta \in Y$ with $\alpha\beta \geq \delta$. Consider the following situations:

      (i) If $a \in Q_\alpha, b \in Q_\beta$ and $ab \in Q_{\alpha\beta}$, then

         (a) $(ab)\theta_{\alpha\beta,\delta} = a\theta_{\alpha,\delta}b\theta_{\beta,\delta}$;

         (b) $\psi_{\alpha\beta,\delta}^{ab} = \psi_{\alpha,\delta}^a \psi_{\beta,\delta}^b$.

      (ii) If $a \in Q_\alpha, b \in Q_\beta$ and $ab \notin Q_{\alpha\beta}$, then write $u = a\theta_{\beta,\alpha\beta}b\theta_{\beta,\alpha\beta}$ and $i = \langle \psi_{\alpha,\alpha\beta}^a \psi_{\beta,\alpha\beta}^b \rangle$. Also, write $\psi_{\alpha,\delta}^a \psi_{\beta,\delta}^b = \psi_{\alpha\beta,\delta}^{(u,i)}$.

      (iii) If $a \in Q_\alpha, (u,j) \in S_\beta^{(1)}$, then write $u = a\theta_{\beta,\alpha\beta}u\theta_{\beta,\alpha\beta}; i = \langle \psi_{\alpha,\alpha\beta}^a \psi_{\beta,\alpha\beta}^{(u,j)} \rangle; \overline{u} = v\theta_{\beta,\alpha\beta}a\theta_{\alpha,\alpha\beta}$ and $\overline{i} = \langle \psi_{\beta,\alpha\beta}^{(u,j)} \psi_{alpha,beta}^a \rangle$. Denote $\psi_{\alpha\beta,\delta}^{(u,i)} = \psi_{\alpha,\delta}^a \psi_{\beta,\delta}^{(v,j)}; \psi_{\alpha\beta,\delta}^{(\overline{u},i)} = \psi_{\beta,\delta}^{(u,j)} \psi_{\alpha,\delta}^a$.

      (iv) If $(u,i) \in S_\alpha^{(1)}$ and $(v,j) \in S_\beta^{(1)}$, then let $k = \langle \psi_{\alpha,\alpha\beta}^{(u,i)} \psi_{\beta,\alpha,\beta}^{(v,j)} \rangle$ and $\psi_{\alpha,\delta}^{(u,i)} \psi_{\beta,\delta}^{(v,j)} = \psi_{\alpha\beta,\alpha}^{(uv,kk)}$, where uv is the semigroup product in T.

Thus, all the possible cases have been considered and have been treated.

(F) Form the triple $\Phi = \{\Phi_{\alpha,\beta}^{(1)}; \Phi_{\alpha,\beta}^{(2)}; \Phi_{\alpha,\beta}^{(3)}; \alpha \geq \beta, \alpha, \beta \in Y\}$. We call this Φ, the three-phase triple, the structure mapping of S_α into S_β.

(G) Let $S = \cup_{\alpha \in Y} S_\alpha$. Define a binary operation "$*$" on S as follows:

   (M1) If $a \in Q_\alpha, b \in Q_\beta$ and $ab \in Q_{\alpha\beta}$, then define $a * b = ab$;

   (M2) If $a \in Q_\alpha, b \in Q_\beta$ and $ab \notin Q_{\alpha\beta}$, then define $a * b = (a\theta_{\alpha,\alpha\beta}b\theta_{\beta,\alpha\beta}, \langle \psi_{\alpha,\alpha\beta}^a \psi_{\beta,\alpha\beta}^b \rangle)$;

(M3)   If $a \in Q_\alpha, (u, j) \in S_\beta^{(1)}$, then define
$$a * (v, j) = (a\theta_{\alpha,\alpha\beta} v \theta_{\beta,\alpha\beta}, \langle \psi_{\alpha,\alpha\beta}^a \psi_{\beta,\alpha\beta}^{(v,j)} \rangle);$$
$$(v, j) * a = (v\theta_{\beta,\alpha\beta} a \theta_{\alpha,\alpha\beta}, \langle \psi_{\beta,\alpha\beta}^{(u,j)} \psi_{\alpha,\alpha\beta}^a \rangle)$$

(M4)   If $(u, i) \in S_\alpha^{(1)}, (v, j) \in S_\beta^{(1)}$, then define
$$(u, i) * (v, j) = (u\theta_{\alpha,\alpha\beta} v \theta_{\beta,\alpha\beta}, \langle \psi_{\alpha,\alpha\beta}^{(u,i)} \psi_{\beta,\alpha\beta}^{(u,j)} \rangle) = (uv, < \psi_{\alpha,\alpha\beta}^{(u,i)} \psi_{\beta,\alpha\beta}^{(v,j)} >)$$

It can be easily verified, after tedious calculations and routine checking, that the operation "$*$" is indeed associative on $S = \cup_{\alpha \in Y} S_\alpha$. Thus, $(S, *)$ is a semigroup.

Definition 6.6 [38]   The semigroup $(S, *)$ constructed above is called the left generalized $\triangle$-product of the partial semigroup Q, the semigroup T and the set I with respect to the semilattice Y together with the structure mapping Φ defined above. Denote this left generalized $\triangle$-product by $S = S(Q, T, I, Y; \Phi)$.

With the aid of the above definition, we can describe the left c-quasiregular semigroups via the left generalized $\triangle$-products. The following theorem is hence established.

Theroem 6.7 [38]   Let Y be a semilattice; Q a power breaking partial semigroup; $T = [Y; G_\alpha; \varphi_{\alpha,\beta}]$ a strong semilattice of groups G_α; $I = \cup_{\alpha \in Y} I_\alpha$ a left regular band expressed by a semilattice of left zero bands I_α on Y and Φ a three phase structure mapping. Then the constructed left generalized $\triangle$-product $S = S(Q, T, I, Y; \Phi)$ is a left c-quasiregular semigroup. Conversely, every left c-quasiregular semigroup can be expressed by a left generalized $\triangle$-product $S(Q, T, I, Y; \Phi)$ described above.

The most difficult part of the above theorem is the sufficiency part. The crucial point is how to construct a structure morphism Φ for $S(Q, T, I, Y; \Phi)$. We sketch the procedures of construction as followings:

Step I : To construct a mapping $\Phi_{\alpha,\beta}^{(2)}$ from $S_\alpha^{(2)}$ to G_β. Since $S_\alpha^{(2)} = Q_\alpha \cup G_\alpha$, so by Lemma 2.5, Reg $S = \cup_{\alpha \in Y} S_\alpha^{(1)}$. For $\alpha, \beta \in Y$ with $\alpha \geq \beta$, consider the following situations:

(i)   Suppose $a \in Q_\alpha, (e_\beta, j) \in S_\beta^{(1)} \cap E$, where e_β is the identity of the group G_β. Then $(e_\beta, j)a \in S_\beta^{(1)} = S_\beta \cap \text{Reg } S$, say $(e_\beta, j)a = (u', j')$ for some $u' \in G_\beta, j' \in I_\beta$. Multiplying (e_β, j) on the left hand side of the above equation, since I is a left zero band, we have

$$(e_\beta, j)a = (e_\beta, j)(u', j')$$
$$= (u', jj') = (u', j). \qquad \ldots\ldots\ldots(1)$$

This means that (1) determines $\theta_{\alpha,\beta} : a \longmapsto a\theta_{\alpha,\beta}$ from Q_α to G_β such that

$$(e_\beta, j)a = (a\theta_{\alpha,\beta}, j). \qquad \ldots\ldots\ldots(2)$$

(ii)   Suppose $a = (u,i) \in S_\alpha^{(1)}$ and $(e_\beta, j) \in S_\beta^{(1)} \cap E$. Then

$$(e_\beta, j)(u, i) = (u'', j) \in S_\beta^{(1)}. \quad \ldots\ldots\ldots(3)$$

On the other hand, if $(e_\alpha, i) \in S_\alpha^{(1)} \cap E$ and $i' \in I_\alpha$, then

$$\begin{aligned}
(e_\beta, j)(u, i') &= [(e_\beta, j)(e_\alpha, i)](u, i') \\
&= (e_\beta, j)[(e_\alpha, i)(u, i')] \\
&= (e_\beta, j)(u, i). \quad \ldots\ldots\ldots\ldots(4)
\end{aligned}$$

Using (3) and (4), we then obtain a mapping $\theta_{\alpha,\beta} : u \longmapsto u\theta_{\alpha,\beta}$ from G_α into G_β:

$$(e_\beta, j)(u, i) = (u\theta_{\alpha,\beta}, j). \quad \ldots\ldots\ldots(5)$$

Combining (i) and (ii), we define a mapping $\Phi_{\alpha,\beta}^{(2)}$ from $S_\alpha^{(2)} = Q_\alpha \cup G_\alpha$ into G_β by $\Phi_{\alpha,\beta}^{(2)} : a \longmapsto a\theta_{\alpha,\beta}$.

Step II.   Now, we verify that $\Phi_{\alpha,\beta}^{(2)}$ satisfies conditions (C3) and (C4).

We simply choose any $(e_\beta, j) \in S_\beta^{(1)} \cap E$ and any $(e_\gamma, k) \in S_\gamma^{(1)} \cap E$ with $\beta \geq \gamma$. Then, since E is a subsemigroup, $(e_\gamma, k)(e_\beta, j) = (e_\gamma, k) \in S_\gamma^{(1)}$. Also, by (2) and (5) in step I, we have, for any $a \in Q_\alpha, \alpha \geq \beta \geq \gamma$,

$$(e_\gamma, k)[(e_\beta, j)a] = (e_\gamma, k)(a\theta_{\alpha,\beta}, j) = (a\theta_{\alpha,\beta}\theta_{\beta,\gamma}, k) \text{ and } (e_\gamma, k)a = (a\theta_{\alpha,\gamma}, k).$$

Hence, $a\theta_{\alpha,\beta}\theta_{\beta,\gamma} = a\theta_{\alpha,\gamma}$. This shows that $\Phi_{\alpha,\beta}^{(2)}$ satisfies condition (C3) (ii).

Likewise, we can show that for any $(u, i) \in S_\alpha^{(1)}, \alpha \geq \beta \geq \gamma$,

$$u\theta_{\alpha,\beta}\theta_{\beta,\gamma} = u\theta_{\alpha,\gamma}. \quad \ldots\ldots\ldots(6)$$

Assume that $a \in Q_\alpha, b \in Q_\beta$ and $ab \in Q_{\alpha\beta}$ such that $(e_\delta, l) \in S_\delta^{(1)} \cap E$ when $\alpha\beta \geq \delta$. Then, applying (2) and (5), we get

$$(e_\delta, l)ab = [(ab)\theta_{\alpha\beta,\delta}, l]$$

and

$$\begin{aligned}
[(e_\delta, l)a]b &= (a\theta_{\alpha,\delta}, l)(e_\delta, l)b \\
&= (a\theta_{\alpha,\delta}, l)(b\theta_{\beta,\delta}, l) \\
&= (a\theta_{\alpha,\delta}b\theta_{\beta,\delta}, l).
\end{aligned}$$

This implies that $(ab)\theta_{\alpha\beta,\delta} = a\theta_{\alpha,\delta}b\theta_{\beta,\delta}$. Hence, condition (C4) (i) (a) is satisfied by $\Phi_\alpha^{(2)}$. Moreover, if (u,i) and $(u',i') \in S_\alpha^{(1)}, \alpha \geq \beta$, then we can similarly verify that

$$(uu')\theta_{\alpha,\beta} = u\theta_{\alpha,\beta} \cdot u'\theta_{\alpha,\beta}. \qquad ....(7)$$

Applying (4), we immediately obtain

$$u\theta_{\alpha,\alpha} = u. \qquad ....(8).$$

Denote $\Phi_{\alpha,\beta}^{(2)}|_{G_\alpha}$ by $\varphi_{\alpha,\beta}$. Then it is easy to see that $\varphi_{\alpha,\beta}$ is a structure homomorphism of the strong semilattice $T = [Y; G_\alpha; \varphi_{\alpha,\beta}]$. Hence, condition (C3) (i) is verified.

Step III.   Define a mapping $\Phi_{\alpha,\beta}^{(3)}$ from Q_α to $T(I_\beta)$.

For this purpose, we pick $a \in Q_\alpha$ and any $(e_\beta, j) \in S_\beta^{(1)} \cap E$ with $\alpha \geq \beta$. Write $a(e_\beta, j) \in S_\beta^{(1)}$ by $a(e_\beta, j) = (v, k)$. Multiplying by (e_β, k) from the right, we obtain $a(e_\beta, j)(e_\beta, k) = (v, k)(e_\beta, k)$. This shows that

$$a(e_\beta, j) = (a\theta_{\alpha,\beta}, k) \in S_\beta^{(1)}. \qquad ....(9)$$

In other words, the equality (9) determines a mapping $\Phi_{\alpha,\beta}^{(3)} : a \longmapsto \psi_{\alpha,\beta}^a$ from Q_α to $T(I_\beta)$ if $k = \psi_{\alpha,\beta}^a j$; that is,

$$a(e_\beta, j) = (a\theta_{\alpha,\beta}, \ \psi_{\alpha,\beta}^a j). \qquad ....(10)$$

Clearly, it can be seen that $\Phi_{\alpha,\beta}^{(3)}$ is the required mapping.

Step IV.   To define a mapping $\Phi_{\alpha,\beta}^{(1)}$ from $S_\alpha^{(1)}$ to $T(I_\beta)$.

This can be easily done by picking any $(u,i) \in S_\alpha^{(1)}$. Then, by using the arguments similar to our Step III, we can define $\Phi_{\alpha,\beta}^{(1)} : (u,i) \longmapsto \psi_{\alpha,\beta}^{(u,i)}$ from $S_\alpha^{(1)}$ to $T(I_\beta)$ by

$$(u,i)(e_\beta, j) = (u\theta_{\alpha,\beta}, \psi_{\alpha,\beta}^{(u,i)} \cdot j). \qquad ....(11)$$

Observe that if $(u,i) \in S_\alpha^{(1)}, i' \in I_\alpha$, then, by (8) and (11), we have

$$(u,i)(e_\alpha, i') = (u\theta_{\alpha,\alpha}, \psi_{\alpha,\alpha}^{(u,i)} i')$$
$$= (u, \psi_{\alpha,\alpha}^{(u,i)} i') \qquad ....(12).$$

As $(u,i)(e_\alpha, i') = (u,i)$, we get $\psi_{\alpha,\alpha}^{(u,i)} i' = i$. Thus, $\Phi_{\alpha,\beta}^{(1)}$ satisfies (C1).

Step V. To verify that (C2) holds, we need to verify that if $\alpha, \beta \in Y, a \in S_\alpha, b \in S_\beta$ and $ab \notin Q_{\alpha\beta}$, then $\psi^a_{\alpha,\alpha\beta} \psi^b_{\beta,\alpha\beta}$ is a constant mapping with constant value $\langle \psi^a_{\alpha,\alpha\beta} \psi^b_{\beta,\alpha\beta} \rangle$.

For this purpose, take $ab = (u, i)$. Since $ab \notin Q_{\alpha\beta}$, we have $ab = (u, i) \in S^{(1)}_{\alpha,\beta}$. If $(e_{\alpha\beta}, k) \in S^{(1)}_{\alpha\beta} \cap E$, then by (10), we have

$$a[b(e_{\alpha\beta}, k)] = a(b\theta_{\beta,\alpha\beta}, \psi^b_{\beta,\alpha\beta} k)$$
$$= a(e_{\alpha\beta}, \psi^b_{\beta,\alpha\beta} k)(b\theta_{\beta,\alpha\beta}, \psi^b_{\beta,\alpha\beta} k)$$
$$= (a\theta_{\alpha,\alpha\beta}, \psi^a_{\alpha,\alpha\beta} \psi^b_{\beta,\alpha\beta} k)(b\theta_{\beta,\alpha\beta}, \psi^b_{\beta,\alpha\beta} k)$$
$$= (a\theta_{\alpha,\alpha\beta} b\theta_{\beta,\alpha\beta}, \psi^a_{\alpha,\alpha\beta} \psi^b_{\beta,\alpha\beta} k), \quad \ldots\ldots\ldots(13)$$

and $\quad ab(e_{\alpha\beta}, k) = (u, i)(e_{\alpha\beta}, k) = (u, i). \quad \ldots\ldots\ldots(14)$

This shows that $i = \psi^a_{\alpha,\alpha\beta} \psi^b_{\beta,\alpha\beta} k$. Hence, $\psi^a_{\alpha,\alpha\beta} \psi^b_{\beta,\alpha\beta}$ is a constant mapping acting on $I_{\alpha\beta}$.

Step VI. In order to verify that the semigroup multiplication on S coincides with the semigroup operation "$*$" defined on the left generalized Δ-product $S = (Q, T, I; Y; \Phi)$, we need to show that the conditions (M1) - (M4) are satisfied.

If $a \in Q_\alpha, b \in Q_\beta, ab \notin Q_{\alpha\beta}$, then, by (13) and (14) in Step V, we have

$$ab = (u, i) = (a\theta_{\alpha,\alpha\beta} b\theta_{\beta,\alpha\beta}, \langle \psi^{(a)}_{\alpha,\alpha\beta} \psi^{(b)}_{\beta,\alpha\beta} \rangle)$$
$$= a * b. \quad \ldots\ldots\ldots(15)$$

Also, if $a \in Q_\alpha, b \in Q_\beta, ab \in Q_{\alpha\beta}$, then $ab = a * b$. This shows that the conditions (M1) and (M2) are satisfied. Also, conditions (M3) and (M4) can be verified likewisely.

Step VII. Last, we still need to verify that the mapping

$$\Phi = (\Phi^{(1)}_{\alpha,\beta}; \Phi^{(2)}_{\alpha,\beta}; \Phi^{(3)}_{\alpha,\beta} | \alpha \geq \beta, \alpha, \beta \in Y)$$

is indeed a structure mapping. By our construction shown above, the required conditions (C1), (C2) and (C3) are all fulfilled. (Refer to Step II, IV, V and VI). The only thing left is to verify condition (C4).

Let $a \in Q_\alpha, b \in Q_\beta, ab \notin Q_{\alpha\beta}$ with $\alpha\beta \geq \delta$. Pick any $(e_\delta, l) \in S^{(1)}_\delta \cap E$. Then, by (15), $ab = (a\theta_{\alpha,\alpha\beta} b\theta_{\beta,\alpha\beta}, \langle \psi^a_{\alpha,\alpha\beta} \psi^b_{\beta,\alpha\beta} \rangle)$; denote it by (u, i). Invoking (10),(11) and (C3) (ii), we therefore derive that

$$ab(e_\delta, l) = (u, i)(e_\delta, l) = (u\theta_{\alpha\beta,\delta}, \psi^{(u,i)}_{\alpha\beta,\delta}l)$$

$$= [(a\theta_{\alpha,\alpha\beta}b\theta_{\beta,\alpha\beta})\theta_{\alpha\beta,\delta}, \psi^{(u,i)}_{\alpha\beta,\delta}l]$$

$$= (a\theta_{\alpha,\delta}b\theta_{\beta,\delta}, \psi^{(u,i)}_{\alpha\beta,\delta}l), \quad \dots\dots\dots(16)$$

and
$$a[b(e_\delta, l)] = a(b\theta_{\beta,\delta}, \psi^b_{\beta,\delta}l)$$

$$= a(e_\delta, \psi^b_{\beta,\delta}l)(b\theta_{\beta,\delta}, \psi^b_{\beta,\delta}l)$$

$$= [(a\theta_{\alpha,\delta}, \psi^a_{\alpha,\delta}\psi^b_{\beta,\delta}l)(b\theta_{\beta,\delta}, \psi^b_{\beta,\delta}l)$$

$$= (a\theta_{\alpha,\delta}b\theta_{\beta,\delta}, \psi^a_{\alpha,\delta}\psi^b_{\beta,\delta}l). \quad \dots\dots\dots(17)$$

By (16) and (17), we obtain $\psi^a_{\alpha,\delta}\psi^b_{\beta,\delta} = \psi^{(u,i)}_{\alpha\beta,\delta}$. and so (C4)(ii) holds. The remaining parts of (C4) also hold verbatim. Thus, the converse part of Theorem 6.7 is proved.

It can now be seen that the Δ-product of left c-semigroups is just a particular case of left generalized Δ-product structure of left c-quasiregular semigroups. In fact, if $T = [Y; G_\alpha; \varphi_{\alpha,\beta}]$ is a strong semilattice of groups G_α; $I = \bigcup_{\alpha \in Y} I_\alpha$ is a semilattice of left zero bands I_α; and $Q = \phi$, then pick $\psi_{\alpha,\beta} = \Phi^{(1)}_{\alpha,\beta}$ which is the phase I structure map of $S = S(T, I, Y; \Phi)$. Then form the Δ-product $T\Delta_{Y,\Psi}I$. Since $\Psi = \Phi^{(1)}_{\alpha,\beta}$ satisfies the conditions C(1), C(2) and C(4)(iv) in the definition of left generalized Δ-product, Ψ also satisfies the required conditions (P1) and (P2) of the Δ-product. As a consequence, the structure theorem for left c-semigroups becomes an immediate corollary of theorem 6.7. Similarly, the θ-product for c-quasiregular semigroups is also a special case of the left generalized Δ-product for left c-quasiregular semigroups.

At the end of this paper, we construct an example of a non-trivial left c-quasiregular semigroup. The procedures of the construction are displayed as follows:

Step I :   Let $Y = \{\alpha, \beta, \alpha\beta\}$ be a basic semilattice.

Step II :   We contruct a group on each vertex of Y. For instance, let $G_\alpha = \{e_0\}$ and $G_\beta = \{g_0\}$ be trivial groups and $G_{\alpha\beta} = \{w_0, a_0, b_0 | a_0^2 = b_0, a_0^3 = w_0\}$ a cyclic group with three elements. Mount all the above groups on its corresponding vertex of Y. Thus, $T = [Y; G_\alpha; \varphi_{\alpha,\beta}]$ is obvious a strong semilattice of groups.

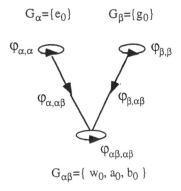

$$G_\alpha = \{e_0\} \qquad G_\beta = \{g_0\}$$

$$\varphi_{\alpha,\alpha} \qquad \qquad \varphi_{\beta,\beta}$$

$$\varphi_{\alpha,\alpha\beta} \qquad \varphi_{\beta,\alpha\beta}$$

$$\varphi_{\alpha\beta,\alpha\beta}$$

$$G_{\alpha\beta} = \{ w_0, a_0, b_0 \}$$

Step III :   Construct a left regular band on the semilattice Y by letting $I_\alpha = \{i, j\}$, $I_\beta = \{k, l\}$ and $I_{\alpha\beta} = \{m\}$ that are left zero bands. Form $I = \bigcup_{\alpha \in Y} I_\alpha$. Then I is clearly a left regular band.

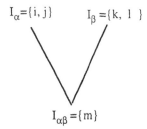

$$I_\alpha = \{i, j\} \qquad I_\beta = \{k, l\}$$

$$I_{\alpha\beta} = \{m\}$$

Step IV :   On each vertex of the semilattice Y, we form the Catersian product of the groups and the left zero bands, namely $S_\alpha^{(1)} = G_\alpha \times I_\alpha$; $S_\beta^{(1)} = G_\beta \times I_\beta$ and $S_{\alpha\beta}^{(1)} = G_{\alpha\beta} \times I_{\alpha\beta}$.   Let $(e_0, i) = e$; $(e_0, j) = f$; $(g_0, k) = g$; $(g_0, l) = h$; $(w_0, m) = w$; $(a_0, m) = u$ and $(b_0, m) = v$ respectively. Then, we have $S_\alpha^{(1)} = \{e, f\}$; $S_\beta^{(1)} = \{g, h\}$ and $S_{\alpha\beta}^{(1)} = \{w, u, v\}$. Mount the above these three semigroups on the semilattice Y, we obtain the following figure.

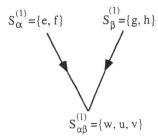

$$S_\alpha^{(1)} = \{e, f\} \qquad S_\beta^{(1)} = \{g, h\}$$

$$S_{\alpha\beta}^{(1)} = \{w, u, v\}$$

Step V :   Construct the phase I structure map of S, that is, $\Phi^{(1)}_{\gamma,\delta} : S^{(1)}_\gamma \longmapsto \mathcal{T}(I_\delta)$ as shown below :

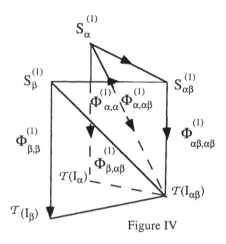

Figure IV

where

$$\Phi^{(1)}_{\alpha,\alpha} : e = (e_0, i) \longmapsto \begin{pmatrix} ij \\ ii \end{pmatrix} ; \quad f = (e_0, j) \longmapsto \begin{pmatrix} ij \\ jj \end{pmatrix}$$

$$\Phi^{(1)}_{\beta,\beta} : g = (g_0, l) \longmapsto \begin{pmatrix} kl \\ kk \end{pmatrix} ; \quad h = (g_0, k) \longmapsto \begin{pmatrix} kl \\ ll \end{pmatrix},$$

and the mappings $\Phi^{(1)}_{\alpha,\alpha\beta}$ and $\Phi^{(1)}_{\beta,\alpha\beta}$ are the trivial mappings which map respectively S_α and S_β onto $\mathcal{T}(I_{\alpha\beta})$.

Step VI :   Let $Q = Q_\alpha \cup Q_\beta \cup Q_{\alpha\beta}$, where $Q_\alpha = \{a\}$; $Q_\beta = \{b\}$ and $Q_{\alpha\beta} = \phi$. Suppose that a^2, b^2, ab and ba are not in Q. Then form the phase II component of the left generalized Δ-product S, say $S^{(2)}_\alpha = Q_\alpha \cup G_\alpha$; $S^{(2)}_\beta = Q_\beta \cup G_\beta$ and $S^{(2)}_{\alpha\beta} = Q_{\alpha\beta} \cup G_{\alpha\beta}$. Thus, $S^{(2)}_\alpha = \{a, e_0\}$; $S^{(2)}_\beta = \{b, g_0\}$ and $S^{(2)}_{\alpha\beta} = G_{\alpha\beta} = \{w_0, a_0, b_0\}$.

Construct the phase II structure maps of S by considering $\Phi^{(2)}_{\gamma,\delta} : S^{(2)}_\gamma \to G_\delta$ shown below :

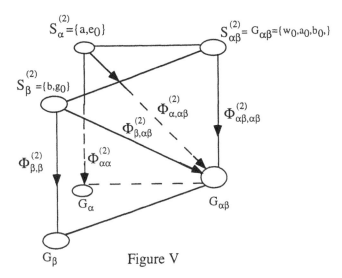

Figure V

Now the mappings $\Phi^{(2)}_{\alpha,\alpha}$, $\Phi^{(2)}_{\beta,\beta}$ and $\Phi^{(2)}_{\alpha\beta,\alpha\beta}$ are clearly displayed in the above figure; where $\Phi^{(2)}_{\alpha\beta,\alpha\beta}$ is the identity map and

$$\Phi^{(2)}_{\alpha,\alpha\beta} : x \longmapsto w_0, x \in S^{(2)}_\alpha, \quad \Phi^{(2)}_{\beta,\alpha\beta} : x \longmapsto w_0, x \in S^{(2)}_\beta$$

Step VII : Construct the phase III structure mappings of S by $\Phi^{(3)}_{\gamma,\delta} : Q_\gamma \longmapsto \mathcal{T}(I_\delta)$ shown below:

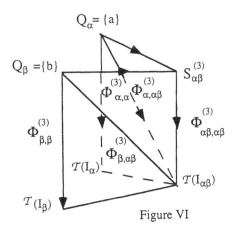

Figure VI

where

$$\Phi^{(3)}_{\alpha,\alpha} : a \longmapsto \begin{pmatrix} ij \\ ii \end{pmatrix}, \quad \Phi^{(3)}_{\beta,\beta} : b \longmapsto \begin{pmatrix} kl \\ kk \end{pmatrix},$$

The other mappings are the identity maps.

Step VIII : Finally, form the left generalized Δ-product $S = \bigcup_{\alpha \in Y} S_\alpha$ on the semilattice Y, where $S_\alpha = Q_\alpha \cup S_\alpha^{(1)}$; $S_\beta = Q_\beta \cup S_\beta^{(1)}$ and $S_{\alpha\beta} = Q_{\alpha\beta} \cup S_{\alpha\beta}^{(1)}$, together with the 3-phase structure mapping $\Phi = (\Phi_{\gamma,\delta}^{(1)}; \ \Phi_{\gamma,\delta}^{(2)}; \ \Phi_{\gamma,\delta}^{(3)})$. Then, $S = \{a, b, e, f, g, h, w, u, v\}$ is the required semigroup on the basic semilattice Y.

Summing up all the above steps, we obtain the following Cayley table for the left generalized Δ-product S :

*	a	b	e	f	g	h	w	u	v
a	e	w	e	e	w	w	w	u	v
b	w	g	w	w	g	g	w	u	v
e	e	w	e	e	w	w	w	u	v
f	f	w	f	f	w	w	w	u	v
g	g	w	w	w	g	g	w	u	v
h	h	w	w	w	h	h	w	u	v
w	w	w	w	w	w	w	w	u	v
u	u	u	u	u	u	u	u	v	w
v	v	v	v	v	v	v	v	w	u

Step IX : We show how to glue up the above steps altogether.

(I) Form a strong semilattice $T = [G_\alpha; Y; \varphi_{\alpha, \beta}]$ on the semilattice Y.

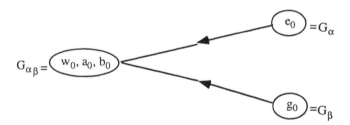

(II)   Construct a power breaking partial semigroup $Q = \bigcup_{\alpha \in Y} Q_\alpha$ on Y.

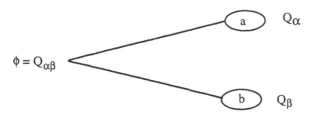

(III)   Form the phase II component of the left generalized Δ-product.

(IV)   Form $\bigcup_{\alpha \in Y} (G_\alpha \times I_\alpha)$ on Y.

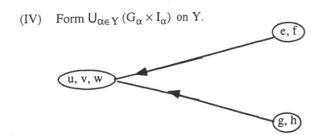

(V)   Form the left generalized Δ-product on Y.

(VI) Finally, form the left generalized Δ-product $S(T,I,Y,Q; \Phi)$, with $\Phi=(\Phi^{(1)};\Phi^{(2)};\Phi^{(3)})$

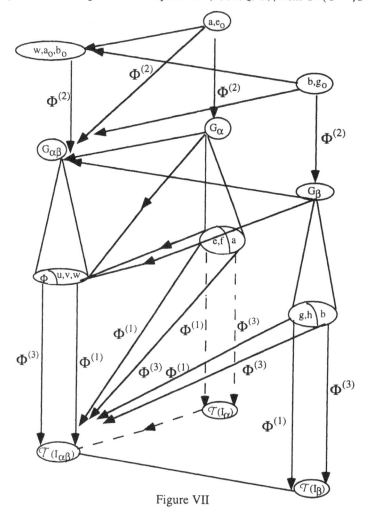

Figure VII

In closing this paper, we remark reiterately that both the Δ-products and the θ-products are particular cases of the left generalized Δ-products. This fact is not hard to observe. Thus, by taking the Clifford semigroups as the starting point for research, we can develope a Clifford hierarchy which contains the class of regular semigroups as its core. We expect that more research can be done and some more generalized classes of regular semigroups can be explored along this direction. We also believe that if some more structures are assigned to the base semilattices, then the structure of semigroups on such semilattices may be more fascinating.

References

[1] Bogdanović S., Semigroups With a System of Subsemigroups; Inst. of Math,. Novi Sad, Yugoslavia, 1985.

[2] Clifford A.H., Bands of Semigroups, Proc. Amer. Math. Soc. 5 (1954), 499-504.

[3] Clifford A.H., The Structure of Orthodox Union of Groups, Semigroup Forum 3 (1992), 283-337.

[4] Clifford A.H. and Petrich M., Some Classes of Completely Regular Semigroups, J. Algebra 46 (1997), 462-480.

[5] Clifford A.H. and G.B. Preston, The Algebraic Theory of Semigroups, Vol I, Math. Survey of the American Math. Soc. 7, Providence, R.I., (1961).

[6] Clifford A.H. and G.B. Preston, The Algebraic Theory of Semigroups, Vol II, Math. Surveys of the American Math. Soc. I, Providence R.I., (1967).

[7] Fountain J.B., Right pp Monoids with Central Idempotents, Semigroup Forum 13, (1977), 103-129.

[8] Guillet P.A., Semigroups: An Introduction to the Structure Theory (1995), Marcel Dekker, INC., New York. Basel. Hong Kong.

[9] Guo Y.Q., Weakly Left C-rpp Semigroup, Science Bulletin (in Chinese), Vol.40, No. 19 (1995), 1744-1747

[10] Guo Y.Q., On Right C-rpp Semigroups, to appear in Science Bulletin, 1996

[11] Guo Y.Q., and P. Trotter, On (ℓ)-Green Relations on A Semigroup, to apppear in Scinence in China, 1996.

[12] Guo Y.Q., Ren X.M., & K.P. Shum, On Clifford Quasi-regular Semigroups, Chinese Annals of Math., series A. (1994), 319-325.

[13] Guo Y.Q., Ren X.M. & K.P. Shum, A New structure of Left C-semigroups, Advance in Mathematics, ♯ 2, 1993.

[14] Guo Y.Q., Shum K.P. & P.Y. Zhu, The Structure of Left C-rpp Semigroups, Semigroup Forum, Vol 50 (1995), 9-23.

[15] Guo Y.Q, Shum K.P., & P.Y. Zhu, A Note on Left C-rpp Semigroups, Bulletin of Science, China, Vol 34 ♯ 4, (1992) 292-294.

[16] Guo Y.Q., Zhu P.Y., and K.P. Shum, The Structure of Quasi-c-Semigroups, to appear in "Southeast Asian Buttelin of Mathematics", 1997.

[17] Hall T.E., On Regular Semigroups whose Idempotents form a Semigroup, Bull. Austral. Math. Soc. (1969), 195-208.

[18] Howie J.M., An Introduction to Semigroup Theory, Academic Press, London (1976).

[19] Howie J.M. & G. Lallement, Certain Fundamental Congruences on Regular Semigroups; Prod. Glascow Math Asso. 7 (1966), 145-149.

[20] Jarek P., Commutative Regular Semigroups, Colloq. Math. 12 (1964), 195-208.

[21] Kimura N., The Strurcture of Idempotent Semigroups (I), Pacific J. Math. 8 (1958), 257-275.

[22] Kilp M., Commutative Monoids all whose Principal Ideals are Projective, Semigroup Forum 6, (1973), 334-339.

[23] Lajos S., A Note on Semilattice Of Groups, Acta Sci. Math. Szeged 33 (1972), 315-317.

[24] Lallement G., Structure Theorem For Regular Semigroups, Semigroup Forum 4 (1972), 95-123.

[25] Lallement G., Demi-groupes régulies, Ann. Mat. Prue Appl.(4) 77 (1967), 47-129.

[26] Latorre D.R., On Semigroups that are Semilattices of Groups, Czechoslovak J. Math. 21 (1971), 369-370.

[27] Li Lide & B.M. Schein, Strongly Regular Rings, Semigroup Forum, 32 (1985), 145-161.

[28] Mclean D., Idempotent Semigroups, Amer. Math. Monthly, 6 (1954), 110-113.

[29] Nambooripad K.S.S. & F. Pastijn, V-Regular Semigroups, Proceeding of Royal Soc. of Edingburgh, 88A (1981), 275-291.

[30] O' Carroll L., Strongly E-Refluxive Inverse Semigroup II, Proc. Edinburgh Math. Soc. (2) 21 (1978), 1-10.

[31] O' Carroll L., A Note on Strongly E-Reflexive Inverse Semigroups, Proc. Amer. Math. Soc. 79 (1980), 133-138.

[32] Petrich M., The Maximal Semilattice Decomposition of a Semigroup, Math Zeit., 85 (1964), 68-82.

[33] Petrich M., Introduction to Semigroups, Charles E. Merill Publishing Co., (1973).

[34] Petrich M., Inverse Semigroups, John Wiley & Sons, (1984).

[35] Reilly N.R. & H.E. Scheiblich, Congruence on Regular Semigroups, Pacific J. Math. 23 (1967), 349-360.

[36] Ren X.M., Guo Y.Q., & P. Zhu, E-Rectangular Quasi-Regular Semigroups, Acta Math. Sinica (1988), 396-413.

[37] Ren X.M. & Y.Q., Guo, E-Ideal Quasi-Regular Semigroups, Science in China, Series A, ♯ 12 (1989), 1437-1446.

[38] Ren X.M., Guo Y.Q., & K.P Shum, On the Structure of Left C-Quasiregular Semigroups, Prco. of International Colloqium on Words, Languages & Combinatorics at Kyoto, 1991, World Scientific, (1994), 466-493.

[39] Schein B.M., Homomorphism and Subdirect Decompositions of Semigroups, Pacific J. Math. 17 (1966), 529-547.

[40] Shum K.P., Ren X.M., & Y.Q. Guo, Some Remarks on $\triangle$-Products and θ- Products of Semigroups. (Submitted)

[41] Shum K.P., Ren X.M., & Y.Q. Guo, On Quasi-Left Groups, Groups Korea 1994,(Walter de Gruyter & Co., Berlin-New York)(1995) 285-288.

[42] Shum K.P., Ideals with Non-Addmissible Idempotents, SEA Math. Bulletin 13.1 (1989), 35-42.

[43] Yamada M., On Regular Semigroups in which the Idempotents form a Band, Pacific J. Math. 33 (1970).

[44] Yamada M. & M.K. Sen, P-Regularity in Semigroups, Memo. Fac. Sci., Shimane Univ. Vol 12 (1987), 47-54.

[45] Yamada M., On P-Regularity in Semigroups and the Structure of P-Regular Semigroups, Proc. of International Conference in Algebraic Structure and Number Theory, Hong Kong, 1988. (World Scientific INC, 1990), 297-331.

[46] Yu B.J., Idempotent Methods in Algebraic Theory of Semigroups, Ph.D Thesis, Lanzhou University, (1989).

[47] Yu B.J., A Generalization Theorem from Regular Semigroups to Quasi-Regular Semigroups, Science Communication, China. Vol 33 ♯ 14 (1988), 117-118.

[48] Zhang L., Guo Y.Q., & K.P. Shum, On Perfect Bands, Proc. of the International Conference on Ordered Structure and Algebra of Computer Languages, Hong Kong, 1991. (World Scientific INC.), (1992), 300-309.

[49] Zheng H.W, On the Research of P-Regular Semigroups, Ph.D. Thesis, Lanzhou University,(1994)

[50] Zhu P., Guo Y.Q., & X.M. Ren, The Semilattice (Matrix)-Matrix (Semilattice) Decomposition of the Quasi-Completely Orthodox Semigroups, Chinese J. of Contemporary Math. ♯ 4 (1989), 425-438.

[51] Zhu P.Y., Guo Y.Q., & K.P. Shum, Characterization and Structure of Left C-Semigroups, Science in China, Series A ♯ 6 (1991), 582-590.

[52] Zhu P.Y., On π-Regular $\mathcal{L}$-Trivial Semigroups, J. of Qinghai Normal University (Natural Sciences), No 4. (1987), 1-4.

Primitive Rings, Infinite Matrix Rings, and Morita-like Equivalences

Shum Kar-Ping Department of Mathematics, The Chinese University of Hong Kong, Shatin, Hong Kong

Xu Yonghua Institute of Mathematics, Fudan University, Shanghai, People Republic of China

Introduction

It is well known that the density theory of Jacobson plays a central role in the theory of primitive rings. In the last two or three decades, much effort has been spent to extend the theory to more general classes of rings. In a sequence of papers, most notably [19] and [25], Koh and Mewborn achieved a generalized density theorem for the type of rings with weak transitivity. Later on, Zelmanowitz [52] studied the rings which possess faithful critically compressible modules and a generalized density theory for the weakly primitive rings has been established. Another version of density theory was given by Amitsur [2] and Zelmanowitz [53], [54]. The subclass of weakly dense rings which possesses some non-singular uniform one-sided ideals was investigated by them. These rings were described by a module-theoretical dense theory akin to that for primitive rings with minimal one-sided ideals [17]. In addition, there are some other types of weakly dense rings mentioned by other authors such as K-primitive rings [21], [30]; k-primitive rings [4] ,[5] and endo-primitive rings [10]. However, a general structure theory for primitive rings is still not established;

This research is partially supported by Tian Yuan Foundation of China and UGC (HK) grant 221.600.370

only little information can be found in the literarture about primitive rings with non-finite fold of transitivity.

Concerning primitive rings with infinite-fold of transitivity, Y.H. Xu has written a number of papers on this topic since 1979 (see [36]-[43]). As most of Xu's papers were written in Chinese and published in the mathematical journals in China, some of his work may not familiar to the western colleagues. In this expository article, we first summarize the work done by Xu on primitive rings and discuss the presentation and amalgamation of rings with infinite matrices. The relationships between free modules and their endomorphism rings will be particularly displayed. Also, Bolla's theorem [3] which gave a categorical description for the isomorphism between endomorphism rings of free modules will be sketched and generalized to endomorphism rings of quasi-progenerators (see [24] and [45]). Moreover, some theorems of Jacobson on primitive rings will be extended and amplified. The representation of primitive rings by rings of matrices will be discussed in detail.

About infinite matrix rings, the problem "What kind of infinite matrix rings would determine the Morita equivalence of rings?" has been tackled by Xu, Shum and Turner-Smith [46]. In this report, we sketch how to unify and extend some of the previous results on infinite matrix rings obtained by Stephenson [34], Camillo [8], Abrams [1] and Garcia [13] by using the replacement technique of modules with ring structure. Finally, we introduce the concept of Morita-like equivalence of rings by Xu, Shum and Turner-Smith [46]. Some recent results obtained by Xu, Shum and Fong [47] on Morita-like equivalence of rings will be included in this report. The concept of Morita-like equivalence of rings is a generalization of the Morita equivalence developed by Morita in 1954. This new concept brings the old theory with some fresh ideas.

The reader is always referred to Jacobson [17], [18] for terminology and definitions not mentioned in this article.

1   Notation and Basic Results

We first enunciate some notations and basic results on primitive rings that will be frequently used throughout this paper.

Let $(\mathcal{M}, \mathcal{M}')$ be a pair of dual vector spaces over a division ring F. Then $\mathcal{M}$ becomes a topological space if $\mathcal{M}$ is topologized by the well known $\mathcal{M}'$-topology. Denote the complete ring of continuous linear transformations of $\mathcal{M}$ by $\mathcal{L}_{\mathcal{M}'}(\mathcal{M})$ and let $\mathcal{F}_{\mathcal{M}'}(\mathcal{M})$ be the socle of $\mathcal{L}_{\mathcal{M}'}(\mathcal{M})$.

The following theorem of Jacobson [17] contains the core results on primitive rings as much research evolved from this well known theorem.

Theorem 1.1 (i) A ring R is primitive with non-zero socle $\mathcal{G}$ if and only if there exists an $\mathcal{M}'$-topology in a vector space $\mathcal{M}$ such that

$$\mathcal{G} = \mathcal{F}_{\mathcal{M}'}(\mathcal{M}) \subset R \subset \mathcal{L}_{\mathcal{M}'}(\mathcal{M})$$

(ii) Let $(\mathcal{N}, \mathcal{N}')$ be another pair of dual spaces over a division ring G and T be a ring such that

$$\mathcal{F}_{\mathcal{N}'}(\mathcal{N}) \subset T \subset \mathcal{L}_{\mathcal{N}'}(\mathcal{N})$$

Assume also that there exists a ring isomorphism σ of R onto T; then there exists a semi-linear homeomorphism S of the space $\mathcal{M}$ onto the space $\mathcal{N}$ such that

$$r^\sigma = S^{-1} r S \text{ for all } r \epsilon R$$

$$\mathcal{L}_{\mathcal{M}'}(\mathcal{M})^\sigma = \mathcal{L}_{\mathcal{N}'}(\mathcal{N});$$

$$\text{and } \mathcal{F}_{\mathcal{M}'}(\mathcal{M})^\sigma = \mathcal{F}_{\mathcal{N}'}(\mathcal{N})$$

In view of the above results, one would naturally ask about the structures of the rings $\mathcal{L}_{\mathcal{M}'}(\mathcal{M})$ and $\mathcal{F}_{\mathcal{M}'}(\mathcal{M})$. Also, as there are primitive rings which are not finite-fold transitive, the method of finite topology is hence not adequate for us to deal with rings of this type. In order to consolidate the content of Jacobson's theorem 1.1 and to provide for further investigation of primitive rings with infinite-fold transitivity, Xu subsequently developed some methods and techniques. His main idea is to replace the pair of the given dual vector spaces $(\mathcal{M}, \mathcal{M}')$ without ring structures by another dual pair of spaces equipped with ring structures. The essential steps are the following (see [36] and [37]):

Step I Let $\mathcal{M} = \sum_{i\epsilon\Gamma} F\mu_i$ be a vector space over a division ring F with an arbitrary index set Γ. Let $\mathcal{M}^* = \sum_{j\epsilon\Gamma'} v_j F$ be the conjugate space of $\mathcal{M}$. Form the endomorphism ring $\Omega = \text{End}_F\mathcal{M}$.

Step II. To show that every minimal right ideal $A = E\Omega$ of the ring Ω can be expressed as a vector space over the division ring $K = E\Omega E$, where $E^2 = E\epsilon\Omega$ such that

$$A = \sum_{i\epsilon\Gamma} K\alpha_i \text{ and } \text{End}_K A = \Omega.$$

Step III. To show that the conjugate space A^* of A is a minimal left ideal of Ω which can be written as

$$A^* = \Omega E = \sum_{j\epsilon\Gamma'} \beta_j K.$$

Step IV. To show that there exists a semi-linear isomorphism S from the space $\mathcal{M}$ to the space A. Also, to show that there exists a semi-linear isomorphism S' from the space A^* to the space $\mathcal{M}^*$ such that S' is the adjoint of S and $\omega = S\omega S^{-1}$; $\omega = S'^{-1}\omega S'$ for all $\omega\epsilon\Omega$.

After getting through the above four steps, we can observe that the spaces $\mathcal{M}$ and $\mathcal{M}^*$ have been replaced by the spaces A and A^* respectively. As $A = E\Omega$ and $A^* = \Omega E$ are respectively the minimal right and minimal left ideals of the ring Ω, they possess ring structures while the spaces $\mathcal{M}$ and $\mathcal{M}'$ do not possess such a nice property.

Now let A' be the total subspace (i.e., dense subspace) of $A^* = \Omega E$ and $\mathcal{M}' = A'S'$. Then $(\mathcal{M}, \mathcal{M}')$ and (A, A') become the corresponding pairs of dual vector spaces over the division rings F and $K = E\Omega E$ respectively.

Step V. To show that

$$(\text{i}) \mathcal{L}_{\mathcal{M}'}(\mathcal{M}) = \mathcal{L}_{A'}(A) = \{\omega\epsilon\Omega | \omega A' \subset A\}$$

$$(\text{ii}) \mathcal{F}_{\mathcal{M}'}(\mathcal{M}) = \mathcal{F}_{A'}(A) = A'\Omega \text{ and } A' = \mathcal{F}_{A'}(A)K.$$

The result in Step V is called the **replacement theorem** as the dual pair of spaces $(\mathcal{M}, \mathcal{M}')$ can be replaced by another dual pair (A, A'). One of the main advantages of replacing $\mathcal{M}'$ by A' is that A' possesses a ring structure while $\mathcal{M}'$ does not . Also, the relation between the socle $\mathcal{F}_{A'}(A)$ and the total subspace A' of A^* can be clearly displayed.

By using the replacement technique, primitive rings with infinite-fold transitivity can be easily studied. Moreover, the proofs of some well-known theorems on ring theory in the literature can perhaps be shortened. We enumerate here some of the main results obtained by Xu in [36]-[42] in following sections.

2   Primitive rings with ν-socles

Throughout this section, primitive rings with infinite-fold transitivity will be considered.

Definition 2.1 An ideal $\mathcal{G}_\nu$ of a primitive ring R is called a $\underline{\nu\text{-socle}}$ if and only if the following conditions are satisfied:

(i) $\mathcal{G}_\nu = \{r\epsilon R| \text{ rank } r < \aleph_\nu \}$

(ii) $\mathcal{G}_\nu$ is $\aleph_\nu$-fold transitive. (This means, for any subset $\{x_i\}_{i\epsilon I}$ of linearly independent elements of space $\mathcal{M}$ with cardinal $I < \aleph_\nu$ and $\{ y_i\}_{i\epsilon I} \subset \mathcal{M}$, there exists an element $r\epsilon\mathcal{G}_\nu$ such that $x_i r = y_i$ for all $i\epsilon I$).

Clearly, the concept of ν-socle $\mathcal{G}_\nu$ generalizes the usual concept of socle, in fact, when $\nu = 0$, $\mathcal{G}_0$ is nothing but just the usual socle of the ring R.

The following theorem obtained by Xu in [38] outlines the features of primitive rings with ν-socles.

Theorem 2.2 [38]. Let R be a primitive ring with ν-socle $\mathcal{G}_\nu$. Then the following statements hold:

(i) $\mathcal{G}_0 \subsetneq \mathcal{G}_1 \subsetneq \cdots \subsetneq \mathcal{G}_\mu \subsetneq \cdots \subsetneq \mathcal{G}_\nu$, where $\mathcal{G}_\mu$ is the μ-socle of R for any $\mu < \nu$.

(ii) if μ is a non-limit ordinal, then there is an idempotent $\ell_\mu \epsilon \mathcal{G}_\mu$ with rank $\ell_\mu = \aleph_{\mu-1}$ such that $\mathcal{G}_\mu = R\ell_\mu R$.

(iii) if μ is a limit ordinal then $\mathcal{G}_\mu = \cup_{\lambda<\mu}\mathcal{G}_\lambda$.

(iv) if L is an ideal of R such that $L \subset \mathcal{G}_\nu$ then $L = \mathcal{G}_\lambda$ for some $\lambda < \nu$.

3 ν-Normal primitive rings

Before we give a full description for the rings $\mathcal{L}_{\mathcal{M}'}(\mathcal{M})$ and $\mathcal{F}_{\mathcal{M}'}(\mathcal{M})$, we first consider a special case, namely the ν-normal primitive rings.

Same as before, $\mathcal{M} = \sum_{i\epsilon\Gamma} F\mu_i$ is the vector space over a division ring F and $\Omega = \mathrm{End}_F\mathcal{M}$. Let R be a subring of Ω. In order to describe the ν-normal primitive rings, the following definitions are needed:

Definition 3.1 A subset $\{E_i\}_{i\epsilon\Gamma}$ of Ω is called a <u>corresponding basis</u> of Ω if there exists a basis $\{\mu_i\}_{i\epsilon\Gamma}$ of $\mathcal{M}$ such that $\mu_i E_i = \mu_i$, $\mu_j E_i = 0$ for $j \neq i$. Occasionallly, we call $\{E_i\}_{i\epsilon\Gamma}$ the corresponding basis with respect to the basis $\{\mu_i\}_{i\epsilon\Gamma}$ of $\mathcal{M}$.

Definition 3.2 An idempotent $\ell\epsilon\Omega$ is called <u>$\{E_i\}_{i\epsilon\Gamma}$ type</u> if and only if $\{E_i\}_{i\epsilon\Gamma}$ is a corresponding basis of Ω and there exists a subset $I \subset \Gamma$ such that $E_i\ell = \ell E_i = E_i$ for all $i\epsilon I$; $E_j\ell = \ell E_j = 0$ for all $j\epsilon\Gamma\backslash I$.

A primitive ring R with ν-socle $\mathcal{G}_\nu$ is called ν-normal with rank $\aleph$ if and only if there exists a corresponding basis $\{E_i\}_{i\epsilon\Gamma}$ of Ω in R such that the cardinal number of $\{E_i\}_{i\epsilon\Gamma} = \aleph$, and the ν-socle

$$\mathcal{G}_\nu = \sum_{\ell\epsilon L_\nu} \ell R = \sum_{\ell\epsilon L_\nu} R\ell.$$

where L_ν is the set of all $\{E_i\}_{i\epsilon\Gamma}$-type idempotent elements of Ω with rank $< \aleph$.

Invoking the idea of Jacobson (see theorem 1.1), we give here the following definition of complete ν-normal rings.

Definition 3.3 Let R be a ν-normal primitive ring of rank $\aleph$. Let $A = ER$ and $A' = RE$ be minimal right and left ideal of R respectively. Then the complete ring of all continuous linear transformaitons of the space $A = \sum_{i\epsilon\Gamma} K\alpha_i$, topologized by the A'-topology, is called the complete ν-normal primitive ring of R, denote it by $L_{A'}(A)$.

For the presentation of the complete ν-normal primitive rings by infinite matrix rings, the following class of matrices is considered:

Definition 3.4. A matrix $M_{\Gamma\times\Gamma} = (f_{ij})_{\Gamma\times\Gamma}$ over a ring F is called ν-normal with order $\aleph$ if and only if the following conditions are satisfied:

(a) $|\Gamma| = \aleph$.

(b) every row of $M_{\Gamma\times\Gamma} = (f_{ij})_{\Gamma\times\Gamma}$ is finite and the cardinal number of the set of all non-zero entries of every columns in $M_{\Gamma\times\Gamma}$ is smaller than $\aleph$.

The following structure theorem for ν-normal primitive rings is found in [42].

Theorem 3.5 Let R be a ν-normal primitive rings of rank $\aleph$ with $\mathcal{G}_\nu$ being its ν-socle. Denote the complete ν-normal primitive ring of R by $L_{A'}(A)$. (Note: if $\nu = 0$, then the ring $L_{A'}(A)$ is simply called the complete normal primitive ring) Let $\mathbb{M}$ be the collection of all ν-normal matrices $(k_{ij})_{\Gamma\times\Gamma}$ with order $|\Gamma| = \aleph$ over the division ring $K = ERE$. Then the following statements hold:

(i) $L_{A'}(A) \cong \mathbb{M}$.

(ii) the socle $\mathcal{G}_\nu$ of R is also the ν-socle of $L_{A'}(A)$.

(iii) $L_{A'}(A)$ is just $\aleph_\nu$-fold transitive (this means that $L_{A'}(A)$ is precisely $\aleph_\nu$-fold but not $\aleph_{\nu+1}$-fold transitive anymore).

(iv) $\mathcal{F}_{A'}(A) \cong \{(k_{ij})_{\Gamma \times \Gamma} \epsilon \mathbb{M}$, where the matrix $(k_{ij})_{\Gamma \times \Gamma}$ has only a finite member of non-zero columns $\}$.

Conversely, let F be a division ring and $\mathbb{M}_\Gamma$ the matrix ring consisting of all row finite $\Gamma \times \Gamma$-matrices over F. Also, let $\mathbb{N}_\Gamma$ be the subring of $\mathbb{M}_\Gamma$ consisting of all matrices $(f_{ij})_{\Gamma \times \Gamma}$ such that the cardinal number of the set of non-zero entries of every columns of $(f_{ij})_{\Gamma \times \Gamma}$ is smaller than $\aleph_\nu$. Then, $\mathbb{N}_\Gamma$ is a complete ν-normal primitive ring with exactly $\aleph_\nu$-fold transitivity.

Because theorem 3.5 associates the ν-normal primitive rings with matrix rings, it can be used as a tool to generalize some well-known theorems of rings with matrix representations. In the following section, we shall state the generalized versions of some well-known theorems on this topic.

4   Generalized Wedderburn-Artin theorem and Generalized Faith-Utumi theorem

The Wedderburn-Artin theorem is among one of the most elegant theorems in ring theory because it leads to a matrix presentation of the simple Artinean rings. By using this theorem, one can give a complete characterization for a semi-simple ring by means of finitely many division rings and finitely many natural numbers. However, in virture of theorem 3.5 in the last section, the Wedderburn-Artin theorem can be further extended and can be used to deal with some rings with infinite cardinality.

Let us call a ring R a <u>normal</u> (that is, 0-normal) <u>primitive ring</u> of rank $\aleph$ if and only if R has a corresponding basis $\{E_i\}_{i \epsilon \Gamma}$ with card $\Gamma = \aleph$ such that the socle $\mathcal{G}_0$ of R is of the form $\mathcal{G}_0 = \sum_{i \epsilon \Gamma} RE_i = \sum_{i \epsilon \Gamma} E_i R$. We then let $L_{A'}(A)$ be the complete normal primitive ring described in the last section, where, $A = ER$ and $A' = RE$ are a pair of minimal right and minimal left ideals of R respectively. By using theorem 3.5, Xu established the following generalized Wedderburn-Artin theorem.

Theorem 4.1 [42] (Wedderburn-Artin theorem for primitive rings).

Let $\mathcal{G}_0$ be the socle of a normal primitive ring R with rank $\aleph$ and let $L_{A'}(A)$ be the complete normal primitive ring of R described above. Denote the set of all matrices $(k_{ij})_{\Gamma \times \Gamma}$ over the division ring $K = ERE$ with card $\Gamma = \aleph$ by $\mathbb{M}$. If every row and column of the matrix $(k_{ij})_{\Gamma \times \Gamma}$ is finite, then $L_{A'}(A) \cong \mathbb{M}$. Moreover, if $L_{A'}(A)$ is isomorphic to another matrix ring consisting of all matrices $(k'_{ij})_{\Gamma' \times \Gamma'}$ over a division ring K' with card $\Gamma' = \aleph'$ such that every row and every column are finite, then $\aleph = \aleph'$ and $K \cong K'$.

It is not difficult for us to see that if R is a simple Artinean ring, then theorem 4.1 is just the well-known Wedderburn-Artin theorem. In fact, if $\aleph = n$ is finite, then the complete normal primitive ring $L_{A'}(A)$ is a simple Artinean ring with rank n. Thus, our theorem 4.1 is indeed a generalization of the usual Wedderburn-Artin theorem.

In connection with the matrix representation of rings, we have the Faith-Utumi theorem. This theorem is one of the most important theorems concering the matrix representation of rings. A modified version of the Faith-Utumi theorem recorded in [49] says that a ring R is a prime right Goldie ring with Goldie dimension n if and only if R is isomorphic to a matrix subring M_n such that $D_n \subseteq M_n \subseteq K_n$, where K is a division ring and D is a right order of K. In view of this theorem, one would naturally ask whether the "Goldie dimension n" in the Faith-Utumi theorem can be relaxed to "Goldie dimension $\aleph$." This extension can be done only after several relevent defintions have been overhauled.

Definition 4.2. A ring R is called a <u>generalized right Goldie ring</u> of rank $\aleph$ if and only if R has a right quotient ring which is a complete normal primitive ring of rank $\aleph$.

It has been noticed in [42] that all maximal independent sets of uniform right ideals of R have the same cardinality $\aleph$, which is the rank of the generalized right Goldie ring. We call this number the <u>right Goldie dimension of R</u>.

After introducing the above theorem, the following general version of Faith-Utumi theorem can be obtained by applying theorem 3.5.

Theorem 4.3 [42]. (Generalized Faith-Utumi theorem)

A ring R is a generalized prime Goldie ring with right Goldie dimension $\aleph$ if and only if R is a ring isomorphic to a ring R' such that Nor.$D_\Gamma \subset R' \subset$ Nor.K_Γ, where D is a right order of a division ring K with card $\Gamma = \aleph$; Nor. D_Γ and Nor.K_Γ are the $\Gamma \times \Gamma$ matrix rings over D and K consisting of all row and column finite matrices over D and K respectively.

Obviously, if card $\Gamma = n$, then theorem 4.3 is just the original version of the Faith-Utumi theorem. There is nothing new.

We conclude this section with a remark on the annihilators of the semi-prime Goldie rings.

Let G be a subset of a ring R. Let $G^\perp = \{x \epsilon R | Gx = 0\}$ be the right annihilator of G in R. Then we can easily obtain a necessary and sufficient condition for the equivalence of right and left semi-prime Goldie rings.

Theorem 4.4. [31]

Let R be a semi-prime (right) Goldie ring. Then the following conditions are equivalent.

(i) R is a left Goldie ring;

(ii) if $x_i^\perp = y_i^\perp$ for some $x_i, y_i \epsilon R, i = 1, 2, ..., t$, then there exist two subsets $\{d, r_1, ..., r_t\}$ and $\{d', r'_1, ..., r'_t\}$ of R such that $dx_i = r_i y_i \neq 0, d'y_i = r'_i x_i \neq 0$ for $i = 1, 2, ..., t$.

It was noticed by Xu in [49] that if R is a right generalized prime Goldie ring with right Goldie dimension $\aleph$, then a composition series of right (left) annihilators of R with length $\aleph$ can be found; moreover every left annihilator $^{\perp}S$ of R for $S \subset R$ can be expressed by $^{\perp}S =^{\perp} x$ for some singleton $x \epsilon R$.

5 Imbedding theorems.

One of the most interesting imbedding theorems in ring theory was due to Litoff (see [20]). He called a ring R a <u>locally matrix ring</u> over a division ring K if every finite subset of R is contained in a subring of R which is isomorphic to the full matrix ring K_n for some natural number n. Litoff and Ánh then asserted [11] that every non-trivial simple ring having a minimal one-sided ideal is a locally matrix ring over a division ring K. Recently, Shum noticed that the results of Litoff can also be proved by using the Morita context.

In order to extend Litoff's theorem, Xu supplied the following generalized version [42].

Theorem 5.1 (i) Let R be a simple ring with minimal one-sided ideals. If $\{x_i\}_{i \epsilon I}$ is an arbitrary subset of R with $|I| = $ card $I < \aleph_\nu$ then the ring R' generated by $\{x_i\}_{i \epsilon I}$ can be imbedded into an $I' \times I'$-matrix ring over a division ring with $|I'| < \aleph_\nu$.

(ii) Let $\Omega = \text{End}_F \mathcal{M}$, where $\mathcal{M} = \sum_{i \epsilon \Gamma} F \mu_i$ is a vector space over a division ring F. Let $S = \{\omega \epsilon \Omega | \text{ rank } \omega < \aleph_\nu\}$ with $|S| = \text{card} S < \aleph_\nu$. Then the ring R' generated by S can be imbedded into a ν-socle of a ν-normal primitive ring.

It should be noticed here that if $|I| = n < \aleph_0$ in (i), then theorem 5.1 (i) says that the ring R' generated by $\{x_1, x_2, ..., x_n\}$ can be imbedded into a $m \times m$-matrix ring K_m over a division ring K for a suitable integer m; this is precisely the original version of Litoff's theorem. Thus, theorem 5.1 is indeed a generalization of Litoff-Ánh's theorem. An additional version of the Litoff-Ánh theorem was obtained by Ng and Shum [29]. They have shown that a ring R is a simple ring with non-zero socle if and only if there is a division ring D and a cardinal Υ such that R is isomorphic to a left ideal S of $M_\Upsilon^0(D)$ for some left S-unital subring S of $M_\Upsilon^0(D)$, where $M_\Upsilon^0(D)$ is an matrix subring which consists of all row-finite and almost-zero column $\Gamma \times \Gamma$ matrices.

Concerning a primitive ring R with $\aleph_\nu$-fold transitivity, we can also provide the following information:

(a) If there exists an element $r\epsilon R$ such that the rank $r = \aleph_\alpha < \aleph_\nu$ then there exists a subring S_α of R which is isomorphic to the complete ring of linear transformations of an $\aleph_\alpha$-dimensional vector space over a division ring.

(b) For any ordinal $\alpha < \nu$, there exists a subring S_α of R such that S_α is ring homomorphic to the complete ring of linear transformations of an $\aleph_\nu$-dimensional vector space over a division ring.

For the case of m-fold transitivity case, the corresponding results are the following:

(a') If R has non-zero socle then there is a subring S_m of R such that S_m is isomorphic to an $m \times m$-matrix ring $\triangle_m$ over a division ring $\triangle$.

(b') If R has no non-zero socle then there exists a subring S_m of R such that S_m is homomorphic to $\triangle_m$.

Recall that a matrix M is <u>almost zero</u> if and only if its entries are all zero except a finite number of them. Consequently, we call a matrix ring $\mathbb{M}$ almost zero if every matrix of $\mathbb{M}$ is an almost zero matrix.

By applying the concept of normal primitive rings, a global imbedding theorem is established by Xu.

Theorem 5.2 [37, 48]

(i) Every simple ring with its minimal one-sided ideals regarded as additive groups can be imbedded into an almost zero matrix ring over a division ring.

(ii) Let $\Omega^* = \text{End}_F \mathcal{M}^*$ be the conjugate ring of $\Omega = \text{End}_F \mathcal{M}$, where $\mathcal{M}^* =$

$\text{Hom}_F(\mathcal{M}, F)$ and $\mathcal{M} = \sum_{i\epsilon\Gamma} F\mu_i$ is a vector space over the division ring F. Then every complete normal dense subring of Ω can be imbedded into any complete normal dense subring of Ω^*.

6 The structure of the ring $\mathcal{L}_{\mathcal{M}'}(\mathcal{M})$

We are now in a position to determine the general structure of the ring $\mathcal{L}_{\mathcal{M}'}(\mathcal{M})$ which was described in our introduction. In order to proceed with our discussion, we first mention the following relations:

Let K be a ring and $K^{\Gamma\times\Gamma'}$ the set of functions $\sigma : \Gamma \times \Gamma' \to K$. Write $\sigma = \sigma(\Gamma, \Gamma') = (k_{ij})_{\Gamma\times\Gamma'}$, which is a matrix over K. Also, write

$$\sigma(i, \Gamma') = \text{the i-th row of the matrix } \sigma(\Gamma, \Gamma') \text{ for } i\epsilon\Gamma;$$

$$\sigma(\Gamma, j) = \text{the j-th column of the matrix } \sigma(\Gamma, \Gamma') \text{ for } j\epsilon\Gamma'$$

Furthermore, we use $K^{\Gamma 0}$ and $K^{0\Gamma'}$ for the following sets:

$$K^{\Gamma 0} = \text{ the set of the functions } C : \Gamma \to K, \text{ denoted by } C = \begin{pmatrix} \cdot \\ \cdot \\ \cdot \\ k_i \\ \cdot \\ \cdot \end{pmatrix},$$

$$K^{0\Gamma'} = \text{the set of the functions } r : \Gamma' \to K, \text{ denoted by } r = (..., k_i, ...)_{i\epsilon\Gamma'}$$

Definition 6.1. Let S be a subgroup of $K^{\Gamma 0}$ and T a subgroup of $K^{0\Gamma'}$. Then the set

$$W = \{\sigma\epsilon K^{\Gamma\times\Gamma'} | \sigma(i, \Gamma')\epsilon T; \sigma(\Gamma, j)\epsilon S \text{ for } i\epsilon\Gamma, j\epsilon\Gamma'\}$$

is called an <u>ST-matrix group</u>.

Definition 6.2. Let D be the set of row finite matrices of $K^{\Gamma\times\Gamma}$. A map $\lambda : K^{\Gamma\times\Gamma'} \to D$ is called a <u>D-map</u> if and only if for any $\sigma_1, \sigma_2\epsilon K^{\Gamma\times\Gamma'}$, the following conditons are satisfied:

$$\lambda(\sigma_1 + \sigma_2) = \lambda\sigma_1 + \lambda\sigma_2$$

$$\text{and } \lambda(\lambda(\sigma_1)\sigma_2) = \lambda(\sigma_1)\lambda(\sigma_2).$$

We now define the (D, λ)-group that is determined by the D-maps.

Definition 6.3. An additive subgroup M of $K^{\Gamma \times \Gamma'}$ is called a $\underline{(D, \lambda)\text{-group}}$ if and only if there is a D-map λ satisfying the following condition.:

$$\lambda(\sigma_1)\sigma_2 \epsilon M \text{ for any } \sigma_1, \sigma_2 \epsilon M.$$

It is trivial to see that the (D, λ)-group M is a matrix ring under the multiplication $\sigma_1 \cdot \sigma_2 = \lambda(\sigma_1)\sigma_2$ for any elements $\sigma_1, \sigma_2 \epsilon M$.

Defintion 6.4 (i) A (D, λ)-matrix ring is called an $\underline{(ST, D, \lambda)\text{-matrix ring}}$ if and only if it is an ST-matrix group.

(ii)An (ST, D, λ)-matrix ring M is said to be $\underline{\text{complete}}$ if and only if every ST-matrix is contained in M.

Equipped with the above new definitions, we immediately obtain the following theorem for $L_{\mathcal{M}'}(\mathcal{M})$.

Theorem 6.5 (Structure theorem for the ring $L_{\mathcal{M}'}(\mathcal{M})$)

Let $A = \sum_{i \epsilon \Gamma} K\alpha_i$ be a vector space over a division ring K which is topologized by the A'-topology, where $A' = \sum_{j \epsilon \Gamma'} \beta_j K$. Denote an arbitrary element of A' by a'.

For any fixed bases $\{\alpha_i\}_{i \epsilon \Gamma}$ of A and $\{\beta_j\}_{j \epsilon \Gamma'}$ of A', we form the sets:

$$S = \left\{ \left(\begin{array}{c} \cdot \\ \cdot \\ \alpha_{i_1} \\ \cdot \\ \cdot \\ \cdot \\ \alpha_{i_m} \\ \cdot \\ \cdot \\ \cdot \end{array} \right)_\Gamma \middle| a' \right\} = \left\{ \left(\begin{array}{c} \cdot \\ \cdot \\ \alpha_{i_1}a' \\ \cdot \\ \cdot \\ \cdot \\ \alpha_{i_m}a' \\ \cdot \\ \cdot \\ \cdot \end{array} \right) \right\} \text{ for all } a' \epsilon A'$$

and let $T = \{a(..., \beta_j, ...)_{\Gamma'}\} = \{(..., a\beta_j, ...)_{\Gamma'}\}$ for all $a\epsilon A$. Then $L_{A'}(A)$ is a ring which is isomorphic to the complete (ST, D, λ)-matrix ring.

Remark: Recall that in section 1(step V), we have, by using the replacement arguments, identified the ring $\mathcal{L}_{\mathcal{M}'}(\mathcal{M})$ with the ring $L_{A'}(A)$. Now, in view of theorem 6.5, we have further shown that the ring $\mathcal{L}_{\mathcal{M}'}(\mathcal{M})$ is in fact ring isomorphic to the complete (ST, D, λ)-matrix ring. This theorem hence provides us some very useful information concerning the representation of primitive rings with infinite-fold transitivity.

7 Structure theory of general primitive rings with non-zero socles

In this section, some results on primitive rings with ν-socles stated in section 2 will be consolidated and further strengthened. Theorem 2.2 in section 2 will be particularly extended to a more general setting. The results obtained in this section will enable us to get a deeper insight into the structure of the class of primitive rings with non-zero socles. Unless otherwise stated, we adopt the notation and terminology mentioned in §2.

We first let $\ell\epsilon\Omega$ be an idempotent element such that rank $\ell = \aleph_\nu$. Write $A_\nu = \ell\Omega, A_\nu^* = \Omega\ell$ and $K_\nu = \ell\Omega\ell$. Also, let A_ν' be a submodule of the K_ν-module of A_ν^*. Then call the pair (A_ν, A_ν') a __dual pair__ if and only if $\omega A_\nu' = 0$ implies $\omega = 0$ for any $\omega\epsilon A_\nu$.

Next, we make the following definition for the dual modules pair (A_ν, A_ν').

Definition 7.1 A pair of dual modules (A_ν, A_ν') is called $\aleph_\nu$-typical if and only if for any subset $\{x_i\}_{i\epsilon I}$ of linearly independent elements of $A = \sum_{i\epsilon\Gamma} K\alpha_i$ (where $K = E\Omega E$, Card $I < \aleph_\nu$), and any subset $\{y_i\}_{i\epsilon I} \subset K$, there exists an element $a'\epsilon A_\nu'$ such that $x_i a' = y_i$ for all $i\epsilon I$.

By using definition 7.1, a characterization for primitive rings with ν-socle $\mathcal{G}_\nu$ can now be deduced.

Theorem 7.2 [38, 48] A ring R is a primitive ring with ν-socle $\mathcal{G}_\nu$ if and only if all pairs of dual modules (A_α, A'_α) for $0 \le \alpha \le \nu$ are $\aleph_\alpha$-typical and there is an adhesive inclusion chain of subrings of Ω lying between $\mathcal{G}_0$ and Ω:

$$\mathcal{F}_{\mathcal{M}'}(\mathcal{M}) = \mathcal{G}_0 \subsetneqq \mathcal{G}_1 \subsetneqq \cdots \subsetneqq \mathcal{G}_\alpha \subsetneqq \cdots \subsetneqq \mathcal{G}_\nu \subset R \subset L_{A'_\nu}(A_\nu) \subset \cdots$$
$$\subset L_{A'_\alpha}(A_\alpha) \subset \cdots \subset L_{A'_1}(A_1) \subset \mathcal{L}_{A'_0}(A_0) = L_{\mathcal{M}'}(\mathcal{M}) \subset \Omega.$$

where $\mathcal{G}_\alpha$ is an α-socle and $L_{A'_\alpha}(A_\alpha) = \{\omega \in \Omega \mid \omega A'_\alpha \subseteq A'_\alpha\}$.

The following theorem is a generalized version of the well known isomorphism theorem of primitive rings (refer to theorem 1.1.).

Theorem 7.3 [37,48] (Extention theorem of isomorphism)

Let (N, N') be an another pair of vector spaces over a division ring G and T a ring with $\mathcal{F}_{N'}(N) \subset T \subset L_{N'}(N)$. Write $\Omega = \text{End}_F \mathcal{M}$ and $\overline{\Omega} = \text{End}_G N$. For any non-limit ordinal $\alpha < \nu$, let ℓ_α be an idempotent of ring R with rank $\ell_\alpha = \aleph_{\alpha-1}$. Write $A_\alpha = \ell_\alpha R$ and $A'_\alpha = R\ell_\alpha$. Then the following statements hold:

(i) If σ is a ring isomorphism of the right ideal A_α of R onto a right ideal of the ring T, then σ can be extended to a ring isomorphism σ' of Ω onto $\overline{\Omega}$;

(ii) If the extended ring isomorphism σ' is also a ring isomorphism of the left ideal A'_α of R onto a left ideal of T, then we have the following situations:

(1) T is a dense sub-ring of $\overline{\Omega}$ with α-socle $\overline{\mathcal{G}}_\alpha$;

(2) For the dual pair (A_μ, A'_μ), when μ is a non-limit ordinal such that $\mu < \alpha$, form $L_\mu = L_{A'_\mu}(A_\mu) = \{\omega \epsilon \Omega \mid \omega A'_\mu \subset A'_\mu\}$. When $\mu < \alpha$ is a limit ordinal μ, form $L_\mu = \cap_{\Lambda < \mu} L_{A'_\Lambda}(A_\Lambda)$, where Λ is a non-limit ordinal. Then the following two chains of subrings of Ω and $\overline{\Omega}$ are isomorphic:

$$R \subset L_\alpha \subset \cdots \subset L_\mu \subset \cdots \subset L_1 \subset L_0 \subset \Omega$$
$$T \subset \overline{L}_\alpha \subset \cdots \subset \overline{L}_\mu \subset \cdots \subset \overline{L}_1 \subset \overline{L}_0 \subset \overline{\Omega}$$

such that $\Omega^{\sigma'} = \overline{\Omega}$ and $L_\mu^{\sigma'} = \overline{L}_\mu$ for all $\alpha \geqslant \mu \geqslant 0$.

(3) Denote the μ-socle of R by $\mathcal{G}_\mu$ and the arbitrary direct sum of n distinct minimal left (right) ideals of R by $\mathcal{G}_0^{(n)}$. Then we obtain the following strict ring isomorphic chains of left (right) ideals of R and T:

$$(0) \subsetneq \mathcal{G}_o^{(1)} \subsetneq \cdots \subsetneq \mathcal{G}_0^{(n)} \subsetneq \cdots \subsetneq \mathcal{G}_0 \subsetneq \mathcal{G}_1 \subsetneq \cdots \subsetneq \mathcal{G}_\mu \subsetneq \cdots \subsetneq \mathcal{G}_\alpha \subset R$$

$$(0) \subsetneq \overline{\mathcal{G}}_o^{(1)} \subsetneq \cdots \subsetneq \overline{\mathcal{G}}_0^{(n)} \subsetneq \cdots \subsetneq \overline{\mathcal{G}}_0 \subsetneq \overline{\mathcal{G}}_1 \subsetneq \cdots \subsetneq \overline{\mathcal{G}}_\mu \subsetneq \cdots \subsetneq \overline{\mathcal{G}}_\alpha \subset T$$

in the sense that $\mathcal{G}_0^{(n)} \cong \overline{\mathcal{G}}_0^{(n)}$ for each integer $0 < n < \infty$ and $\mathcal{G}_\mu^{\sigma'} = \overline{\mathcal{G}}_\mu$ for every $0 \leqslant \mu \leqslant \alpha$.

(4) there exists a semi-linear isomorphism S from $\mathcal{M}$ onto N such that $\omega^{\sigma'} = S^{-1}\omega S$ for any $\omega \epsilon \Omega$ and also the adjoint S' of S is a semi-linear isomorphism of N' onto $\mathcal{M}'$ such that

$$\omega^{\sigma'} = S'^{-1}\omega S' \text{ for all } \omega \epsilon L_{\mathcal{M}'}(\mathcal{M}).$$

As a corollary of the above theorem, we know immediately that if R is a primitive ring with non-zero socle, then any pair of minimal right (left) ideals of R are themselves ring isomorphic.

8 Galois theory of division rings

We are now going to demonstrate how to of build up the Galois theory for finite division rings. For this purpose, we consider the base division ring F of the vector space $\mathcal{M} = \sum F\mu_i$. Call a subring P of F a <u>Galois subring</u> if there exists a subgroup G of Aut (F) such that $P = \{x \epsilon F | x^g = x, \forall g \epsilon G\} = I(G)$, i.e., the fixed subring of F under the action of G. Denote the group of all inner automorphisms of F belonging to G by G_0 and denote the inner automorphism $r \to xrx^{-1}$ by I_r for all $r \epsilon F$. Then, express the algebra of the group G_0 by $E' = \sum_{I_{r_j} \epsilon G_0} \Phi r_j$, where Φ is the center of

F. Write $P' = C_F(E')$, which is the centralizer of E' in F. All the above notations are taken from the text of N. Jacobson [17].

Denote the set $\{\omega \epsilon L(\mathcal{M}, F) |$ rank $\omega < \aleph_\nu\}$ by $T_\nu(\mathcal{M}, F)$, where $L(\mathcal{M}, F)$ is the complete ring of F-linear transformations of $\mathcal{M}$ over F. Then, the set $T_\nu(\mathcal{M}, \mathcal{A})$ can be written verbatim when $\mathcal{A}$ is a sub-division ring of F. After the concept $T_\nu(\mathcal{M}, F)$ is introduced, a theorem of Galois type which associated the set $T_\nu(\mathcal{M}, \mathcal{A})$ with the dimension of Galois ring extension in L can be established. The theorem reads as follows:

Theorem 8.1 [50]

(i) $[F : P']_L = n < \infty$ if and only if $T_\nu(\mathcal{M}, P') = \sum_{j=1}^{n} \oplus r_{jL} T_\nu(\mathcal{M}, F)$, where $r_j \epsilon E'$ and r_{jL} is just the scalar left multiplication of r_j.

(ii) $[P' : P]_L = m < \infty$ if and only if $T_\nu(\mathcal{M}, P) = \sum_{j=1}^{m} \oplus S_j T_\nu(\mathcal{M}, P')$, where S_j is an F-semi-linear automorphism of $\mathcal{M} = \sum F\mu_i$, with an associates isomorphism $\psi_j \epsilon G$.

(iii) If there exists $T_\nu(\mathcal{M}, P), T_\nu(\mathcal{M}, P')$ and $T_\nu(\mathcal{M}, F)$ satisfying the above prepositions (i) and (ii), then the same relationships hold for any suitably chosen $T_\mu(\mathcal{M}, P), T_\mu(\mathcal{M}, P')$ and $T_\mu(\mathcal{M}, F)$.

(iv) If $[F : P]_L < \infty$ then we have $C_F(C_F(E')) = E'$, $[F : P']_L = \dim E'$ and $[P' : P]_L = [G/G_0]$ where $\dim E'$ is the dimension of E' over Φ and $[G/G_0]$ is the index of G_0 in G. In particular, if G is a Gaolis group then $C_F(P') = C_F(P) = E'$.

(v) If $\tilde{G}$ is another group of automorphisms of F such that $I(\tilde{G}) = I(G) = P$, then $[G/G_0] = [\tilde{G}/\tilde{G}_0]$ and $\dim E' = \dim \tilde{E}'$, where $\tilde{E}'$ is the algebra of the group $\tilde{G}$.

Perhaps, we should emphasize here that if the subring P is the center Φ of F then we can deduce from theorem 8.1 that $L(\mathcal{M}, \Phi) = L(\mathcal{M}, F) \otimes_\Phi F_L$ if and only if $[F : \Phi]_L < \infty$. This is actually the result exhibited by Jacobson in [17].

For a more general situation, we consider a division subring P of a division ring F such that P is Galois in F and $[F : P]_L < \infty$. Let G be the Galois group of P in F and let K be a division subring of F containing P, that is, $P \subset K \subset F$. Also, let the group of F-semi-linear automorphisms associated with G be Θ. Then, the following theorem for the ring $L(\mathcal{M}, K)$ was obtained by Xu [49]; it describes the finite extension of the division ring F over K and its associated Galois group H.

Theorem 8.2 If $[F : K]_L = n$, then $L(\mathcal{M}, K) = \sum_{j=1}^n \oplus S_j L(\mathcal{M}, F) = BL(\mathcal{M}, F)$ and $T_\nu(\mathcal{M}, K) = BT_\nu(\mathcal{M}, F)$, where $B = \Theta \cap L(\mathcal{M}, K)$ and $S_j \epsilon B$. Moreover, if $H = \{\psi | S \epsilon B$ and ψ is associated with $S\}$ then H is a Galois group of K over F.

In view of theorem 8.1 and theorem 8.2, the finite Galois theory for division rings can be established. Also, by using theorem 8.1, we can obtain a fundamental theorem for infinite almost inner Galois theory for division rings. (see [50]).

9   Tensor products of algebras [41, 42]

Let ∂ and ∂^* be a pair of algebras over a field Φ. It is natural to ask: when will their tensor product, namely, $\partial \otimes_\Phi \partial^*$ be a primitive ring? The first result generated by this problem was due to Jacobson [17]. He considered the tensor product of primitve rings with non-zero socles. Later on, Lanski, Resco and Small [21] all considered the case that ∂ is a primitive algebra with non-zero socle while ∂^* is just an arbitrary Φ-algebra. On the other hand, Resco [31], by relaxing Φ to a commutative domain, established a correspondence theorem for tensor products. This result is significant as there are a number of results in the literature which are more or less closely related to the correspondence theorem of Resco, in particular the famous Azumaya-

Nakayama theorem [17]. However, the correspondence theorem describes only the one-sided correspondence between the right modules and the right ideals; the picture is not yet complete. In this section, we will strengthen the correspondence theorem by extending the result of Azumaya-Nakayama to tensor product of simple modules along the lines of the Morita theorem [17].

Let ∂_i be an irreducible algebra of linear transformations of the vector space $\mathcal{M}_i$ over a field Φ; $i = 1, 2$. Denote the centralizer and the socle of ∂_i by $\triangle_i$ and $\mathcal{G}_i$ respectively. Let $(\mathcal{M}_i, \mathcal{M}'_i)$ be a dual pair of vector spaces associated with ∂_i. Then, form the tensor products $\mathcal{M} = \mathcal{M}_1 \otimes \mathcal{M}_2$ and $\mathcal{M}' = \mathcal{M}'_1 \otimes \mathcal{M}'_2$ relative to the field Φ.

The following results are the main results on tensor products of algebras:

Theorem 9.1 [43]

(i) (Azumaya-Nakayama Theorem). The lattice of left (right) $\partial_1 \otimes \partial_2$-submodules of $\mathcal{M}$ is isomorphic to the lattice of left (right) ideals of $\triangle_1 \otimes \triangle_2$.

(ii) The lattice of right (resp. left) $\triangle_1 \otimes \triangle_2$- submodules of $\mathcal{M}'$ (resp. $\mathcal{M}$) is isomorphic to the lattice of all right (resp.left) ideals of the form $\partial_1 \otimes \partial_2$ which is contained in $\mathcal{G}_1 \otimes \mathcal{G}_2$.

(iii) Let $\mathcal{C}$ be a full subcategory of mod-$\partial_1 \otimes \partial_2$, where Obj $\mathcal{C}$ consists of all objects N with $N(\mathcal{G}_1 \otimes \mathcal{G}_2) = N$. Then the category $\mathcal{C}$ and the category mod-$\triangle_1 \otimes \triangle_2$ are equivalent. This equivalence is actually determined by the pair of functors $\otimes_{\triangle_1 \otimes \triangle_2} \mathcal{M}$ and $\otimes_{\partial_1 \otimes \partial_2} \mathcal{M}'$ between these two categories. The equivalence of $\triangle_1 \otimes \triangle_2$-mod and a full sub-category of $\partial_1 \otimes \partial_2$-mod can be similarly determined by the pair of functors $\mathcal{M}'_{\triangle_1 \otimes \triangle_2} \otimes$ and $\mathcal{M}_{\partial_1 \otimes \partial_2} \otimes$.

(iv) Let H_{∂_i} be a full subcategory of the category mod-∂_i whose objects are homogeneous completely reducible ∂_i-modules. Then the category mod-$\triangle_i$ and the full subcategory H_{∂_i} are equivalent.

(V) Denote the $n_i \times n_i$-matrix ring over $\triangle_i$ by $\triangle_{in_i}$. Then the categories mod-$\triangle_1 \otimes \triangle_2$; mod-$\triangle_{1n_1} \otimes \triangle_{2n_2}$ and the full subcategory $C_{\mathcal{G}_1 \otimes \mathcal{G}_2}^{\partial_1 \otimes \partial_2}$ of mod-$\partial_1 \otimes \partial_2$ are all equivalent, where the objects N of $C_{\mathcal{G}_1 \otimes \mathcal{G}_2}^{\partial_1 \otimes \partial_2}$ all satisfy the condition $N(\mathcal{G}_1 \otimes \mathcal{G}_2) = N$.

In extending our study on tensor products of algebras, let ∂^* be an algebra over the field Φ. We assume that ∂^* always contains an identity element. In case if ∂^* does not contain an identity, then we just write $\partial_1^* = \{(m, a^*)|m\epsilon\Phi, a^*\epsilon\partial^*\}$ in the usual way so that ∂_1^* turns out to be a Φ-algebra with identity and ∂^* becomes an ideal of ∂_1^*.

In building up the correspondence theorem for tensor products, the following definitions are required.

Definition 9.2. An $\partial \otimes_\Phi \partial^*$-submodule S of $\mathcal{M} \otimes \partial^*$ is called an $\underline{\partial \otimes \partial_1^*}$ $\underline{\text{-submodule}}$ if and only if $S(\partial \otimes \partial_1^*) \subset S$.

A $\triangle \otimes_\Phi \partial^*$-submodule S of $\mathcal{M} \otimes \partial^*$ is called a $\underline{\triangle \otimes \partial_1^*\text{-submodule}}$ if and only if $(\triangle \otimes \partial_1^*)S \subset S$.

A similar definition can be given for the module $\partial^* \otimes \mathcal{M}'$.

Definition 9.3. A left ideal L (resp. right ideal I) of $\partial \otimes \partial^*$ is called a $\underline{\mathcal{G} \otimes \partial_1^*\text{-unitary}}$ left (resp. right) ideal if and only if $(\mathcal{G}\otimes\partial_1^*)L = L$ (resp. $I(\mathcal{G}\otimes\partial_1^*) = I$). An ideal of $\partial \otimes \partial^*$ is called a $\mathcal{G} \otimes \partial_1^*$-unitary ideal if and only if it is both a $\mathcal{G} \otimes \partial_1^*$-unitary left and right ideal of $\partial \otimes \partial^*$.

A similar definition can be given for the algebra $\triangle \otimes \partial^*$. The following theorem is an improved version of the correspondence theorem.

Theorem 9.4 (Correspondence theorem) [41, 43]

Let $\mathcal{M}$ be a vector space over a field Φ, ∂ an irreducible algebra of linear transformation in $\mathcal{M}$, $\triangle$ the centralizer of $\mathcal{M}$ and $\mathcal{G}$ the socle of ∂. Let ∂^* be a Φ-algebra which is faithful as an ∂^*-bimodule over Φ. Denote the dual pair of vector spaces over $\triangle$ associated with ∂ by $(\mathcal{M}, \mathcal{M}')$. Then the following statements hold:

(i) The lattice of right $\partial \otimes \partial_1^*$-submodules of $\mathcal{M} \otimes \partial^*$ is isomorphic to the lattice of $\triangle \otimes \partial_1^*$-unitary right ideals of $\triangle \otimes \partial^*$; the lattice of left $\partial \otimes \partial_1^*$-submodules of $\mathcal{M}' \otimes \partial^*$ is isomorphic to the lattice of $\triangle \otimes \partial_1^*$-unitary left ideals of $\triangle \otimes \partial^*$.

(ii) The lattice of left (resp. right) $\triangle \otimes \partial_1^*$-submodules of $\mathcal{M} \otimes \partial^*$ (resp. $\mathcal{M}' \otimes \partial^*$) is isomorphic to the lattice of $\mathcal{G} \otimes \partial_1^*$-unitary left (resp. right) ideals of $\partial \otimes \partial^*$.

(iii) The lattice of $\triangle \otimes \partial_1^*$-unitary ideals of $\triangle \otimes \partial^*$ is isomorphic to the lattice of $\mathcal{G} \otimes \partial_1^*$-unitary ideals of $\partial \otimes \partial^*$.

By using our theorem 9.4, we can deduce the following theorem for tensor products of primitive rings and the results obtained by Lanski, Resco and Small [22] will hence become its immediate corollary:

Theorem 9.5 [41,43] Let ∂ be a (left) primitive algebra over a field Φ with an associated division algebra $\triangle$ and ∂^* be any algebra. Then $\partial \otimes_\Phi \partial^*$ is (left) primitive if $\triangle \otimes \partial^*$ is (left) primitive.

Corollary 9.6 (Lanski, Resco and Small). Let ∂ be a primitive algebra with non-zero socle and associated division algebra $\triangle$. Let ∂^* be an arbitrary algebra. Then $\triangle \otimes_\Phi \partial^*$ is primitive if and only if $\partial \otimes_\Phi \partial^*$ is primitive.

The following theorems elucidate various situations of the tensor algebra $\partial \otimes \partial^*$, where ∂ is an algebra over the field Φ.

Theorem 9.7[43]. Let ∂ be an irreducible algebra of linear transformations on the vector space $\mathcal{M}$ over Φ. Let $\triangle$ be the centralizer of $\mathcal{M}$ and $\mathcal{G}$ the socle of ∂. If ∂^* is an algebra over Φ, then $\triangle \otimes \partial^*$ is a simple algebra having some minimal one-sided ideals if and only if $\partial \otimes \partial^*$ is a primitive algebra with socle $\mathcal{G} \otimes \partial^*$.

Theorem 9.8 [43] Let (∂, ∂^*) be a pair of primitive algebras over Φ with non-zero socles $\mathcal{G}$ and $\mathcal{G}^*$ respectively. Let $(\triangle, \triangle^*)$ be its corresponding pair of associated division algebras with respect to (∂, ∂^*). Then $\triangle \otimes_\Phi \triangle^*$ is simple Artinean if and only if $\partial \otimes_\Phi \partial^*$ is a primitive algebra with non-zero socle $\mathcal{G} \otimes_\Phi \mathcal{G}^*$.

Theorem 9.9[43]. Let ∂ be a primitive algebra with non-zero socle $\mathcal{G}$ and be associated with a division algebra $\triangle$. Let ∂^* be a Φ-algebra which is faithful as an ∂^*-bimodule. Then

(i) $\partial \otimes_\Phi \partial^*$ is semi-prime (prime) if and only if $\triangle \otimes_\Phi \partial^*$ is semi-prime (prime);

(ii) if H is the intersection of all ideals of $\partial \otimes \partial^*$ and H' is the intersection of all ideals of $\triangle \otimes \partial^*$, then $H \neq 0$ if and only if $H' \neq 0$.

(iii) the nil radical of $\partial \otimes \partial^*$ is non-zero if and only if the nil radical of $\triangle \otimes \partial^* \neq 0$.

10   Free modules and endomorphism rings

We have already noticed in §7 that any two vector spaces $_F\mathcal{M}$ and $_G\mathcal{N}$ over the division rings F and G are semi-linear isomorphic if and only if there exists a ring isomorphism σ of $\mathrm{End}_F\mathcal{M}$ onto $\mathrm{End}_G\mathcal{N}$. It would be interesting if this result can be extended to a more general form, that is, extending to a module which is determined by its endomorphism ring. In fact, Wolfson [35] showed in 1962 that any isomorphism between $\mathrm{End}_F\mathcal{M}$ and $\mathrm{End}_G\mathcal{N}$ can be induced by a semi-linear isomorphism S of $_F\mathcal{M}$ onto $_G\mathcal{N}$ such that $\omega^\sigma = S^{-1}\omega S$ for all $\omega\epsilon\mathrm{End}_F\mathcal{M}$, subject to F and G both being principal ideal domains. We state here a more general result concerning free modules which contains Wolfson's theorem as its corollary.

Before stating our result, we need the following crucial lemma:

Lemma 10.1 Let $_F\mathcal{M}$ be a free module over an arbitrary ring F with identity and let $\Omega = \mathrm{End}_F\mathcal{M}$. Let $\{\mu_i\}_{i\epsilon\Gamma}$ be an arbitrary basis of $_F\mathcal{M}$ and $\{E_i\}_{i\epsilon\Gamma}$ a subset of idempotents of Ω satisfying $\mu_i E_i = \mu_i, \mu_j E_i = 0$, $i \neq j$ for all $i, j\epsilon\Gamma$. (Note: we call such $\{E_i\}_{i\epsilon\Gamma}$ a corresponding basis in Ω.) Write $\mathcal{G} = \sum_{i\epsilon\Gamma} \Omega E_i$. Then we have the following statments:

(i) For any element $E_1\epsilon\{E_i\}_{i\epsilon\Gamma}$, we there is a free module $E_1\Omega = E_1\mathcal{G} = \sum_{i\epsilon\Gamma} K\alpha_i$

over $K = E_1 \Omega E_1$ with a basis $\{\alpha_i\}_{i \epsilon \Gamma}$ such that $\mu_i = \mu_1 \alpha_i$, $\alpha_i = E_1 \alpha_i = \alpha_i E_i$ for all $i \epsilon \Gamma$.

(ii) Denote $A = \sum_{i \epsilon \Gamma} K \alpha_i$. Then $\mathrm{End}_K A = \Omega$.

(iii) There exists a semi-linear isomorphism S of $_F \mathcal{M}$ onto $_K A$ such that $\omega = S^{-1} \omega S$ for all $\omega \epsilon \Omega$.

Definition 10.2 An isomorphism σ of $\Omega = \mathrm{End}_F \mathcal{M}$ onto $\Omega' = \mathrm{End}_G \mathcal{N}$ is called <u>strict</u> if and only if there exists a corresponding basis $\{E_i\}_{i \epsilon \Gamma}$ in Ω and a corresponding basis $\{e_j\}_{j \epsilon \Gamma'}$ in Ω' such that $e E^\sigma \Omega' = e \Omega'$, $\Omega' e E^\sigma = \Omega' E^\sigma$ for some elements $E \epsilon \{E_i\}_{i \epsilon \Gamma}$ and $e \epsilon \{e_j\}_{j \epsilon \Gamma'}$.

The strict isomorphism can now be related with the semi-linear isomorphism between $_F \mathcal{M}$ and $_G \mathcal{N}$, by using lemma 10.1.

Theorem 10.3. Let $_F \mathcal{M}$ and $_G \mathcal{N}$ be free modules over arbitrary rings F and G with identity. Then we have

(i) if σ is a strict isomomorphism of Ω onto Ω', then there exists a semilinear isomorphism S of $_F \mathcal{M}$ onto $_G \mathcal{N}$ such that $\omega^\sigma = S^{-1} \omega S$ for all $\omega \epsilon \Omega$.

(ii) Conversely, if S^* is a semi-linear isomorphism of $_F \mathcal{M}$ onto $_G \mathcal{N}$, then there exists a strict isomorphism σ of Ω onto Ω' such that $\omega^\sigma = S^{*-1} \omega S^*$ for all $\omega \epsilon \Omega$.

We call a free module $_F \mathcal{M}$ <u>direct</u> if and only if every direct summand of $_F \mathcal{M}$ is still free. By using this terminology, the generalized version of Wolfson's theorem can be stated as follows:

Theorem 10.4 (Generalized Wolfson's theorem) [45].

Let $_F \mathcal{M}$ and $_G \mathcal{N}$ be direct free modules over arbitrary rings F and G respectively. Assume that these two rings have identities but are without proper zero divisors. If σ is an isomorphism from the ring $\Omega = \mathrm{End}_F \mathcal{M}$ onto the ring $\Omega' = \mathrm{End}_G \mathcal{N}$, then there exists a semi-linear isomorphism S of $_F \mathcal{M}$ onto $_G \mathcal{N}$ such that $\omega^\sigma = S^{-1}\omega S$ for all $\omega\epsilon\Omega$.

11 Infinite matrix rings and Morita equivalence

In the previos section, a necessary and sufficient condition has been found for the isomorphism σ between the ring $\Omega = \mathrm{End}_F\mathcal{M}$ and the ring $\overline{\Omega} = \mathrm{End}_G\mathcal{N}$ which is induced by a semi-linear isomorphism S between the free module $_F\mathcal{M}$ and the free module $_G\mathcal{N}$. But in general, not all isomorphisms between free modules can be induced by semi-linear isomorphisms. It was noticed by Bolla [3] in 1985 that isomorphism between endomorphism rings of free modules can be described by the categorical equivalence of rings. In fact, the work of Bolla indicated that such isomorphism between endomorphism rings of free modules is closely related to the Morita equivalence of their corresponding base rings. On this aspect, Stephenson [32], dating back to 1965, proved that the Morita equivalence of rings can be determined by the isomorphism of matrix rings with finitely many non-zero entries in their matrices. Camillo [8], in 1984, made significant progress on this result by proving that the Morita equivalence of rings can also be determined by the isomorphism of some matrix rings which are composed of the $\mathbb{N} \times \mathbb{N}$ row finite matrices. His method of approach is somewhat different from the method of Stephenson. Later on, the infinite matrix rings and their relationships with Morita equivalence of rings were investigated by Abrams [1] and Garcia [13]. Just recently, Xu, Shum and Fong [47] have extended the study to the matrices with an arbitrary index set Γ. By using the techniques of rings developed by Xu described in §2, some further results on infinite matrix rings are obtained. These results are in fact to unify and extend all the results previously obtained by Stephenson, Camillo, Abrams and Garcia, in particular, all matrix rings with a finite or countable infinite indexed set are swathed together as a special case of the matrix rings with an arbitrary index set

Γ. Moreover, the well known Bolla's theorem can be generalized by using the above results. In addition, Bolla's theorem has been recently extended from progenerators to quasi-progenerators by Lok and Shum in [24]

For the sake of convenience to the reader, we restate here the definition of Morita equivalence of rings for reference.

Definition 11.1 Let R and R' be two rings with identity. Let $_R\mathcal{M}$ and $_{R'}\mathcal{M}$ be the categories of left R-modules and left R'-modules respectively. Then R and R' are said to be <u>Morita equivalent</u>, denote by $R \approx R'$, if $_R\mathcal{M}$ and $_{R'}\mathcal{M}$ are isomorphic.

Throughout this section, the following notation will be adopted:

Notation 11.2 Let $\mathbb{N}$ be the set of all natural numbers and Γ an arbitrary index set.

(i) Denote the ring of all row-finite $\mathbb{N} \times \mathbb{N}$-matrices over the ring R by $\mathrm{RFM}_\mathbb{N}(R)$, and the ring of all row-finite $\Gamma \times \Gamma$-matrices over R by $\mathrm{RFM}_\Gamma(R)$;

(ii) The subring of $\mathrm{RFM}_\mathbb{N}(R)$ consisting of all matrices with a finite number of non-zero entries over R is denoted by $FM_\mathbb{N}(R)$; the subring $\mathrm{FM}_\Gamma(R)$ of $\mathrm{RFM}_\Gamma(R)$ is similarly defined.

(iii) $\mathrm{FC}_\mathbb{N}(R)$ means the subring consisting of all matrices in $RFM_\mathbb{N}(R)$ which have at most a finite number of non-zero columns; the subring $FC_\Gamma(R)$ is similarly defined.

(iv) $CRFM_\Gamma(R)$ means the subring of $RFM_\Gamma(R)$ which consists of all the column-finite matrices over R.

(v) $\mathcal{M} = \sum_{i\epsilon\Gamma} F\mu_i$ and $\mathcal{N} = \sum_{j\epsilon\Gamma'} Gv_j$ are free modules over the rings F and G with identity respectively, where the index sets Γ and Γ' are arbitrary.

Concerning ring isomorphisms between endomorphism rings, it was noticed by Bolla [3] in 1985 that, from the category point of view, a finite dimensional vector space can be regarded as a progenerator of the underlying category of modules. Thus, the isomorphism between matrix rings can be viewed as an isomorphism between the endomorphism rings of progenerators.

The following theorem due to Xu [45] can be regarded as a generalized version of Bolla's theorem because the restricting condition: " countably infinite basis for free modules" has been removed. The theorem now reads as follows:

Theorem 11.3 [45] (A Generalized version of Bolla's theorem)

Consider the endomorphism rings $\Omega = \mathrm{End}_F \mathcal{M}$ and $\Omega' = \mathrm{End}_G \mathcal{N}$. If there exists a ring isomorphism σ of Ω onto Ω' then there exists a unique (up to natural-isomorphism) categorical equivalence $T_\sigma : {}_F\mathcal{M} \to {}_G\mathcal{M}$ such that $T_\sigma(\mathcal{M}) = \mathcal{N}$ and $T_\sigma(\omega) = \omega^\sigma$ for all $\omega \epsilon \Omega$.

Bolla's theorem can also be extended to quasi-progenerators. The following result is due to Fuller [12].

Theorem 11.5 Let R be a ring (not necessarily with identity) and A a ring with identity. If ${}_A U_R$ is a bimodule such that U_R has a spanning set, then the following conditions are equivalent:

(i) The functors

$$\mathrm{Hom}_R(U, \_) : C[U_R] \to \mathcal{M}_A$$

and

$$\_ \otimes_A U : \mathcal{M}_A \to C[U_R]$$

are inverse categorical equivalent, where $C[U_R]$ is the smallest full subcategory of $\mathcal{M}_R$ that contains U_R and it closed under submodules, epimorphic images and direct sums.

(ii) U_R is a quasi-progenerator and A is canonically isomorphic to $\mathrm{End}_R(U)$.

The following theorem is also an extended version of Bolla's theorem, obtained by using the techniques in homological algebra.

Theorem 11.6 [24] Let U_R and V_S be quasi-progenerators of $\mathcal{M}_R$ and $\mathcal{M}_S$ over the rings R and S respectively. Let

$$\theta : \mathrm{End}_R(U) \rightarrow \mathrm{End}_S(V)$$

be a ring isomorphism. Then there exists a categorical equivalence

$$F_\theta : C[U_R] \rightarrow C[V_S]$$

which is unique (up to natural isomorphism) such that $F_\theta(U_R) = V_S$ and $F_\theta(f) = \theta(f)$ for all $f \in \mathrm{End}_R(V)$.

As progenerators are quasi-progenerators, theorem 11.6 is clearly an extension version of Bolla's theorem [3].

Garcia observed in [13] that the Morita equivalence of rings can be determined by the intermediate matrix ring $FC_\mathbb{N}(R)$. His result has been extended to the matrix ring $FC_\Gamma(R)$ by Xu [41] as well. Most of the results on Morita equivalence of rings obtained by Stephenson, Camillo, Abrams and Garcia are hence included in the following theorem.

Theorem 11.7 [47] Let R and R' be two arbitrary rings with identity. Let Γ and Γ' be two arbitrary index sets. The following statements hold:

(i) if $RFM_\Gamma(R) \cong RFM_{\Gamma'}(R')$ under an isomorphism σ, then σ is also an isomorphism restricted to $FC_\Gamma(R) \cong FC_{\Gamma'}(R')$. Conversely, if $FC_\Gamma(R) \cong FC_{\Gamma'}(R')$ under an isomorphism ψ, then ψ can be uniquely extended to an isomorphism from $RFM_\Gamma(R)$ onto $RFM_{\Gamma'}(R)$.

(ii) if $FM_\Gamma(R) \cong FM_{\Gamma'}(R)$ under an isomorphism ψ, then ψ can be uniquely extended to an isomorphism from $RFM_\Gamma(R)$ onto $RFM_{\Gamma'}(R)$.

(iii) if $FM_\Gamma(R) \cong FM_{\Gamma'}(R')$ under an isomorphism ψ, then ψ can be uniquely extended to an isomorphism from $CRFM_{\Gamma'}(R')$ onto $CRFM_{\Gamma'}(R)$. Conversely, if $CRFM_\Gamma(R) \cong CRFM_{\Gamma'}(R')$ under an isomorphism σ then σ is also an isomorphism restricted to $FM_\Gamma(R) \cong FM_{\Gamma'}(R)$ if and only if there exists an element $E_{11} \in \{E_{ii}\}_{j \in \Gamma}$ and element $e_{11} \in \{e_{jj}\}_{j \in \Gamma'}$, such that $E_{11}^\sigma \in FM_{\Gamma'}(K')$ and $e_{11}^{\sigma^{-1}} \in FM_\Gamma(K)$, where $\{E_{ij}\}_{i,j \in \Gamma}$ and $\{e_{ij}\}_{i,j \in \Gamma'}$, are the usual matrix units of $FM_\Gamma(K)$ and $FM_{\Gamma'}(K')$, respectively.

The following question was asked by Abrams in [1]: "Do the intermediate matrix rings, that is, the rings T with $FM_\mathbb{N}(R) \subset T \subset RFM_\mathbb{N}(R)$, also determine the Morita equivalence of rings?". We now attempt to answer this question, by considering those intermediate rings T sitting in between $FM_\Gamma(R)$ and $RFM_\Gamma(R)$ for any arbitrary index set Γ. In other words, we try to solve Abram's problem in a broader sense.

For brevity, we just let $\Omega = RFM_\Gamma(R)$, $\mathcal{G} = FC_\Gamma(R)$ and $\mathcal{F} = FM_\Gamma(R)$. Likewise, let $\Omega' = RFM_{\Gamma'}(R')$, $\mathcal{G}' = FC_{\Gamma'}(R')$ and $\mathcal{F}' = FM_{\Gamma'}(R')$ respectively. Now, form the following sets:

$$\Lambda = \{T | T \text{ is a subring such that } \mathcal{F} \subseteq T \subseteq \Omega\}$$
$$\text{and } \Lambda' = \{T' | T' \text{ is a subring such that } \mathcal{F}' \subseteq T' \subseteq \Omega'\}.$$

The properties of the intermediate ring T are hence given below:

Theorem 11.8 [47].

(i) $T \cap \mathcal{G}$ is a right ideal of Ω for any $T \epsilon \Lambda$.

(ii) Let $T \epsilon \Lambda$ and $T' \epsilon \Lambda'$. Let σ be a ring isomorphism from T to $T' = T^\sigma$. Then σ can be uniquely extended to a ring isomorphism from Ω to Ω' if and only if $(T \cap \mathcal{G})^\sigma \subset \mathcal{G}'$ and $(T' \cap \mathcal{G}')^{\sigma^{-1}} \subset \mathcal{G}$.

By using the above theorem , one can easily establish the following isomorphism theorem for intermediate rings without too much difficulty.

Theorem 11.9 [47]. Let T and T' be intermediate subrings such that $\mathcal{F} \subseteq T \subseteq \mathcal{G}$ and $\mathcal{F}' \subseteq T' \subseteq \mathcal{G}'$. Denote $\mathcal{L}_T = \{\omega\epsilon\Omega|\omega T \subseteq T\}$ and $\mathcal{L}'_{T'} = \{\omega'\epsilon\Omega'|\omega'T' \subset T'\}$. Then the following statements hold:

(i) If σ is a ring isomorphism from T to T' then σ can be uniquely extended to a ring isomorphism σ' from Ω to Ω' such that $\mathcal{L}_T^{\sigma'} = \mathcal{L}'_{T'}$.

(ii) If σ is a ring isomorphism from $\mathcal{L}_T$ to $\mathcal{L}'_{T'}$, then $T^\sigma = T'$ if and only if there exists an element $E_{11} \in \{E_{jj}\}_{j\in\Gamma}$ and an element $e_{11} \in \{e_{jj}\}_{j\in\Gamma'}$, such that $E_{11}^\sigma \in T'$ and $e_{11}^{\sigma^{-1}} \in T$, where $\{E_{ij}\}_{i,j\in\Gamma}$ and $\{e_{ij}\}_{ij\in\Gamma'}$ are the usual matrix units of $FM_\Gamma(K)$ and $FM_{\Gamma'}(K')$, respectively.

Hence, by combining theorems 11.7 and 11.8 and also theorem 11.3 (the generalized version of Bolla's theorem), the question raised by Abrams in [1] can be easily answered. We omit the details.

In closing this section, we consider the $\aleph_\nu$-matrices.

Definition 11.10 Let $(r_{ij})_{\Gamma\times\Gamma}$ be a $\Gamma \times \Gamma$-matrix. A column of $(r_{ij})_{\Gamma\times\Gamma}$ is called a $\aleph_\nu$-column if the cardinal number of the set of its non-zero entries is less than $\aleph_\nu$.

A matrix $(r_{ij})_{\Gamma\times\Gamma}$ is called a $\underline{\aleph_\nu\text{-matrix}}$ if each column of $(r_{ij})_{\Gamma\times\Gamma}$ is a $\aleph_\nu$-column; a matrix subring M of $RFM_\Gamma(R)$ is called a $\underline{\aleph_\nu\text{-matrix subring}}$ if every matrix of M is $\aleph_\nu$-matrix; A $\aleph_\nu$-matrix subring is called $\underline{\aleph_\nu\text{-maximal}}$, denoted by $\aleph_\nu - RFM_\Gamma(R)$, if every $\aleph_\nu$-matrix of $RFM_\Gamma(R)$ is contained in this matrix subring.

Using the above definition, the following theorem can be formulated.

Theorem 11.11 [47] (Isomorphism theorem for $\aleph_\nu$-maximal subrings)

Let $T_\nu = \mathcal{G} \cap \aleph_\nu\text{-RFM}_\Gamma(R)$ and $T'_\nu = \mathcal{G}' \cap \aleph_\nu\text{-RFM}_{\Gamma'}(R')$. Then $\aleph_\nu\text{-RFM}_\Gamma(R) = \{\omega\epsilon\Omega|\omega T_\nu \subseteq T_\nu\}$, $\aleph_\nu\text{-RFM}_{\Gamma'}(R') = \{\omega'\epsilon\Omega'|\omega'T'_\nu \subset T'_\nu\}$. Furthermore, if $T_\nu \cong T'_\nu$ under σ, then σ can be uniquely extended to a ring isomorphism σ' such that $\Omega \cong \Omega'$ under σ' and $(\aleph_\nu\text{-RFM}_\Gamma(R))^{\sigma'} = \aleph_\nu\text{-RFM}_{\Gamma'}(R')$.

12   Morita-like equivalence of infinite matrix subrings

In this section, we will extend the concept of Morita equivalence of rings to the concept of Morita-like equivalence. Let $\mathcal{M}_R$ and $\mathcal{M}_S$ be the categories of right R-modules over the rings R and S respectively. In order to extend the Morita equivalence to the Morita-like equivalence, the crucial step is to consider the categorical equivalence of the complete additive subcategories $\mathcal{C}_R$ of $\mathcal{M}_R$ and $\mathcal{C}_S$ of $\mathcal{M}_S$ instead of considering the equivalence of the categories of right modules $\mathcal{M}_R$ and $\mathcal{M}_S$ itselves. It is obvious to see that the class of Morita-like equivalent rings includes the class of all Morita equivalent rings as its subclass, but not vice versa.

Morita-like equivalence of infinite matrix subrings has recently been investigated by Xu, Shum and Turner-Smith in [46]. Some results of Garcia [13] on Morita-like equivalence of categories of modules were generalized and strengthened by them in [46]; however, the method introduced there uses mainly the techniques of ring theory and is quite different from the method adopted by Garcia. Some further results on Morita-like equivalence of matrix rings are obtained in a forthcoming paper by Xu, Shum and Fong [47].

Hereafter, all modules over a ring with identity are assumed to be unital. For any ring R (with or without an identity) , we just denote the category of all right modules over R by $\mathcal{M}_R$ and the symbol $\mathcal{C}_R$ stands for the full subcategory of $\mathcal{M}_R$ such that

$$\text{Obj}\,(\mathcal{C}_R) = \{N\epsilon\ \text{Obj}\,(\mathcal{M}_R)|NR = N\}$$

We now formulate the following definition.

Definition 12.1 Two rings R and S are called Morita-like equivalent (written $R \sim S$) if $\mathcal{C}_R$ and $\mathcal{C}_S$ are complete additive subcategories of $\mathcal{M}_R$ and $\mathcal{M}_S$ respectively such that $\mathcal{C}_R$ and $\mathcal{C}_S$ are categorically equivalent, that is, $\mathcal{C}_R \approx \mathcal{C}_S$.

Remark: If R has an identity, since we assume each $N \epsilon \text{Obj} \ (\mathcal{M}_R)$ is unital, then $NR = N$. It hence follows that if R and S both contain an identity, then the Morita-like equivalence of R and S is exactly the same as the Morita equivalence of R and S.

Before we state our theorem concerning Morita-like equivalence on rings, we recall that $_RM = \sum_{i \epsilon \Gamma} R\mu_i$ is a left free module over a ring R with identity and with basis $\{\mu_i\}_{i \epsilon \Gamma}$; Ω in the ring $\text{End}_R M$ and $\{E_i\}_{i \epsilon \Gamma} \subseteq \Omega$ is a set of basic projections (we call such $\{E_i\}_{i \epsilon \Gamma}$ a corresponding basis in Ω, see Lemma 10.1 in §10) with respect to the basis $\{\mu_i\}_{i \epsilon \Gamma}$ such that $\Phi = \sum_{i \epsilon \Gamma} \Omega E_i$. Also, the notations given in section 11 will be retained as before.

Suppose $\ell^2 = \ell \epsilon \Phi$ such that $\Phi \ell \Phi = \Phi$. Then we can establish the following theorem on Morite-like equivalence for $\ell \Phi \ell$ and Φ.

Theorem 12.2 [46]. Let $\mathbb{A} = \ell \Phi$ and $\mathbb{A}' = \Phi \ell$. Then the functor pair $(\otimes_{\ell \Phi \ell} \mathbb{A}, \ \otimes_\Phi \mathbb{A}')$ defines an equivalence of complete additive subcategories between $\mathcal{C}_{\ell \Phi \ell}$ and $\mathcal{C}_\Phi$ (so that $\mathcal{C}_{\ell \Phi \ell} \approx \mathcal{C}_\Phi$), and hence $\ell \Phi \ell$ and Φ are Morita-like equivalent (i.e. $\ell \Phi \ell \sim \Phi$).

In fact, we simply consider the intermediate subring T satisfying:

$$\mathcal{F} = FM_\Gamma(R) \subset T \subset \mathcal{G} = FC_\Gamma(R)$$

and let $\mathcal{L}_T = \{\omega \epsilon \Omega \ | \omega T \subset T\}$. If $\ell^2 = \ell \epsilon \mathcal{L}_T$ and $T\ell T = \ell$, then the result obtained in theorem 12.2 can be carried over verbatim to a given intermediate ring T.

Theorem 12.3 [47] Let $\mathbb{A} = \ell T$ and $\mathbb{A}' = T\ell$. Then the functor pair $(\otimes_{\ell T \ell} \mathbb{A}, \ \otimes_T \mathbb{A}')$ defines a categorical equivalence of the complete additive subcategories $\mathcal{C}_{\ell T \ell}$ and $\mathcal{C}_T$ and hence the rings $\ell T \ell$ and T are Morita-like equivalent, i.e., $\ell T \ell \sim T$.

The following theorem is a matrix representation for rings that are Morita-like equivalence:

Theorem 12.4 [46] Let $M_\Gamma(R)$ be the ring of row-finite $\Gamma \times \Gamma$ matrices of the ring R over an arbitrary index set Γ. Let M_Γ^0 be the subring of $M_\Gamma(R)$ consisting of almost zero-column matrices over R. $\left(\text{i..e. } M_\Gamma^0(R) = FM_\Gamma(R) \text{ by the notation given in 11.2 (see §11)}\right)$. If $\ell^2 = \ell\epsilon M_\Gamma(R)$ is such that $M_\Gamma^0(R)\ell M_\Gamma^0(R) = M_\Gamma^0(R)$ then $\ell M_\Gamma^0(R)\ell \sim M_\Gamma^0(R)$. Moreover, the functor pair $(\otimes_S \ell M_\Gamma^0(R), \otimes_T M_\Gamma^0(R)\ell)$ defines an equivalence between the complete additive subcategories $\mathcal{C}_S$ and $\mathcal{C}_T$, where $S = \ell M_\Gamma^0(R)\ell$ and $T = M_\Gamma^0(R)$.

The statement corresponding to theorem 12.4 for intermediate rings now reads as follows:

Theorem 12.5 [47] Let T and T' be intermediate rings satisfying $\mathcal{F} \subseteq T \subset \mathcal{G}$ and $\mathcal{F}' \subseteq T' \subset \mathcal{G}'$ respectively. Then the following statements hold:

(i) $\ell_n \mathbb{M}_n(R)\ell_n \sim \mathbb{M}_n(R) \sim R \sim \mathcal{F} \sim \ell_\mathcal{F}\mathcal{F}\ell_\mathcal{F} \sim T \sim \ell_T T\ell_T \sim \mathcal{G} \sim \ell_\mathcal{G}\mathcal{G}\ell_\mathcal{G}$ where $\ell_n^2 = \ell_n\epsilon\mathbb{M}_n(R)$, $\ell_\mathcal{F}^2 = \ell_\mathcal{F}\epsilon\mathcal{L}_\mathcal{F}$, $\ell_T^2 = \ell_T\epsilon\mathcal{L}_T$, $T\ell_T T = T$ and $\ell_\mathcal{G}^2 = \ell_\mathcal{G}\epsilon\mathcal{G}$, $\mathcal{G}\ell_\mathcal{G}\mathcal{G} = \mathcal{G}$.

(ii) If there exist T and T' such that $\mathcal{L}_T \cong \mathcal{L}_{T'}$, then $\ell_n\mathbb{M}_n(R)\ell_n \sim \mathbb{M}_n(R) \sim R \sim S \sim \ell S\ell \sim S' \sim \ell'S'\ell' \sim R' \sim \mathbb{M}_n(R') \sim \ell_n'\mathbb{M}_n(R')\ell_n'$ for any intermediate rings S and S' satisfying $\mathcal{F} \subseteq S \subseteq \mathcal{G}$ and $\mathcal{F}' \subseteq S' \subseteq \mathcal{G}'$, where $\ell^2 = \ell\epsilon\mathcal{L}_S$, $S\ell S = S$ and $\ell'^2 = \ell'\epsilon\mathcal{L}_{S'}$, $S'\ell'S' = S'$.

In closing this paper, we remark that not only the results of Morita-like equivalence of matrix rings can be carried over to intermediate rings, but in addition, most of the results concerning the isomorphism of primitive rings obtained by Jacobson [17] can also be sharpened and restricted to intermediate rings. The details will be shown in a forthcoming paper by Xu and Shum in [48]. Finally, we would like to mention that the problem posed by Xu at the end of [43] concerning the Morita-like equivalence of rings has just been answered negatively by Simón in [33]. The problem reads as follows: "Let R be a ring with identity and S another ring, but not necessarily with identity. If $R \sim S$, is there an idempotent matrix ℓ in the ring $M_\Gamma(R)$ such that $S \cong \ell M_\Gamma^0(R)\ell$"?

References

[1] G.D. Abrams, "Infinite matrix type which determine Morita equivalence", Arch. Math. Vol.46 (1986), 33-37.

[2] S.A. Amitsur, "Rings of quotients and Morita contexts", J.Algebra 17 (1971), 273-298.

[3] M.L. Bolla, "Isomorphisms between infinite Matrix Rings", Linear Algebra and its applications, (1985), 239-247.

[4] A.K. Boyle and E.M. Feller, "Semicritical modules and K-Primitive rings in module theory", Springer Verlag Lecture Notes, 700 (1970), 57-74.

[5] A.K. Boyle, M.G. Desh-paude and E.M. Feller, "On nonsingularly K-primitive rings", Pacific J. Math. 68 (1977), 303-311.

[6] G.M. Brokskii, " Endomorphism rings of free modules over perfect rings", Nat. Sb. (N.S.) 88 (1971), 137-147.

[7] G.M. Brokokii, "Endomorphism rings of free modules", Mt. Sb (N.S.) 94 (136) (1974), 226-242.

[8] V. Camillo, "Morita equivalence and infinite matrix rings," Proc. Amer. Math. Soc. (2) 90 (1984) 186-188.

[9] G.M. Cukerman, "Rings of endomorphisms of free modules" Sibirsk Mat Z., 7 (1966), 1161-1167.

[10] G. Deale and W.K. Nicholson, "Endoprimitive rings" J. Algebra 70 (1981), 548-560.

[11] C. Faith and Y. Utumi, "On a new proof of Litoff's theorem", Acta Math. Acad. Sci. Hungar 14 (1963), 369-371.

[12] K.R.Fuller, "Density and Equivalence", J.Algebra 29 (1974), 550-628.

[13] J.L. Garcia, "The finite column matrix ring", Proc. of the first Belgian-Spanish week on Algebra and Geometry (1988), 64-79.

[14] V.I. Geminlern, "Self injective rings of endomorphisms of free modules", Mat. Z. 6 (1969), 533-540.

[15] V.I. Geminlern, "Rings of endomorphisms of the injective hulls of free modules", Uspebi Nat. Nauk. 24(1969) 6, (150) 185-186.

[16] I.N. Herstein, "Noncommutative rings, " Math. Associ. Amer. 1973.

[17] N. Jacobson, "Structure of rings" Amer. Math. Soci. Colloquium Vol. 36, revised edition, 1964, Providence, R.I.

[18] N. Jacobson, "Basic Algebra II" 1980.

[19] K. Koh and A.C. Mewborn, "A class of primitive rings", Canad. Math. Bull. 9 (1966), 63-72.

[20] A. Kertesz, "Lectures on Artinian Rings" (Edit by R. Wiegandt), Akademiai Kiado, Budapest ,1987.

[21] T.R. Kezlan, "On K-primitive rings", Proc Amer. Math. Soc. 74 (1) (1979) 24-28.

[22] C. Lanski, R. Resco and L. Small, "On the primitivity of primitive rings" J. Algebra 59 (1979), 395-398.

[23] J. Lambek, "Lectures on rings and modules", Blaisdell, Toronto, London and Waltham (Mass), 1966.

[24] T.M. Lok and K.P. Shum, Isomorphisms between Endomorphism Rings of Quasi-progenerators, Algebra Colloqium, Chinese Academy of Sciences, vol.2 2 (1995), 157-166.

[25] A.C. Mewborn, Quasi-simple modules and weak transitivity in ring theory, Academic Press, New York, (1972), 241-249.

[26] K. Morita, "Duality of Modules and its applications to the theory of rings with minimum conditions", Sci. Rep, Tokyo Kyoiku Daigaku, Sect A6 (1958), 85-142.

[27] W.K. Nicholson and J.F. Walters, "Normal class of rings and tensor products", Comm. Algebra 9(3) (1981), 299-311.

[28] ——, "Normal radicals and normal class of rings" J. Algebra 59, (1979) 5-15.

[29] S.H.Ng and K.P. Shum, "On simple rings with non-trivial socle" P.U.M.A. vol 4 (1993), No.4, 467-470.

[30] A.M. Ortiz, "On the structure of semiprime rings" Proc. Amer. Math. Soc. 38 (1973), 22-26.

[31] R. Resco, "A reduction theorem for the primitivity of tensor products", Math. Z., 170 (1980), 65-76

[32] R.F. Shanny, "Regular endomorphism of free modules", J. London Math. Soc. (2) 4, (1971), 253-354.

[33] J.J. Simón, "Subrings of $\mathbb{RFM}(R)$ and equivalences", preprint.

[34] W. Stephenson, "Characterization of rings and modules by means of ideals", Ph.D thesis, Bedford College, University of London, (1965).

[35] K.G. Wolfson, "Isomorphisms of the endomorphism ring of a free module over a principal left ideal domain", Michigan Math. J. 9 (1962), 69-75.

[36] Xu Yonghua, "A theory of rings that are isomorphic to the complete rings of linear transformations (I)", Acta Math. Sinica, 22(1979), 204-218.

[37] Xu Yonghua, "On the structure of primitive rings having minimal one-sided ideals", Scientia Sinica Vol. 24 (1981), 1056-1065.

[38] —, "On the structure of primitive rings", Chin. Ann. of Math. 4B (2) (1983), 133-143.

[39] —, "Normal primitive rings", Scientia Sinica, (Series A), Vol.26 (1983) 1156-1166.

[40] —, "On the structure of homogeneous completely reducible modules", Chin. Ann. of Math. 5B(1) (1984), 59-66.

[41] —, "On the structure of Kronecher product of primitive algebras", Scientia Sinica Ser.A 27(5) (1984), 449-456.

[42] —, "On the structure of complete ν-normal primitive rings and its applications", J. Algebra 113 (1988), 19-39.

[43] —, "On the tensor products of algebras", J. Algebra 113 (1988), 40-70.

[44] —, "An equivalence between ring F and infinite matrix subring over F", Chin. Ann. of Math. 11B:1 (1990), 66-69.

[45] —,"Isomorphisms between endomorphism rings of free modules", Science in Chin. (Series A), Vol. 35 No.7 (1992), 769-779.

[46] Xu Yonghua, Kar-Ping Shum and Turner-Smith, R.F., "Morita-like equivalence of infinite matrix subrings", J. Algebra 159, (1993), 425-435.

[47] Xu Yonghua, Kar-Ping Shum and Yuen Fong, "Infinite Matrix Rings and Morita-like Equivalence", (submitted for publication)

[48] Xu Yonghua, Kar-Ping Shum, "Some isomorphism theorems for primitive rings", (submitted for publication)

[49] Xu Yonghua, "A note on semi-prime Goldie rings", Rendiconti del circols Matematics di Palermo, Serie II, Tomo XXXVI (1987), 220-240.

[50] ——, "A finite structure theorem between primitive rings and its application to Galois theory", Chin. Ann. of Math. 1:2 (1980), 183-197.

[51] ——, "On the Faith-Utumi theorem", Acta Math Sinica, New Series, (1990), Vol. 6, No.4, 323-326.

[52] J.M. Zelmanowitz, "Weakly primitive rings", Comm. Algebra 9 (1), (1981), 23-45.

[53] ——, "Representations of rings with faithful monoform modules", J. London Math. Soc. 29 (1984), 237-248.

[54] ——, "The structure of rings with nonsingular modules", Trans. Amer. Math. Soc. 278 (1983), 347-359.

Maximal Subgroups of Finite Groups

YANMING WANG

DEPARTMENT OF MATHEMATICS, ZHONGSHAN UNIVERSITY GUANGZHOU 510275, PEOPLE'S REPUBLIC OF CHINA

E-mail: stswym@zsulink.zsu.edu

ABSTRACT. In this paper, we discuss some problems concerning the maximal subgroups of a finite group and explore some relationships between maximal subgroups and the structure of the group.

0. Introduction

The relationship between the properties of maximal subgroups of a finite group G and the structure of G has been studied by many people. In this paper, we limit our discussion to the following three aspects. (I) Frattini-like subgroups. (II) c-normality of maximal subgroups. (III) Maximal A-invariant subgroups.

Let π be a set of primes and π' the complementary set of primes. Let G be a finite group. Then we write $M < \cdot G$ to indicate that M is a maximal subgroup of G. Also, $|G : M|_\pi$ denotes the π-part of $|G : M|$. Consider the following families of subgroups:

Definition 1:

$\mathcal{F} = \{M : M < \cdot G\}$.

$\mathcal{F}_c = \{M : M < \cdot G\}$ with $|G : M|$ is composite.

$\mathcal{F}_\pi = \{M : M < \cdot G, |G : M|_\pi = 1\}$.

$\mathcal{F}_{\pi c} = \mathcal{F}_\pi \cap \mathcal{F}_c$.

$\mathcal{F}^p = \{M : M < \cdot G, N_G(P) \leq M\}$ for a $P \in Syl_p(G)$.

$\mathcal{F}^s = \bigcup_{p \in \pi(G)} \mathcal{F}^p$.

$\mathcal{F}^{pc} = \mathcal{F}^p \cap \mathcal{F}_c$.

$\mathcal{F}^{sc} = \mathcal{F}^s \cap \mathcal{F}_c$.

Definition 2:

$\Phi_\pi(G) = \bigcap\{M : M \in \mathcal{F}_\pi\}$ if $\mathcal{F}_\pi$ is non-empty, otherwise $\Phi_\pi(G) = G$.

$S_\pi(G) = \bigcap\{M : M \in \mathcal{F}_{\pi c}\}$ if $\mathcal{F}_{\pi c}$ is non-empty, otherwise $S_\pi(G) = G$.

$\Phi^p(G) = \bigcap\{M : M \in \mathcal{F}^p\}$ if $\mathcal{F}_p$ is non-empty, otherwise $\Phi^p(G) = G$.

$\Phi^s(G) = \bigcap\{M : M \in \mathcal{F}^s\}$ if $\mathcal{F}^s$ is non-empty, otherwise $\Phi^s(G) = G$.

$S^p(G) = \bigcap\{M : M \in \mathcal{F}^{pc}\}$ if $\mathcal{F}^{pc}$ is non-empty, otherwise $S^p(G) = G$.

$S^s(G) = \bigcap\{M : M \in \mathcal{F}^{sc}\}$ if $\mathcal{F}^{sc}$ is non-empty, otherwise $S^s(G) = G$.

It is clear that all the above subgroups are characteristic subgroups of G.

Project supported in part by National Natural Science Foundation of China and STFG .

1. On Frattini Subgroups

Of late there has been considerable interest in the study of analogs of Frattini subgroups of a finite group and investigation of their properties, particularly, their influence on the structure of the group. Both $\Phi_\pi(G)$ and $S_\pi(G)$ are examples of Frattini-like subgroups. One can easily check the following properties:

Lemma 1.1. *Let $K \trianglelefteq G$. Then*

(1) $\Phi_\pi(G)/K \leq \Phi_\pi(G/K)$; *consequently, if $K \leq \Phi_\pi(G)$, it follows that $\Phi_\pi(G)K/K$* $= \Phi_\pi(G/K)$.

(2) $S_\pi(G)/K \leq S_\pi(G/K)$; *consequently, if $K \leq \S_\pi(G)$, it follows that $S_\pi(G)K/K$* $= S_\pi(G/K)$.

(3) $O_\pi(G)\Phi(G) \leq \Phi_\pi(G) \leq S_\pi(G)$.

(4) $1 = O_\pi(G/O_\pi(G)) = O_\pi(G/\Phi_\pi(G)) = O_\pi(G/S_\pi(G))$.

Lemma 1.2. *Let G be a π-separable group. Then:*

(1) *Every maximal subgroup of G has index with either π-number or π'-number.*

(2) $\Phi_\pi(G) \cap \Phi_{\pi'}(G) = \Phi(G)$ *and* $S_\pi(G) \cap S_{\pi'}(G) = L(G)$ *are supersolvable.*

(3) *If $K \trianglelefteq\trianglelefteq G$ and K is a π'-subgroup, then $K \leq \Phi_\pi(G)$ if and only if $L \leq \Phi(G)$, and $K \leq S_\pi(G)$ if and only if $L \leq L(G)$.*

Lemma 1.3.

(1) $\Phi_\pi(G) = < x : x \in G, x$ *is a π-non-generator of $G >$. (an element x in G is called a π-non-generator, if for any subset $T \subset G$ with $|G :< T >|_\pi = 1$ and $G = < T, x >$ then $G = < T >$).*

(2) *Suppose that $N \trianglelefteq G$ and $U \leq G$. If $N \leq \Phi_\pi(U)$, then $N \leq \Phi_\pi(G)$.*

(3) *If $N \trianglelefteq G$, then $\Phi_\pi(N) \leq \Phi_\pi(G)$.*

Theorem 1.4. *Let G be a finite π-separable group. Then:*

(1) $\Phi_\pi(G)/O_\pi(G) = \Phi(G/O_\pi(G))$ *is a nilpotent π'-group and $\Phi_\pi(G)$ is π-closed.*

(2) *Let $K \trianglelefteq G$ and $K \leq \Phi_\pi(G)$. Then M is π-closed if and only if MK/K is π-closed for every normal subgroup M of G*

(3) *If $G = G_1 \times \cdots \times G_k$, then $\Phi_\pi(G) = \Phi_\pi(G_1) \times \cdots \times \Phi_\pi(G_k)$.*

Theorem 1.5. *Let G be a finite π-separable group. Then:*

(i) $S_\pi(G)/O_\pi(G)$ *is supersolvable.*

(ii) $G/O_\pi(G)$ *is supersolvable, if and only if $G/S_\pi(G)$ is supersolvable, if and only if $S_\pi(G) = G$.*

(iii) $S_\pi(S_\pi(G)) = S_\pi(G)$.

We give only the proof of Theorem 1.5 (i).

Proof. Assume that the result is false and consider a counterexample G with minimal order. Then

(1) $O_\pi(G) = 1$

In fact, if $O_\pi(G) \neq 1$, consider $\overline{G} = G/O_\pi(G)$, by basic properties, $\Phi(G/O_\pi(G)) = \Phi_\pi(G)/O_\pi(G)$ and $O_\pi(G/O_\pi(G)) = 1$. Hence $\Phi_\pi(G)/O_\pi(G) \cong \Phi_\pi(\overline{G})/O_\pi(\overline{G}) = \Phi(\overline{G}/O_\pi(\overline{G})) \cong \Phi(G/\Phi_\pi(G))$, a contradiction.

(2) $\Phi(G) = 1$

If $\Phi(G) \neq 1$, we consider $\overline{G} = G/\Phi(G)$. Since $\Phi(G)$ is nilpotent and $O_\pi(\Phi(G)) \leq \Phi_\pi(G) = 1$ by (1), we have that $\Phi(G)$ is a π'-subgroup. Let $L/\Phi(G) = O_\pi(G/\Phi(G)) \trianglelefteq G/\Phi(G)$. Then $L \trianglelefteq G$ and $\Phi(G)$ is a Hall π'-subgroup of L. By the Schur-Zassenhaus Theorem, ([G] Theorem 6.2.1), there is Hall π-subgroup L_1 of L such that $L = L_1\Phi(G)$ and all the Hall π-subgroups of L are conjugated to L_1 in L. A direct generalization of a Frattini argument yields that $G = N_G(L_1)\Phi(G) = N_G(L_1)$. i.e. $L_1 \leq \Phi_\pi(G) = 1$. This implies that $O_\pi(\overline{G}) = 1$. Since $S_\pi(\overline{G}) = S_\pi(\overline{G})/O_\pi(\overline{G}) \cong S_\pi(G)/O_\pi(G)$. By the choice of G, $S_\pi(G)/O_\pi(G) \cong S_\pi(\overline{G})/O_\pi(\overline{G}) = S_\pi(G)/\Phi(G)$ is supersolvable. By [B-M 1] Theorem 9, $S_\pi(G)$ is supersolvable and so is $S_\pi(G)/O_\pi(G)$, contrary to our assumption.

For simplicity, we denote $S_\pi(G)$ by S and the Fitting subgroup of S by F.

(3) $F = N_1 \times \cdots \times N_k$ where N_i are minimal normal subgroup of G. $G = FL$ with $L \cap F = 1$ for a subgroup of G.

Let N be a minimal normal subgroup of G which is contained in $S_\pi(G)$. Since G is π-separable and $O_\pi(N) \leq O_\pi(G) = 1$, we have that N is a π'-group and so $N \leq S_\pi(G) \cap O_{\pi'}(G)$ is solvable by Lemma 1. Hence N is a solvable minimal normal subgroup of G and so N is an elementary abelian p-subgroup with $p \in \pi'$. Certainly $N \leq F$. Since F is a nilpotent normal subgroup of G, $O_\pi(F) = 1 = \Phi(F)$ by (1) and (2). F is abelian. Let H be maximal among all subgroups of F that can be expressed as the direct product of minimal normal subgroups of G. Then $H \trianglelefteq G$ and H is abelian. Let $L = \min\{T : T \leq G, HT = G\}$. Then $H \cap L = 1$. In fact, $H \cap L \trianglelefteq G$. If $H \cap L \neq 1$, since $\Phi(G) = 1$, $\exists M < \cdot G$ such that $H \cap L \not\leq M$ and so $G = (L \cap H)M$. Note that $L = L \cap G = (L \cap M)(L \cap M)$ and $L \cap M \neq L$. However, $G = LH = (L \cap M)H$, contrary to the minimal choice of L. Since $F = H(F \cap L)$ and $F \cap L \triangleleft FL = G$, if $F \cap L \neq 1$, then there is a minimal normal subgroup N of G with $N \leq F \cap L$. As $H \cap L = 1$, we conclude that $H < N \times H$. This contradicts the maximal choice of H and therefore $F \cap L = 1$. Then $F = H$ and the result follows.

(4) $|N_i|$ is a prime for all $i \in \{1 \cdots k\}$, $S' \leq C_S(F) = F$, S/F is abelian group.

In fact, (3) implies that $|N_i| = p_i^{\alpha_i}$ with $p_i \in \pi'$. Since $\Phi(G) = 1$ and $1 \neq N_i \trianglelefteq G$, $\exists M_i < \cdot G$ with $N_1 \not\leq M$ and $G = N_iM_i$. It is clear that $N_i \cap M_i = 1$ and $|G : M_i| = |N_i| = p_i^{\alpha_i}$ is a π'-number. If $|G : M_i|$ is composite, then $N_i \leq S \leq M_i$, a contradiction. Hence $|N_i|$ is a prime. $G/C_G(N_i) = N_G(N_i)/C_G(N_i) \lesssim Aut(N_i)$ is cyclic. Therefore $G' \leq C_G(N_i)$ and $G' \leq \bigcap_{i=1}^k C_G(N_i) = C_G(F)$. $S' \leq G' \cap S \leq C_S(F)$. Since F is abelian, $S = F(S \cap F)$, $C_S(F) = F(L \cap C_S(F) =: FL_1$, where $L_1 = L \cap C_S(F) \triangleleft L$ and commutes with F, hence $L_1 \trianglelefteq S$. If $L_1 \neq 1$, then there is a minimal normal subgroup N of S which is contained in L_1. N is a subnormal subgroup of G and $O_\pi(N) \leq O_\pi(G) = 1$. This yields that N is a π'-group. $N \leq O_{\pi'}(G)$ and N is solvable by Lemma 1.2, hence N is a solvable minimal normal subgroup of S and so $N \leq S$. Then $1 \neq N \leq F \cap L = 1$, a contradiction. The result now follows.

(5) $S_\pi(G)$ is supersolvable.

Let $M < \cdot S_\pi(G)$. We prove that $|S_\pi(G) : M|$ is a prime. If $F \leq M$, then $|S_\pi(G) : M| = |S_\pi(G)/F : M/F|$ is a prime since $S_\pi(G)/F$ is an abelian group by (4). Suppose that $F \not\leq M$. By (3), $\exists N_i \leq F$ such that $N_i \not\leq M$. Hence $S_\pi(G) = N_iM, N_i \cap M = 1$. By (4), $|S_\pi(G) : M| = |N_i| = p_i$ is a prime. Now, a well known Huppert's theorem yields that $S_\pi(G)$ is supersolvable.

This contradicts the choice of G and completes the proof. $\quad\square$

2. On C-Normality of Maximal Subgroups

The normality of subgroups of a finite group plays an important role in the study of finite groups. It is well known that a finite group G is nilpotent if and only if every maximal subgroup of G is normal in G. As for the class of supersolvable groups, B. Huppert's well known theorem shows that a finite group G is supersolvable if and only if every maximal subgroup of G has prime index in G. In terms of normality, we have that G is supersolvable if and only if every maximal subgroup of G is weakly normal in G. (Refer to [We] Theorem 1.8.7.) In this section, we show that G is solvable if and only if M is c-normal in G for every maximal subgroup M of G. By use of the concept "c-normal", we can make things simpler. Also, some people try to characterize the group structure, by using as few as possible maximal subgroups with certain properties [B-B][B-E][Wa]. We can also easily infer and generalize some known theorems by use of the concept "c-normal".

Definition 2.1: Let G be a group. We call a subgroup H c-normal in G if there exists a normal subgroup N of G such that $HN = G$ and $H \cap N \leq H_G$.

It is clear that a normal subgroup of G is a c-normal subgroup of G but the converse is not true. For example, $S_3 = C_3 \rtimes C_2$, $C_2 \ntrianglelefteq S_3$ but C_2 is c-normal in S_3.

Definition 2.2: We call a group G c-simple if G has no c-normal subgroup except the identity group 1 and G.

We can easily show that G is c-simple if and only if G is simple.

In this section, we give some analogous properties of normal subgroups for c-normal subgroups. We prove that a finite group G is solvable if and only if every maximal subgroup of G is c-normal in G. We also try to minimize the number of the maximal subgroups needed to characterize the structure of G. As applications, we replace "normal subgroups" by the weaker condition "c-normal subgroups" to generalize some known theorems.

It is clear that all the above subgroups are characteristic subgroups of G.

Elementary Properties

Lemma 2.1. *Let G be a group. Then*

    (1) *If H is normal in G, then H is c-normal in G;*

    (2) *G is c-simple if and only if G is simple;*

    (3) *If H is c-normal in G, $H \leq K \leq G$, then H is c-normal in K;*

    (4) *Let $K \trianglelefteq G$ and $K \leq H$. Then H is c-normal in G if and only if H/K is c-normal in G/K.*

Lemma 2.2. *Let G be a finite group. Then*

    (1) *$\Phi^p(G)$ is p-closed for every $p \in \pi(G)$;*

    (2) *$\Phi^s(G)$ is nilpotent;*

    (3) *$S^p(G)$ is p-closed if p is maximal in $\pi(S^p(G))$;*

    (4) *$S^s(G)$ has a Sylow tower.*

Lemma 2.3. *Let G be a finite group. Then*

(a) G *is nilpotent if and only if* $G = \Phi^s(G)$.
(b) G *is nilpotent if and only if* M *is normal in* G *for every* $M \in \mathcal{F}^s$.
(c) G *is nilpotent if and only if* G/N *is nilpotent for a normal subgroup* N *of* G *which is contained in* $\Phi^s(G)$.

Lemma 2.4. *Let G be a finite group. Then*

(a) G *is supersolvable if and only if* $|G; M|$ *is a prime for every* $M \in \mathcal{F}^s$.
(b) G *is supersolvable if and only if* $G = S^s(G)$. G *is nilpotent if and only if* M *is normal in* G *for every* $M \in \mathcal{F}^s$.
(c) G *is supersolvable if and only if* G/N *is supersolvable for a normal subgroup* N *of* G *which is contained in* $S^s(G)$.

In [De], Deskins introduced $\eta(G : M)$, the normal index of a maximal subgroup M in a group G. It is defined as the order of a principal factor that supplements M in G. That is, if K/H is a principal factor of G with the property $MK = G$ and $H \leq M$, then $\eta(G : M) := |K/H|$. For convenient, we assume that $\eta(G : M)$ is well defined. By definition, we can easily prove the following:

Lemma 2.5. *If N is a normal subgroup of a group G and M is a maximal subgroup of a finite group G such that $N \leq M$, then $\eta(G/N : M/N) = \eta(G : M)$.*

Theorem 2.6. *Let G be a finite group. Then G is solvable if and only if every maximal subgroup of G is c-normal in G.*

Proof. Suppose that every maximal subgroup M of G is c-normal in G. We prove that G is solvable. Assume this is false and let G be a minimal counterexample. If G is simple, then by Lemma 2.1 (2), G is c-simple, it follows that $M = 1$ and G is a group of prime order, a contradiction. Hence we assume that G is not simple. It is clear that the hypothesis of the theorem are satisfied by any quotient group G/K of G. A trivial argument shows that G has unique minimal normal subgroup K with $K \not\leq \Phi(G)$. Then there exists a maximal subgroup $M < \cdot G$ such that $K \not\leq M$, i.e., $G = KM$. Since M is c-normal in G, there exists $N \trianglelefteq G$ such that $G = MN$ and $N \cap M \leq M_G = 1$. Then $1 \neq N$. Hence $K \leq N$ and so $K \cap M = 1$. Hence $|N| = |G : M| = |K|$, $K = N$. For any maximal subgroup $L < \cdot G$ with $L_G = 1$, we have $KL = G$. Since L is c-normal in G, the same argument shows that $|G : L| = |K|$. By a result of Baer ([Ba], Lemma 3), K is solvable. It is clear that G/K satisfies the hypotheses on G. The minimal choice of G implies that G/K is solvable. Now that both K and G/K are solvable, it follows that G is solvable, a contradiction.

Conversely, suppose that G is solvable and $M < \cdot G$. If $M_G \neq 1$; consider G/M_G and use induction on $|G|$. We get M/M_G is c-normal in G/M_G. Lemma 2.1 (4) shows that M is c-normal in G. Assume $M_G = 1$. Let N be a minimal normal subgroup of G which is certainly abelian. Then $G = NM$ and $N \cap M \leq M_G = 1$. By definition, M is c-normal in G. $\square$

Theorem 2.7. *Let G be a finite group and M be a maximal subgroup of G. Then M is c-normal in G if and only if $\eta(G : M) = |G : M|$.*

Proof. Suppose M is c-normal in G. We prove that $\eta(G : M) = |G : M|$ by induction on $|G|$. If $M_G \neq 1$, then M/M_G is c-normal in G/M_G. By induction, $\eta(G/M_G : M/M_G) = |G/M_G : M/M_G|$. By Lemma 2.5 we have that $\eta(G : M) = \eta(G/M_G : M/M_G) = |G/M_G : M/M_G| = |G : M|$. If $M_G = 1$, by definition, there exists a normal subgroup N of G such that $MN = G$ and $M \cap N \leq M_G = 1$. Since $M <\cdot G$, N is in fact a minimal normal subgroup of G. By the definition of normal index, $\eta(G : M) = |N| = |G : M|$.

Conversely, suppose that $\eta(G : M) = |G : M|$. If G is simple, then $\eta(G : M) = |G|$. Therefore $M = 1$ and so M is c-normal in G. Hence we can assume that G is non-simple. If $M_G \neq 1$, we can use induction on $|G|$. It follows that M/M_G is c-normal in G/M_G and so M is c-normal in G. Hence we can assume that G is a non-simple group with $M_G = 1$. Let N be a minimal normal subgroup of G, then $MN = G$ and $\eta(G : M) = |N|$. By assumption, $|G : M| = \eta(G : M) = |N|$. It yields that $M \cap N = 1$. Therefore M is c-normal in G. □

Corollary 2.8. *Let G be a finite group. Then G is solvable if and only if $\eta(G : M) = |G : M|$ for every maximal subgroup M of G.*

This is the main theorem of [De], and it follows directly from Theorem 2.6 and 2.7

In order to minimize the number of restricted maximal subgroups, we establish the following theorem.

Theorem 2.9. *Let G be a finite group. Then G is solvable if and only if there exists a solvable c-normal maximal subgroup M of G.*

Theorem 3.5. *Let G be a finite group. Then G is solvable if and only if M is c-normal in G for every maximal subgroup M in $\mathcal{F}_c$.*

We can also discuss p-solvability in terms of c-normality.

Theorem 2.10. *Let G be a finite group and p be the maximal prime divisor of $|G|$. If M is c-normal in G for every non-nilpotent maximal subgroup $M \in \mathcal{F}^{pc}$, then G is p-solvable.*

As applications, we generalize theorems of Srnivasan [S] and Buckley [Bu] by replacing the "normality" condition by the "c-normality" condition.

Theorem 2.11. *Let G be a finite group. Suppose P_1 is c-normal in G for every Sylow subgroup P of G and every maximal subgroup P_1 of P. Then G is supersolvable.*

Theorem 2.12. *Let G be a finite group. Suppose that $< x >$ is c-normal in G for every element x of G with prime order or order 4. Then G is supersolvable.*

3. On Maximal A-invariant Subgroups

When we consider a group A acting on a group G. It is useful to investigate the maximal A-invariant subgroups.

First, we consider the following critical case that A acts trivially on every maximal A-invariant subgroup of G but A acts non-trivially on G. It is not surprising that G has a special structure.

Theorem 3.1. *Suppose that G is a group and A is an operator group of G. Suppose A acts trivially on every maximal A-invariant subgroup of G but A acts non-trivially on G. Then, $C = C_G(A) = N_G(A)$ is the unique maximal A-invariant subgroup of G, which is a normal abelian subgroup of G with $[G, A]C \leq C_G(C)$. Moreover, G must be one of the following cases:*

(I) $[G, A] \neq G$. $|G/C| = p$, $H/C_H(G)$ is isomorphic to a subgroup of C and A' acts trivially on G. $F(G) \neq C$ if and only if G is a p-group.

(II) $[G,A]=G$. $C \leq Z(G)$. Furthermore,

(1) If $R_S(G) \neq G$, then G is a p-group in $\mathcal{A}\mathcal{A}_e \cap \mathcal{A}_e\mathcal{A}$ with class ≤ 2. H acts trivially on $\Phi(G)$ and irreducibly on $G/Z(G)$. G' is an elementary p-subgroup. If $p \neq 2$, then $x^p = 1$ for every element x of G.

(2) If $R_S(G) = G$, then $C = Z(G) = F(G) = \Phi(G)$. $G/Z(G) = G_1 \times \cdots \times G_k$ is a direct product of isomorphic nonabelian simple groups. For every $i \in \{1, \cdots k\}$, there exists $A_i \leq A$ such that $|A : A_i| = k$, $G_i \cong Inn(G_i) \leq A_i/C_{A_i}(G_i)\widetilde{\leq}Aut(G_i)$.

This is a generalization of Hall-Higman's reduction theorem [G-H]

We also have the following structure theorem which is useful for minimal counterexample analysis.

Theorem 3.2. *Suppose that G is a group and A is an operator group of G. Suppose every maximal A-invariant subgroup of G is nilpotent but G is not nilpotent. If there exists an A-subgroup $R \in Syl_p(G)$ for every prime $r \in \pi(G)$, then $G = P \rtimes Q$ is a semidirect product of two Sylow subgroups of G. Every proper A-subgroup of Q is contained in $C_Q(P)$ and every proper QA-subgroup of P is contained in $C_P(Q)$.*

Proof. (1). G is solvable.

Note that there exists an A-subgroup $P \in Syl_p(G)$ with odd prime $p \in \pi(G)$. Assume $N_G(ZJ(P)) < G$. Then G is solvable, as proved in Theorem 2.1. Assume $N_G(ZJ(P)) = G$. We consider $\overline{G} = G/ZJ(P)$. If $\overline{G}$ is nilpotent, then G is solvable. If $\overline{G}$ is non-nilpotent, then $\overline{G}$ is also an inner $A - \mathcal{N}$ group. By induction on $|G|$, we have that $\overline{G}$ is solvable, and so is G.

(2). There exists $P \in Syl_p(G)$ with $P \, char \, G$, for a $p \in \pi(G)$.

Since G is solvable, there exists a characteristic series of groups, $1 < O_{p_1}(G) < O_{p_1,p_2}(G) < \cdots < G$, where $p_i \in \pi(G)$. Hence there is a characteristic subgroup G_1 of G such that $|G : G_1| = p_i^k$ for some k with $p_i \in \pi(G)$. Since G_1 is nilpotent, there exists $p \neq p_i$, $p \in \pi(G_1)$, such that $P \in Syl_p(G_1) \cap Syl_p(G)$. Since P char G_1 and G_1 char G, we have P char G. $\forall q \in \pi(G)$ with $q \neq p$, there is $Q \in Syl_q(G)$ such that Q is A-invariant. Note that PQ is an A-subgroup of G. If $PQ < G$, then $PQ = P \times Q$. Obviously, G is nilpotent if $|\pi(G)| \geq 3$. Now we assume that $|\pi(G)| \leq 2$ and $G = P \rtimes Q$. $\forall A$-subgroup Q_1 with $Q_1 < Q$. We have $PQ_1 = P \times Q_1$ and so $Q_1 \leq C_Q(P)$. $\forall$ proper QA-subgroup P_1 of P. We have $P_1Q = P_1 \times Q$ and so $P_1 \leq C_P(Q)$.

This completes our proof. □

We can also apply this method to p-nilpotent groups and 2-closed groups to get the same structure theorem.

As an application, we have the following corollary:

Corollary 3.3. *Suppose that A is an operator group of a finite group G with $C_G(A) = 1$. Suppose that there is no element of order r^2 in A for every prime divisor r of $|A|$. Suppose that A is cyclic or $(|G|, |A|) = 1$. If $C_G(\alpha)$ is a 2-subgroup of G for every nontrivial element α of A, then G is nilpotent.*

Theorem 3.4. *Let G be a finite group and H be an operator group of G. Suppose that H is p-solvable for every prime in $\pi(G)$. If there exists a maximal H-invariant subgroup M of G such that M is nilpotent, and M_2 has no quotient group isomorphic to $Z_2 \wr Z_2$, then G is solvable.*

Proof. Assume that this is false and let G be a counterexample with minimal order. Then

(1). M contains no nontrivial H-invariant normal subgroup N.

Assume this is false and let $1 \neq N$ be a nontrivial H-invariant normal subgroup N contained in M. Then H acts on G/N in a natural way, and M/N is a nilpotent maximal H-invariant subgroup of G/N; $(G/N, H)$ satisfies the hypotheses of (G, H). Our minimal choice implies that G/N is solvable. Since $N \leq M$ and M is nilpotent, G is solvable, a contradiction.

(2). M is a Hall subgroup of G.

In fact, let $p \in \pi(M)$ and $P \in Syl_p(M)$. Then both P and $N_G(P)$ are H-invariant subgroups of G. $M \leq N_G(P)$, since M is nilpotent. If $M < N_G(P)$ then $N_G(P) = G$; this forces $P \trianglelefteq G$, contrary to (1). Hence $N_G(P) = M$. It is easy to show that $P \in Syl_p(G)$, by a basic property of p-groups.

(3). G is p-nilpotent for every prime $p \in \pi(M)$.

Let $p \in \pi(M)$ and $P \in Syl_p(M)$. Consider the H-invariant subgroups $1 \neq ZJ(P)$ and $N_G(ZJ(P))$. It is clear that $M \leq N_G(P) \leq N_G(ZJ(P))$; $N_G(ZJ(P)) < G$ by (1). Hence $N_G(ZJ(P)) = M$ is p-nilpotent. If p is odd, [G] Theorem 8.3.1 implies that G is p-nilpotent. If $p = 2$, (2) yields that $P \in Syl_2(M) \cap Syl_2(G)$ and so $N_G(P) = M$ is 2-nilpotent. Yoshida's Theorem [Y] 4.2 implies that G is 2-nilpotent, and (3) holds.

By (3), $G = K \rtimes M$, where $K = \bigcap_{p \in \pi(M)} O_{p'}(G)$ is the normal $\pi(M)$-complement of M in G. Surely, K is a characteristic subgroup of G, and hence K is an $M \rtimes H$-invariant subgroup of G. Since M is nilpotent, $M \rtimes H$ is p-solvable for every prime $p \in \pi(MH)$. We prove that MH acts irreducibly on K. In fact, if K_1 is an MH-invariant subgroup of K with $1 \neq K_1 < K$, then $K_1 M$ is an H-invariant subgroup of G with $M < K_1 M < KM = G$, contrary to our assumption. Hence MH acts irreducibly on K. We can prove that K is an elementary abelian group. Now both K and M are nilpotent; this yields that G is solvable, contrary to our choice. This shows that there exists no counterexample, and the theorem is proved. $\square$

4. Summary

In 1, we generalize a series of works of [B-M 1], [Bh]. In 2, we define c-normality of a subgroup, and use it to generalize the work of [De], [B-B], [B-M], as well as [B-E]. In 3, we consider the general cases of Hall-Higman reduction theorem, Thompson's theorem on a group with a maximal odd order nilpotent, and the minimal non-nilpotent group. Some of the detailed proofs and applications can be found in [Wa], [Wa 1], [Wa 2] and [Wa 3].

REFERENCES

[Ba].R.Baer, *Classes of finite groups and their properties*. Illinois J. of Math., **Vol 1** (1957), 115-187

[Bh] H. C. Bhatia, *A generalized Frattini subgroup of a finite group*, Ph. D. Thesis, Michigan State University, East Lansing, 1972.

[Bu]. J.Buckley, *Finite groups whose minimal subgroups are normal.* Math.Z. **116** (1970), 15-17.

[B-B]. A.Ballester-Boliches, *On the normal index of maximal subgroups in finite groups.* J. Pure and Applied Algebra 64 (1990), 113-118.

[B-E].A.Bolinches, L.Ezquerro, *On the Deskins index complex of minimal subgroup of a finite group.* Proc. Amer. Math. Soc. **114** (1992), 325-330.

[B-M]. P.Bhattacharya and N.Mukherjee, *The normal index of a finite group.* Pacific J. of Math.**132 No. 1** (1988), 143-149.

[B-M 1]. P.Bhattacharya and N.Mukherjee, *On the intersection of a class of maximal subgroups of a finite group II*, J. Pure and Applied Algebra **42** (1986), 117-124.

[De]. W.E.Deskins, *On maximal subgroups.* Proc. Symp. Pure Math., Amer. Math. Soc., **1** (1959), 100-104.

[Do]. K.Doerk *Minimal nicht uber auflosbbare endliche Gruppen*, Math. Z. (1966), 198–205.

[G].D.Gorenstein, Finite Groups. Chelsea Publishing Company, New York, 1980.

[R].R.Rose, *On finite insolvable groups with nilpotent maximal subgroups*, Journal of Algebra **48** (1977), 182–196.

[S].S.Srinivasan, *Two sufficient conditions for supersolvability of finite groups*, Israel J. Math., **35** (1980), 210-214.

[Wa].Y.Wang, *A class of Frattini-like subgroups of finite group.* J. Pure and Applied Algebra **78** (1992), 101-108.

[Wa 1]. Y.Wang, *A non-coprime Hall-Higman reduction theorem*, J. Austral. Math. Soc. (series A) **57** (1994), 129-137.

[Wa 3]. Y.Wang, *Critical structure of the formation admitting an operator group*, Acta. Math. Sinica, **57** (1994)

[Wa 2]. Y.Wang, *C-normality of groups and its properties,*, Accepted by J. of Algebra. (1995)

[We].M.Weinstein etc. Between Nilpotent and Solvable, Polygonal Publishing House, New Jersey 1982.

[Y].T.Yoshida, *Character-theoretic transfer*, J. of Algebra, **52** (1978) 1–38.

Recent Developments in Morita Duality

WEIMIN XUE   Department of Mathematics, Fujian Normal University, Fuzhou, Fujian 350007, People's Republic of China

ABSTRACT

Morita duality was established by Azumaya and Morita in the late 1950's. Here most of the recent developments in this theory are reviewed and some new results are presented.

1. INTRODUCTION

Morita duality, as a generalization of the duality of vector spaces over division rings, was established by Azumaya [7] and Morita [39] in the late 1950's. Such a duality is an additive contravariant category equivalence between two categories of left R-modules and right S-modules, which are both closed under submodules and factor modules and contain all finitely generated modules. Azumaya [7] and Morita [39] have shown that: (1) These dualities are precisely those equivalent to the functors Hom(-,E) induced by bimodules $_RE_S$ that are injective cogenerators on both

---

This research is supported by the National Education Commission of China

sides and there are canonical isomorphisms $S \cong End(_RE)$ and
$R \cong End(E_S)$, and (2) The natural domain and range of such a
duality are the categories of E-reflexive modules, i.e., linearly
compact modules by a result of Müller [41]. In this paper we first
give a brief introduction to Morita duality, then we review recent
developments in Morita duality. A presentation of Morita duality
can be found in Anderson and Fuller [2, §23, §24], Faith [20,
Chapter 23], and Xue [52]. We freely use terminologies and
notations in [2].

All rings are associative rings with identity $1 \neq 0$, and
modules are unitary. Let R and S be two rings, and let R-Mod
(Mod-S) denote the category of left R- (right S-) modules.

Let $_RE_S$ be a bimodule. Recall that $_RE_S$ is a __balanced bimodule__
(__faithfully balanced bimodule__) if the canonical ring
homomorphisms $R \longrightarrow End(E_S)$ and $S \longrightarrow End(_RE)$ are epimorphisms
(isomorphisms).

If $_RM$ (N_S) is a left R-module (right S-module), then
$Hom_R(M,E)$ ($Hom_S(N,E)$) is a right S-module (left R-module). So
there is a pair of contravariant additive functors $Hom_R(-,_RE_S)$:
R-Mod $\longrightarrow$ Mod-S and $Hom_S(-,_RE_S)$: Mod-S $\longrightarrow$ R-Mod. For brevity
we write

$$(\quad)^* = Hom(-,_RE_S)$$

to denote either of these functors. For each M in R-Mod or Mod-S

$$[\sigma_M(m)](f) = f(m) \qquad (m \in M, \quad f \in M^*)$$

defines the __evaluation homomorphism__

$$\sigma_M: M \longrightarrow M^{**}.$$

A module M is called __E-reflexive__ in case σ_M is an isomorphism. It
is known that (1) If M is E-reflexive, then M^* is also
E-reflexive; (2) If $M \cong M_1 \oplus ... \oplus M_n$, then M is E-reflexive if and
only if each M_i is E-reflexive; and (3) $_RE_S$ is a faithfully
balanced bimodule if and only if both $_RR$ and S_S are E-reflexive.

If $_RC$ and D_S are two full subcategories of R-Mod and Mod-S,
respectively, then there is a __duality__ between $_RC$ and D_S in case
there are contravariant additive functors $F : _RC \longrightarrow D_S$ and
$G : D_S \longrightarrow _RC$ such that $GF \cong 1_C$ and $FG \cong 1_D$; moreover this

duality is called a <u>Morita duality</u> in case $_RC$ and D_S are closed under submodules and factor modules, and contain all finitely generated modules. Unlike Morita equivalence, for no rings R and S is there a duality between R-Mod and Mod-S.

THEOREM 1.1 (Morita [39]). Let $_RC$ and D_S be full subcategories of R-Mod and Mod-S such that $_RR \in {}_RC$ and $S_S \in D_S$, and such that every module in R-Mod (respectively, Mod-S) isomorphic to one in $_RC$ (D_S) is in $_RC$ (D_S). If there is a duality between $_RC$ and D_S induced by $F : {}_RC \longrightarrow D_S$ and $G : D_S \longrightarrow {}_RC$, then there is a bimodule $_RE_S$ such that (1) $_RE \cong G(S_S)$ and $E_S \cong F(_RR)$; (2) There are natural isomorphisms $F \cong \text{Hom}_R(-,E)$ and $G \cong \text{Hom}_S(-,E)$; and (3) All modules M in $_RC$ and all modules N in D_S are E-reflexive.

In view of the Morita theorem, we have an equivalent definition. A bimodule $_RE_S$ defines a <u>Morita duality</u> in case (1) $_RR$ and S_S are E-reflexive; and (2) Every submodule and every factor module of an E-reflexive module is E-reflexive. In this case R has a <u>Morita duality</u> and S has a <u>right Morita duality</u>. Each factor ring of a ring with a (right) Morita duality has a (right) Morita duality.

THEOREM 1.2. The following statements are equivalent for a bimodule $_RE_S$:

(1) $_RE_S$ defines a Morita duality;

(2) Every factor module of $_RR$, S_S, $_RE$ and E_S is E-reflexive;

(3) $_RE_S$ is a balanced bimodule such that both $_RE$ and E_S are injective cogenerators.

Colby and Fuller [17] have proved that if $_RE_S$ is any balanced bimodule such that $_RE$ and E_S are cogenerators, then $_RE$ and E_S are injective so $_RE_S$ defines a Morita duality. This shows that the "injective" condition in Theorem 1.2(3) is redundant.

For convenience, we write module homomorphisms on the opposite side to the scalars. Let $_RM$ be a left R-module. For each $K \leq M$ the <u>right annihilator</u> of K in M^* is

$$r_{M^*}(K) = \{ f \in M^* \mid (K)f = 0 \}$$

and for each $L \leq M^*$ the <u>left annihilator</u> of L in M is

$$l_M(L) = \{ m \in M \mid (m)f = 0 \text{ for all } f \in L \}.$$

Let N_S be a right S-module. Then for $L \subseteq N$ and $K \subseteq N^*$, one similarly defines

$$1_{N^*}(L) = \{ g \in N^* \mid g(L) = 0 \}$$

and

$$r_N(K) = \{ n \in N \mid g(n) = 0 \text{ for all } g \in K \}.$$

THEOREM 1.3. Let $_R E_S$ define a Morita duality. If $_R M$ and N_S are E-reflexive modules, then

(1) $\text{End}(_R M) \cong \text{End}(\text{Hom}_R(M, E)_S)$;

(2) For each submodule K of M and each submodule L of M^*,

$$1_M(r_{M^*}(K)) = K \quad \text{and} \quad r_{M^*}(1_M(L)) = L;$$

(3) For each submodule L of N and each submodule K of N^*,

$$r_N(1_{N^*}(L)) = L \quad \text{and} \quad 1_{N^*}(r_N(K)) = K;$$

(4) The lattices of submodules of M and M^* are anti-isomorphic via the mapping $K \longmapsto r_{M^*}(K)$;

(5) The lattices of submodules of N and N^* are anti-isomorphic via the mapping $L \longmapsto 1_{N^*}(L)$;

(6) The lattices of ideals of R and S are isomorphic, and the centers of R and S are isomorphic rings;

(7) Every finitely generated or finitely cogenerated left R- (right S-) module is E-reflexive;

(8) M is finitely generated (and projective) if and only if M^* is finitely cogenerated (and injective);

(9) M is simple, semisimple, of finite length n, indecomposable, respectively, if and only if M^* is;

(10) M is noetherian if and only if M^* is artinian.

Similar results hold for N in (8), (9), and (10).

Let $_R E_S$ define a Morita duality and let $_R M$ be an E-reflexive module. If $_R M$ is projective, then $_R M$ must be finitely generated and so $(M^*)_S$ is finitely cogenerated and injective. However, if $_R M$ is injective, $_R M$ need not be finitely cogenerated and $(M^*)_S$ need not be projective.

THEOREM 1.4. If $_R E_S$ defines a Morita duality, then (1) $_R E_S$ is faithfully balanced; (2) Both $_R E$ and E_S are finitely cogenerated injective cogenerators; and (3) Both R and S are semiperfect rings.

2. ARTINIAN RINGS WITH MORITA DUALITY

For a Morita duality over artinian rings, we have

THEOREM 2.1. The following statements are equivalent:

(1) There exists a duality between the category R-FMod of finitely generated left R-modules and the category FMod-S of finitely generated right S-modules;

(2) R is left artinian and some bimodule $_R E_S$ defines a Morita duality;

(3) S is right artinian and some bimodule $_R E_S$ defines a Morita duality.

Moreover, if R, S and E satisfy either of the last two conditions, then a left R- (right S-) module is E-reflexive if and only if it is finitely generated if and only if it is finitely cogenerated.

If R is an artinian ring with a Morita duality induced by $_R E_S$ then S is right artinian by Theorem 2.1, but it is an open question whether or not S must be left artinian.

Azumaya [7] and Morita [39] gave several necessary and sufficient conditions on a left artinian ring R, a ring S (will be right artinian), and $_R E_S$ to insure a duality between the category R-FMod and the category FMod-S.

THEOREM 2.2 (Azumaya [7], Morita [39]). Let R be a left artinian ring and let $_R E_S$ be a bimodule. Then the following statements are equivalent:

(1) The $_R E_S$-dual ( )* defines a duality between the category R-FMod and the category FMod-S.

(2) R, E and S satisfy (i) S is right artinian, and (ii) all finitely generated left R-modules and right S-modules are E-reflexive;

(3) R, E and S satisfy (i) S is right artinian, (ii) $_R E$ and E_S are faithful, and (iii) all simple left R-modules and right

S-modules are E-reflexive;

(4) R, E and S satisfy (i) S is right artinian ($_R$E is finitely generated), (ii) $_R$E and E_S are faithful, and (iii) the $_R E_S$-dual takes simples to simples;

(5) R, E and S satisfy (i) S_S is E-reflexive, and (ii) $_R$E is a finitely generated injective cogenerator;

(6) R, E and S satisfy (i) S_S is E-reflexive, and (ii) $_R$E and E_S are injective cogenerators;

(7) R, E and S satisfy (i) S is right artinian ($_R$E is finitely generated), (ii) for each I $\leq$ $_R$R and each V $\leq$ E_S,

$$1_R(r_E(I)) = I \quad \text{and} \quad r_E(1_R(V)) = V,$$

and (iii) for each K $\leq$ S_S and each W $\leq$ $_R$E,

$$r_S(1_E(K)) = K \quad \text{and} \quad 1_E(r_S(W)) = W.$$

By a result of Rosenberg and Zelinsky [44], a left artinian ring R has a Morita duality if and only if $R/J(R)^2$ has a Morita duality. A ring is <u>left (right) duo</u> in case each left (right) ideal is two-sided. It is easy to see that each idempotent of a left (right) duo ring is central. Hence a semiperfect left (right) duo ring is a finite product of local left (right) duo rings. A complete characterization of artinian left duo rings is given in Xue [50].

A ring R has <u>(Morita) self-duality</u> in case there exists an R-bimodule $_R E_R$ that defines a Morita duality. Hence a ring R has a self-duality if and only if there is a bimodule $_R E_S$ ($_S E_R$) that defines a Morita duality such that R $\cong$ S.

Using Cohn's division ring extensions [14], one constructs an artinian ring without a Morita duality. Using Schofield's division ring extensions [45], one constrcuts an artinian ring with a Morita duality but without self-duality.

If R is a two-sided artinian ring such that R/J(R) is commutative, does R have a Morita duality? If "Yes", does R have self-duality?

Azumaya [8] introduced exact rings, an important class of artinian rings. (Recently, Xue [56] dropped the "artinian" assumption and obtained a natural generalization of Azumaya's exact rings.) An artinian ring R is <u>exact</u> in case there is a composition series of ideals

$$0 = I_0 < I_1 < \ldots < I_n = R$$

such that each bimodule $_R(I_i/I_{i-1})_R$ is balanced. Azumaya [8] proved that each exact ring has a Morita duality, and conjectured that it has self-duality. Although this conjecture is still open, some partial answers have been obtained.

The following result was proved in Habeb [30] and Xue [51].

THEOREM 2.3. Let R be an exact ring. If $_R E$ is a finitely cogenerated injective cogenerator, then $S = End(_R E)$ is an exact ring.

An artinian ring R is <u>serial</u> in case each of indecomposable projective left and right R-modules has a unique composition series. Serial rings are exact rings [8]. The following theorem was observed in Amdal and Ringdal [1], and proved by Dischinger and Muller [19] and Waschbüsch [48], independently.

THEOREM 2.4. Every serial ring has self-duality.

A module is <u>distributive</u> in case its lattice of submodules is distributive. As a generalization of serial rings, one calls an artinian ring R <u>locally distributive</u> in case each indecomposable projective left and right R-module is distributive. Locally distributive rings are also exact rings (Camillo, Fuller and Haack [12]). Although the self-duality of locally distributive rings is still open, the follosing partial answer was given by Fuller and Xue [23], where the first assertion was also proved by Belzner [11], independently.

THEOREM 2.5. Let R be a basic locally distributive ring. If $_R E$ is the minimal injective cogenerator in R-Mod then $S = End(_R E)$ is also locally distributive. Moreover R and S have the same left and right diagrams in the sense of Fuller [21].

If R is locally distributive, Belzner [11] constructed an ideal I such that every factor ring of R/I has self-duality, where I is fairly small in the sense that for any two primitive idempotents e and f of R the eRe-fRf-bimodule eIf is simple on each side. A <u>duo</u> ring is a ring which is both left and right duo.

Habeb [29] observed that artinian duo rings are exact. Xue [49] proved the following

THEOREM 2.6. Let R be an artinian duo ring with minimal injective cogenerator $_R E$. Then (1) S = End($_R E$) is an artinian duo ring; and (2) If J(R) is a direct sum of ideals with simple socles then R has a self-duality.

COROLLARY 2.7. If R is an artinian duo ring with $J(R)^2 = 0$ then R has self-duality.

3. MORITA DUALITY AND LINEARLY COMPACT MODULES

The notion of linearly compact modules plays an important role in Morita duality. Müller [41] has demonstrated the connections between linearly compact modules and Morita duality.

Let M be a module. A family $\{m_i, M_i\}_{i \in I}$ (where $m_i \in M$ and $M_i \leq M$, $i \in I$) is called <u>solvable</u> in case there is an $m \in M$ such that $m - m_i \in M_i$ for all $i \in I$, and it is called <u>finitely solvable</u> if $\{m_i, M_i\}_{i \in F}$ is solvable for any finite subset $F \subseteq I$. The module M is called <u>linearly compact</u> (l.c.) in case any finitely solvable family of M is solvable. One notes that a family $\{m_i, M_i\}_{i \in I}$ is solvable if and only if $\cap_{i \in I}(m_i + M_i)$ is non-empty. Then a module M is linearly compact if and only if for any family $\{m_i, M_i\}_{i \in I}$, $\cap_{i \in I}(m_i + M_i)$ is non-empty whenever $\cap_{i \in F}(m_i + M_i)$ is non-empty for all finite subsets $F \subseteq I$. A ring R is called <u>left (right) linearly compact</u> if the module $_R R$ (R_R) is linearly compact.

If $N \leq M$ then M is l.c. if and only if both N and M/N are l.c.. Every artinian module is l.c.; in particular, a left (right) artinian ring is left (right) l.c.. A left (right) l.c. ring is semiperfect.

THEOREM 3.1 (Müller [41]). If $_R E_S$ defines a Morita duality, then the E-reflexive left R-modules (right S-modules) are precisely the l.c. left R-modules (right S-modules).

If $_R E_S$ defines a Morita duality, then R $\cong$ End(E_S) and S $\cong$ End($_R E$) canonically. For convenience, we say that R has a Morita

duality (induced by $_R$E) or that $_R$E defines a Morita duality. The
following Müller's theorem is a criterion for an arbitrary ring to
have a Morita duality.

THEOREM 3.2 (Müller [41]). A ring R has a Morita duality if
and only if both $_R$R and the minimal injective cogenerator $_R$U in
R-Mod are l.c., and then $_R$U induces a Morita duality.

COROLLARY 3.3 (Azumaya [7], Morita [39]) A left artinian ring
R has a Morita duality if and only if every indecomposable
injective left R-module (i.e., the injective envelope of a simple
left R-module) is finitely generated.

The next result was mentioned in Vámos [47] as a slightly
modified version of Müller's Theorem 3.2.

THEOREM 3.4. A left R-module $_R$E defines a Morita duality if
and only if $_R$R is l.c. and $_R$E is a l.c. finitely cogenerated
injective cogenerator.

An immediate consequence of Theorem 3.4 is that Morita duality
and self-duality are preserved by Morita equivalence. Since a
ring with a Morita duality must be semiperfect and any
semiperfect ring is Morita equivalent to its basic ring, we may
always assume R is a basic semiperfect ring in order to show R
has a Morita (self-) duality.

Recently, we have partially generalized the Azumaya-Morita
Theorem 2.2 in Xue [57].

THEOREM 3.5. Let R be a left l.c. ring and let $_R$E$_S$ be a
bimodule. Then the following are equivalent:

(1) $_R$E$_S$ defines a Morita duality;

(2) R, E and S satisfy (i) S is right l.c., (ii) $_R$E and E$_S$
are faithful, l.c. and finitely cogenerated, (iii) the $_R$E$_S$-dual
( )* takes simples to simples;

(3) R, E and S satisfy (i) S$_S$ is E-reflexive, (ii) $_R$E and
E$_S$ are injective cogenerators;

(4) R, E and S satisfy (i) S is right l.c., (ii) for each
I $\leq$ $_R$R and each V $\leq$ E$_S$,

$$l_R(r_E(I)) = I \text{ and } r_E(l_R(V)) = V,$$
(iii) for each $K \leq S_S$ and each $W \leq {}_R E$,
$$r_S(l_E(K)) = K \text{ and } l_E(r_S(W)) = W.$$

In [57], we give two examples, which show that (a) the condition that S is right l.c. in (2) of Theorem 3.5 can not be dropped, i.e., in a left l.c. ring R, we do not have an analogous result of the parenthetical version of (4) of Theorem 2.2; (b) the condition that ${}_R E$ and E_S are finitely cogenerated in (2) of Theorem 3.5 can not be dropped, either, i.e., in a left l.c. ring R, we do not have an analogous result of the non-parenthetical version of (4) of Theorem 2.2; and (c) the condition that S is right l.c. in (4) of Theorem 3.5 can not be replaced by the condition that ${}_R E$ is l.c. and finitely cogenerated, i.e., in a left l.c. ring R, we do not have an analogous result of the parenthetical version of (7) of Theorem 2.2.

If R is a left and right l.c. ring with a Morita duality induced by ${}_R E_S$ then S is right l.c. by Theorem 3.5, but we do not know whether or not S is also left l.c..

Müller [41] proved that a commutative ring with a Morita duality must have self-duality. The next theorem of Anh [3] answers a question of [41] in the affirmative.

THEOREM 3.6. Every l.c. commutative ring has self-duality.

We prove in [56] that each l.c. duo ring R has a Morita duality. Does R have self-duality? In particular, does an artinian duo ring have self-duality? Partial answers have been given in Xue [49] and [56].

Let R be a semiperfect ring with a complete set of (primitive) idempotents $e_1, \ldots, e_n$. Then R is naturally isomorphic to the matrix ring $(e_i R e_j)_{i,j}$ via $r \longmapsto (e_i r e_j)_{i,j}$. Müller [42] characterized a Morita duality of R via Morita dualities of $e_i R e_i$ (the local rings $e_i R e_i$).

THEOREM 3.7. The semiperfect ring R has a Morita duality if and only if each $e_i R e_i$ has a Morita duality (with S_i induced by bimodule E_i) and each $e_i R e_j$ and $e_i R e_j^* = \text{Hom}_{e_i R e_i}(e_i R e_j, E_i)$ is linearly compact as a left module; moreover, the corresponding second ring is the matrix ring $S = (e_i Re_j^{**})_{ij}$.

The following result is established in Gomez Pardo and Rodriguez Gonzalez [26].

THEOREM 3.8. Let R have a Morita duality. Then (1) if $_RP$ is a finitely generated quasi-projective R-module, then $End(_RP)$ has a Morita duality; and (2) if $_RU$ is a finitely cogenerated quasi-injective R-module, then $End(_RU)$ has a right Morita duality.

4. MORITA DUALITY AND RING EXTENSIONS

In this section, we consider Morita duality and some kinds of ring extensions, namely subdirect products, trivial extensions, finite triangular extensions, finite normalizing extensions, power series rings, and group graded rings. All ring extensions share the same identity.

Let $R_1, R_2, \ldots, R_n$ be a finite number of rings. A ring R is called a <u>subdirect product</u> of R_i s in case there is a ring monomorphism $f : R \longrightarrow R_1 \times \ldots \times R_n$ such that each $p_i f : R \longrightarrow R_i$ is a ring epimorphism, where $p_i : R_1 \times \ldots \times R_n \longrightarrow R_i$ is the i-th projection.

THEOREM 4.1 (Haack [28]). Let R be a subdirect product of the rings $R_1, \ldots, R_n$. Then R has a Morita duality if and only if each R_i has a Morita duality.

A ring S with ideal M and subring R such that $M^2 = 0$ and $S = R \oplus M$ is called a <u>trivial extension</u> of R by M, and is denoted by $S = R \propto M$. If $_RM_R$ is a bimodule then one constructs such a ring extension from R × M with multiplication

$$(r_1, m_1)(r_2, m_2) = (r_1 r_2, \; r_1 m_2 + m_1 r_2).$$

The following theorem is due to Muller [40].

THEOREM 4.2. Let $_RE_S$ define a Morita duality and let $_RM_R$ be an R-bimodule. Then $R \propto M$ has a finitely cogenerated injective cogenerator $Hom_R(M,E) \propto E$, with multiplication given by

$(r,m)(f,x)= (rf, rx+(m)f)$, and its endomorphism ring is
$S \propto \text{Hom}_R(\text{Hom}_R(M,E),E)$. We have a Morita duality between these
two rings if and only if both $_RM$ and $_R\text{Hom}_R(M,E)$ are l.c..

COROLLARY 4.3. Let $_RE_R$ define self-duality and let $_RM_R$ be an
R-bimodule. If both $_RM$ and $_R\text{Hom}_R(M,E)$ are l.c., and there is an
isomorphism f of R and an f-semilinear R-R-bimodule isomorphism

$$g : {}_R\text{Hom}_R({}_RM,{}_RE)_R \cong {}_R\text{Hom}_R(M_R,E_R)_R$$

i.e., $g(r_1hr_2) = f(r_1)g(h)f(r_2)$, then $S = R \propto M$ has self-duality
induced by $\text{Hom}_R({}_RM,{}_RE) \propto E$.

Let $_RM_S$ be a bimodule and $\begin{bmatrix} R & M \\ 0 & S \end{bmatrix}$ denote the formal
triangular matrix ring.

COROLLARY 4.4. Let R and S have Morita dualities induced by
$_RE$ and $_SF$, respectively. Then the ring $\begin{bmatrix} R & M \\ 0 & S \end{bmatrix}$ has a Morita
duality if and only if both $_RM$ and $_S\text{Hom}_R(M,E)$ are linearly
compact. In this case $\begin{bmatrix} E & 0 \\ \text{Hom}_R(M,E) & F \end{bmatrix}$ defines a Morita
duality between left $\begin{bmatrix} R & M \\ 0 & S \end{bmatrix}$-modules and right
$\begin{bmatrix} \text{End}(_RE) & \text{Hom}_S(\text{Hom}_R(M,E),F)) \\ 0 & \text{End}(_SF) \end{bmatrix}$-modules.

Cohn [14] has proved that there is a division ring D with a
division subring C such that $\dim(D_C)$ is finite and $\dim(_CD)$ is not.
Then the artinian ring $\begin{bmatrix} D & D \\ 0 & C \end{bmatrix}$ does not have a Morita duality.

COROLLARY 4.5. Let R and S have self-dualities induced by $_RE_R$
and $_SF_S$, respectively. Suppose that both $_RM$ and $_S\text{Hom}_R(M,E)$ are
linearly compact, and there are ring isomorphisms f_1 of R and
f_2 of S such that there is an f_1-f_2-similinear R-S-bimodule
isomorphism $g: {}_S\text{Hom}_R(M,E)_R \cong {}_S\text{Hom}_S(M,F)_R$, i.e.,

$g(shr) = f_2(s)g(h)f_1(r)$. Then $\begin{bmatrix} R & M \\ 0 & S \end{bmatrix}$ has self-duality.

A ring extension $S \geq R$ is called a <u>finite extension</u> in case both $_RS$ and S_R are finitely generated R-modules. The following two examples show that, for a finite extension $S \geq R$, there is no connection between the Morita duality of R and that of S. For this reason, we study finite triangular extensions which bahave much better than finite extensions.

EXAMPLE 4.6. Let $R = \begin{bmatrix} Q & Q & Q \\ 0 & Z & Q \\ 0 & 0 & Q \end{bmatrix}$ and $S = M_3(Q)$, the ring

of 3 by 3 matrices over Q. Then $S \geq R$ is a finite extension.

In fact $S = \Sigma^4_{i=1} Rs_i = \Sigma^4_{i=1} s_i R$ where $s_1 = 1_S$, $s_2 = \begin{bmatrix} 0 & 0 & 0 \\ 1 & 0 & 0 \\ 0 & 0 & 0 \end{bmatrix}$,

$s_3 = \begin{bmatrix} 0 & 0 & 0 \\ 0 & 0 & 0 \\ 1 & 0 & 0 \end{bmatrix}$, and $s_4 = \begin{bmatrix} 0 & 0 & 0 \\ 0 & 0 & 0 \\ 0 & 1 & 0 \end{bmatrix}$. Now S is semisimple hence

has a Morita duality which can be induced by $_SS_S$. But R is not even semilocal, so R is not semiperfect and R does not have a Morita duality.

EXAMPLE 4.7. The artinian ring $S = \begin{bmatrix} D & D \\ 0 & C \end{bmatrix}$ preceding

Corollary 4.5 does not have a Morita duality, but S is a finite

extension over the semisimple ring $R = \begin{bmatrix} D & 0 \\ 0 & C \end{bmatrix}$ since both $_DD$ and

D_C are finitely generated.

A ring extension $S \geq R$ is called a <u>finite triangular extension</u> in case there are a finite number of elements $s_1, \ldots, s_n$ in S such that $S = \Sigma^n_{i=1} Rs_i$ and $\Sigma^j_{i=1} Rs_i = \Sigma^j_{i=1} s_i R$ $(j = 1, \ldots, n)$. The latter n equalities hold if and only if for each $r \in R$, there are upper triangular matrices $[a_{ij}]_{i,j}$ and $[b_{ij}]_{i,j}$ over R such that $rs_j = \Sigma^j_{i=1} s_i a_{ij}$ and $s_j r = \Sigma^j_{i=1} b_{ij} s_i$ $(j = 1, \ldots, n)$. Finite triangular extensions arise from Azumaya's exact rings which are such over a finite direct products of local exact rings. The next result was due to Lemonnier [33],[34].

THEOREM 4.8. Let $S \geq T \geq R$ be ring extensions such that $S \geq R$ is a finite triangular extension. If R has a Morita duality induced by $_RE$ then T has a Morita duality induced by $_THom_R(T,E)$. In particular, S has a Morita duality induced by $_SHom_R(S,E)$.

The following question arises naturally from the above theorem: If $S \geq R$ is a finite triangular extension and S has a Morita duality, does R have a Morita duality? We are unable to settle this problem, but we do have two partial solutions from Xue [52, §8].

THEOREM 4.9. Let $S \geq R$ be a finite triangular extension and let S have a Morita duality induced by ${}_S U$. If both ${}_R S$ and S_R are progenerators, then R has a Morita duality induced by ${}_R U$.

THEOREM 4.10. Let $S \geq R$ be a finite triangular extension. If S is a left artinian ring with a Morita duality, then so is R.

A finite triangular extension $S = \Sigma_{i=1}^n Rs_i \geq R$ is called a <u>finite normalizing extension</u> in case $Rs_i = s_i R$ $(i = 1,\ldots,n)$. In view of Theorem 4.8, we consider finite normalizing extensions with self-duality.

THEOREM 4.11 (Mano [35]). Let $S = \Sigma_{i=1}^n Rs_i \geq R$ be a finite normalizing extension satisfying the followung:
   (1) ${}_R S$ is a free R-module with basis $s_1,\ldots,s_n$ such that each s_i centralizes the elements of R, i.e., $rs_i = s_i r$ for all $r \in R$;
   (2) R has self-duality induced by ${}_R E_R$ such that $b_{ijp} x = x b_{ijp}$ for all $x \in E$ and all i, j, p; where we put $s_i s_j = \Sigma_{p=1}^n b_{ijp} s_p$.
   Then S has self-duality induced by ${}_S \mathrm{Hom}_R(S,E)$.

COROLLARY 4.12. Every finite dimensional algebra over a field has self-duality.

If R is a ring and G is a finite semigroup (with an identity), then the <u>semigroup ring</u> RG has a free basis G which centralizes the elements of R, and the multiplication is defined as

$$(\Sigma_G r_g g)(\Sigma_G r'_g g) = \Sigma_G (\sum_{hk=g} r_h r'_k) g.$$

We may regard R as a subring of RG under the embedding $R \longhookrightarrow RG$ via $r \longmapsto r1_G$. Hence the following result follows from Mano's Theorem.

COROLLARY 4.13 (Fuller and Haack [22]). Let G be a finite

semigroup. If R has self-duality, then the semigroup ring RG has self-duality.

Fuller and Haack [22] actually proved more: If R is a ring and an R-module $_R E$ defines a Morita duality between R and S = End($_R E$), then $_{RG}$Hom$_R$(RG, E) defines a Morita duality between RG and SG. Hence, if R has self-duality, then so does RG. However, it is an open question wthether or not the converse of Corollary 4.13 is true, even for a finite group G.

Kraemer [32] proved the following result.

THEOREM 4.14. If S is a finite normalizing extension over a division ring D, then S has self-duality induced by $_S$Hom$_D$($_D S, _D D$).

Finite normalizing extensions of division rings occur quite frequently. The most common examples are the finite dimensional algebras over fields.

Let R[[x]] be the ring of all formal power series in x with coefficients in R. If $_R E$ is a left R-module, we let $E[x^{-1}]$ consist of all polynomials in x^{-1} with coefficients in E. Thus a typical element of $E[x^{-1}]$ is an expression

$$a_0 + a_1 x^{-1} + a_2 x^{-2} + \ldots + a_n x^{-n}$$

where each $a_i \in E$. Now $E[x^{-1}]$ can be turned into a left R[[x]]-module. The addition in $E[x^{-1}]$ is componentwise and the scalar multiplication is defined as follows

$$(\Sigma_{i \geq 0} r_i x^i)(\Sigma_{j \geq 0} a_j x^{-j}) = \Sigma_{j \geq 0}(\Sigma_{i \geq 0} r_i a_{i+j}) x^{-j}$$

where $\Sigma_{i \geq 0} r_i x^i \in R[[x]]$ and $\Sigma_{j \geq 0} a_j x^{-j} \in E[x^{-1}]$. Then $E[x^{-1}]$ becomes a left R[[x]]-module. Similarly, if E_S is a right S-module, the $E[x^{-1}]$ is a right S[[x]]-module. Hence if $_R E_S$ is an R-S-bimodule, $E[x^{-1}]$ becomes a left R[[x]]- and right S[[x]]-bimodule. Recently, we have proved the following theorem in Xue [54], [58].

THEOREM 4.15. (1) The power series ring R[[x]] is left l.c. if and only if the ring R is left l.c. and left noetherian.

(2) The power series ring R[[x]] has a Morita duality if and

only if R is a left noetherian ring with a Morita duality induced by a bimodule $_R E_S$ such that S is a right noetherian ring. In this case, the bimodule $_{R[[x]]} E[x^{-1}]_{S[[x]]}$ defines a Morita duality.

If we take a commutative l.c. non-noetherian ring R (e.g., the ring R in [52, Example 10.9]), then R has a Morita duality by Anh's Theorem 3.6, but R[[x]] is not l.c. by Theorem 4.15(1) so R[[x]] does not have a Morita duality. This shows that both [52, Questions 3.7 and 4.16] have negative answers, which was also confirmed by Vamos.

Let $_R E_S$ define a Morita duality. If R is left artinian then S is right artinian. If R is left noetherian then S need not be right noetherian, but if R is (two-sided) noetherian then S must be right noetherian.

COROLLARY 4.16. Let R have a Morita duality. If R is a left artinian ring (resp., a netherian ring) then the left noetherian ring (resp., the noetherian ring) R[[x]] also has a Morita duality.

COROLLARY 4.17. If R is a noetherian ring with self-duality induced by $_R E_R$ then the noetherian ring R[[x]] has self-duality induced by $_{R[[x]]} E[x^{-1}]_{R[[x]]}$.

Let G be a group with identity e. Let $S = \oplus_{g \in G} S_g$ be a graded ring of type G, i.e., each S_g is an abelian group and $S = \oplus_{g \in G} S_g$ as abelian groups satisfying $S_g S_h \subseteq S_{gh}$. In this case, $R = S_e$ is a subring of S and each S_g is an R-bimodule. $S = \oplus_{g \in G} S_g$ is called <u>finitely graded</u> if $S_g = 0$ for almost every $g \in G$. Menini [37] proved the following generalization of Theorem 4.2.

THEOREM 4.18. Let $S = \oplus_{g \in G} S_g$ be a finitely graded ring. Let $_R E$ be the minimal injective cogenerator in R-Mod. Then S has a Morita duality if and only if (1) R has a Morita duality; (2) for each $g \in G \backslash \{e\}$, the left R-module S_g is l.c.; and (3) for each $g \in G \backslash \{e\}$, the left R-module $Hom_R(S_g, E)$ is l.c..

5. OTHER TYPES OF RINGS WITH MORITA DUALITY

In this section, we consider noetherian rings, perfect rings and quasi-perfect rings with a Morita duality. The first result was due to Müller [40].

THEOREM 5.1. Suppose $\cap_{n=1}^{\infty} J(R)^n = 0$ and R has a Morita duality induced by $_RE_S$. Then R is left noetherian, S is right noetherian, both $_RE$ and E_S are artinian modules, and $E = \cup_{n=1}^{\infty} r_E(J(R)^n) = \cup_{n=1}^{\infty} l_E(J(S)^n)$.

Recall that if R is a left artinian ring with a Morita duality induced by $_RE_S$ then S is right artinian. An analogous statement for left noetherian rings is not true, since Menini [36] showed that the left noetherian ring $\begin{bmatrix} F((x)) & F((x)) \\ 0 & F[[x]] \end{bmatrix}$ (which is not right noetherian) has self-duality, where F is a field, F[[x]] is the formal power series ring over F and F((x)) is the quotient field of F[[x]]. Menini [36] also proved that if R is a noetherian ring with a Morita duality (induced by $_RE_S$), then $\cap_{n=1}^{\infty} J(R)^n = 0$ and so S must be right noetherian by Theorem 5.1, but it is an open question whether or not S must be left noetherian.

A ring is called left (right) perfect (Bass [10]) in case each of its left (right) modules has a projective cover. A left or right perfect ring with a Morita duality must be left artinian (Xue [52, Corollary 18.4]). As a generalization of perfect rings, Camillo and Xue [13] called R a left (right) quasi-perfect ring in case each left (right) artinian R-module has a projective cover, and proved that the class of left quasi-perfect rings lies strictly between that of left perfect rings and that of semiperfect rings. The next theorem is given in their paper [13].

THEOREM 5.2. Let $_RE_S$ define a Morita duality. Then R is left quasi-perfect if and only if S is right quasi-perfect.

We do not know whether or not a left quasi-perfect ring with a Morita duality must be left artinian.

6. ON KASCH DUALITY

Dikranjan, Gregorio and Orsatti [18] introduced the notion of
Kasch duality which is a special case of Morita duality. In this
section, we review the characterizations of Kasch duality given
in [18] and [55].

Denote by $_R T$ the full subcategory of R-Mod consisting of all
submodules of the finitely generated modules in R-Mod and by T_S
the analogous subcategory of Mod-S. Following Dikranjan, Gregorio
and Orsatti [18], we call a bimodule $_R K_S$ a Kasch bimodule in case
the functors $\text{Hom}_R(-,K)$ and $\text{Hom}_S(-,K)$ give rise to a duality
between $_R T$ and T_S. Since $_R T$ and T_S are closed under submodules
and factor modules, and contain all finitely generated modules, a
Kasch bimodule $_R K_S$ is a Morita bimodule, i.e., $_R K_S$ induces a
Morita duality. If there is a Kasch bimodule $_R K_S$, the ring R is
said to have a Kasch duality. If R has a Kasch duality, any Morita
bimodule $_R E_S$ is a Kasch bimodule [55].

The following useful characterization of Kasch bimodules was
proved in [18, Proposition 1.4].

THEOREM 6.1. Let $_R K_S$ be a Morita bimodule. The following
conditions are equivalent:
   (a) $_R K_S$ is a Kasch bimodule;
   (b) $_R K$ and K_S are finitely generated;
   (c) $_R R$ and S_S are finitely cogenerated;
   (d) $_R R$ is finitely cogenetrated and $_R K$ is finitely generated.

Our following result in [55] generalizes a theorem of [18],
where R is a Kasch ring if the bimodule $_R R_R$ is Kasch.

THEOREM 6.2. If R is a duo ring, the following conditions are
equivalent:
   (a) R has a Kasch duality;
   (b) R is a l.c. ring with essential socle;
   (c) R is a subdirect product of a finite number of subdirectly
irreducible Kasch rings.

We call an R-module $_R K$ a Kasch module in case $_R K_S$ is a Kasch
bimodule where $S = \text{End}(_R K)$. In [55], we obtain the following

results for Kasch duality with ring extensions.

THEOREM 6.3. Let $_RK$ be a Kasch module and $_RM_R$ be an R-bimodule. Then the trivial extension $S = R \propto M$ has a Kasch duality (which can be induced by $_SHom_R(S,K)$) if and only if $_RM$ is finitely cogenerated and $_RHom_R(M,K)$ is finitely generated.

COROLLARY 6.4. Let $_RK$ be a Kasch module and S have a Kasch duality. If $_RM_S$ is a bimodule, the formal triangular matrix ring $\begin{bmatrix} R & M \\ 0 & S \end{bmatrix}$ has a Kasch duality if and only if $_RM$ is finitely cogenerated and $_SHom_R(M,K)$ is finitely generated.

THEOREM 6.5. Let $S \geq R$ be a finite triangular extension. Suppose $_RS$ is projective and S_R is flat. If R has a Kasch duality induced by $_RK$ then S has a Kasch duality induced by $_SHom_R(S,K)$.

THEOREM 6.6. Let $S \geq R$ be a finite triangular extension. If S has a Kasch duality and R has a Morita duality, then R has a Kasch duality.

COROLLARY 6.7. Let $S \geq R$ be a finite triangular extension. Suppose both $_RS$ and S_R are progenerators. If S has a Kasch duality induced by $_SK$ then R has a Kasch duality induced by $_RK$.

COROLLARY 6.8. The following five conditions are equivalent for a ring R:

(a) R has a Kasch duality;

(b) $\mathbb{U}_n(R)$ has a Kasch duality for each (resp., some) natural number n, where $\mathbb{U}_n(R)$ is the ring of all n by n upper triangular matrices over R;

(c) The semigroup ring RG has a Kasch duality for each (resp., some) finite semigroup G with unity.

Let Aut(R) be the group of the ring automorphisms of R. Let G be a group which acts on R. Then one constructs a <u>skew group ring</u> of G over R, denoted by $R * G$, where the additive group of $R * G$ is the same as the additive group of the group ring RG, and multiplication is the extension of

$$gr = g(r)g$$

where g(r) is the image of r under the action of g. Clearly, we can view R * G as a finite normalizing extension of R; moreover both $_R(R * G)$ and $(R * G)_R$ are progenerators. Hence the next result follows from Theorem 6.5 and Corollary 6.7.

COROLLARY 6.9. Let G be a finite group which acts on R. Then the skew group ring R * G has a Kasch duality if and only if R has a Kasch duality.

Finally, we consider subdirect products related to Kasch duality. Let R be a finite subdirect product of the rings $R_1,\ldots,R_n$. Haack's Theorem 4.1 states that R has a Morita duality if and only if each R_i has a Morita duality. If R has a Kasch duality, R_i need not have a Kasch duality since R_i is a factor ring of R and the Kasch duality is not preserved by factor rings (see Example 2.5 in [55]). However, one direction of Haack's Theorem for Kasch duality is still true.

THEOREM 6.10. Let R be a finite subdirect product of the rings $R_1,\ldots,R_n$. If each R_i has a Kasch duality then R has a Kasch duality.

7. GENERALIZATIONS OF MORITA DUALITY

In this final section, we briefly mention some recent generalizations of Morita duality.

Colby [15] introduced two generalizations of Morita duality in connection with the study of tilting theory. Quasi-duality was introduced in Kraemer [31] and studied in Baccella and Orsatti [9], Gregorio [27] and Xue [53],[58]. Colby and Fuller [16],[17] introduced and studied Morita duality for Grothendieck categories which can be applied to the study of QF-3 and QF-3' rings. Recently, this kind of duality was also studied in Gomez Pardo and Guil Asensio [24],[25], and Anh and Wiegandt [6]. Menini and Rio [38] introduced and studied graded Morita duality for graded rings in connection with the traditional Morita duality for graded rings. Yamagata [59] made a first step in extending Morita duality to rings without identity. He considered rings with enough

idempotents and proved that almost all categorical results on the
traditional Morita duality remain true for this class of rings.
Recently, Anh and Menini [5] studied Morita duality for rings with
local units (a ring with enough idempotents must be a ring with
local units) and obtained a completely analogous characterization
of Morita duality over rings with local units to the classical
case of rings with identity by a suitable modification of the
concept of linear compactness. Morita duality for rings with local
units was also studied in Zhang [60] who proved that this duality
is induced by a bimodule, like the traditional Morita duality
for rings with identity.

ACKNOWLEDGMENT

The author wishes to express his thanks to Professor Fu-an Li
whose friendly recommendation resulted in the writting of this
paper.

REFERENCES

[1] I.K. Amdal and F. Ringdal, Categories uniserielles, C.R.
    Acad. Sci. Paris, Serie A, 267 (1968), 85-87, 247-249.

[2] F.W. Anderson and K.R. Fuller, Rings and Categories of
    Modules, 2nd edition, Springer-Verlag, New York, 1992.

[3] P.N. Anh, Morita duality for commutative rings, Comm. Algebra
    18 (1990), 1781-1788.

[4] P.N. Anh, Characterisation of two-sided PF-rings, J. Algebra
    141 (1991), 316-320.

[5] P.N. Anh and C. Menini, Morita duality for rings with local
    units, J. Algebra 164 (1994), 632-641.

[6] P.N. Anh and R. Wiegandt, Morita duality for Grothendieck
    categories, J. Algebra 167 (1994), 273-293.

[7] G. Azumaya, A duality theory for injective modules, Amer. J.
    Math. 81 (1959), 249-278.

[8] G. Azumaya, Exact and serial rings, J. Algebra 85 (1983),
    477-489.

[9] Baccella and Orsatti, On generalized Morita bimodules and
    their dualities, Rend. Accad. Naz. Sci. XL Mem. Math. 107
    (1989), 323-340.

[10] H. Bass, Finitistic dimension and a homological generalization of semiprimary rings, Trans. Amer. Math. Soc. 95 (1960), 466-488.

[11] Thomas Belzner, Towards self-duality of semidistributive artinian rings, J. Algebra 135 (1990), 74-95.

[12] V.P. Camillo, K.R. Fuller and J.K. Haack, On Azumaya's exact rings, Math. J. Okayama Univ. 28 (1986), 41-51.

[13] V.P. Camillo and Weimin Xue, On quasi-perfect rings, Comm. Algebra 19 (1991), 2841-2850. Addendum, Comm. Algebra 20 (1992), 1839-1840.

[14] P.M. Cohn, On a class of binomial extensions, Illinois J. Math. 10 (1966), 418-424.

[15] R.R. Colby, A generalization of Morita duality and the tilting theorem, Comm. Algebra 17 (1989), 1709-1722.

[16] R.R. Colby and K.R. Fuller, Exactness of the double dual and Morita duality for Grothendieck categories, J. Algebra 82 (1983), 546-558.

[17] R.R. Colby and K.R. Fuller, QF-3´ rings and Morita duality, Tsukuba J. Math. 8 (1984), 183-188.

[18] D.N. Dikranjan, E. Gregorio and A. Orsatti, Kasch bimodules, Rend. Sem. Mat. Univ. Padova 85 (1991), 147-160.

[19] F. Dischinger and W. Müller, Einreihig zerlegbare artinsch ringe sind selbstdual, Arch. Math. 43 (1984), 132-136.

[20] C. Faith, Algebra II Ring Theory, Springer-Verlag, Berlin, 1976.

[21] K.R. Fuller, Algebras from diagrams, J. Pure Appl. Algebra 48 (1987), 23-37.

[22] K.R. Fuller and J.K. Haack, Duality for semigroup rings, J. Pure Appl. Algebra 22 (1981), 113-119.

[23] K.R. Fuller and Weimin Xue, On distributive modules and locally distributive rings, Chinese Ann. Math. Ser.B 12 (1991), 26-32.

[24] J.L. Gomez Pardo and P.A. Guil Asensio, Linear compactness and Morita duality for grothendieck categories, J. Algenra 148 (1992), 53-67.

[25] J.L. Gomez Pardo and P.A. Guil Asensio, Morita duality for grothendieck categories, Pub. Math. 36 (1992), 625-635.

[26] J.L. Gomez Pardo and N. Rodriguez Gonzalez, Endomorphism rings with Morita duality, Comm. Algebra 19 (1991), 2097-2112.

[27] E. Gregorio, Classical Morita equivalence and linear topologies; applications to quasi-dualities, Comm. Algebra 18 (1990), 1137-1146.

[28] J.K. Haack, Finite subdirect products of rings and Morita duality, Comm. Algebra 10 (1982), 2107-2119.

[29] J.M. Habeb, On Azumaya´s exact rings and artinian duo rings, Comm. Algebra 17 (1989), 237-245.

[30] A.V. Jategaonkar, Morita duality and Noetherian rings, J. Algebra 69 (1981), 358-371.

[31] J. Kraemer, Characterizations of the Existence of (Quasi-) Self-Duality for Complete Tensor Rings, Algebra Berichte 56, Verlag Reinhard Fischer, Munchen, 1987.

[32] J. Kraemer, Self-duality for finite normalizing extensions of skew fields, Math. J. Okayama Univ. 32 (1990), 103-109.

[33] B. Lemonnier, Dimension de Krull et dualite de Morita dans les extensions triangulaires, Comm. Algebra 12 (1984), 3071-3110.

[34] B. Lemonnier, Dimension et codimension de Gabriel dans les extensions triangulaires, Comm. Algebra 14 (1986), 941-950.

[35] T. Mano, Self-duality and ring extensions, J. Pure Appl. Algebra 32 (1984), 51-57.

[36] C. Menini, Jacobson´s conjecture, Morita duality and related questions, J. Algebra 103 (1986), 638-655.

[37] C. Menini, Finitely graded rings, Morita duality and self-duality, Comm. Algebra 15 (1987), 1779-1797.

[38] C. Menini and A.N. Rio, Morita duality and graded rings, Comm. Algebra 19 (1991), 1765-1794.

[39] K. Morita, Duality for modules and its applications to the theory of rings with minimum condition, Tokyo Kyoiku Daigaku, Ser A6 (1958), 83-142.

[40] B.J. Müller, On Morita duality, Canad. J. Math. 21 (1969), 1338-1347.

[41] B.J. Müller, Linear compactness and Morita duality, J. Algebra 16 (1970), 60-66.

[42] B.J. Müller, Morita duality-A survey, Abelian Groups and Modules, Proc. Conf., Udine-Italy 1984, CISM Courses Lect. 287 (1984), 395-414.

[43] B.L. Osofsky, A generalization of quasi-Frobenius rings, J. Algebra 4 (1966), 373-387. Erratum, 9 (1968), 120.

[44] A. Rosenberg and D. Zelinsky, Finiteness of the injective hull, Math. Z.· 70 (1959), 372-380.

[45] A.H. Schofield, Artin´s problem for skew field extensions, Math. Proc. Camb. Phil. Soc. 97 (1985), 1-6.

[46] H. Tachikawa, Quasi-Frobenius Rings and Generalizations, Lecture Notes Math. 351, Springer-Verlag, Berlin, 1973.

[47] P. Vámos, Rings with duality, Proc. London Math. Soc. 35 (1977), 275-289.

[48] J. Waschbüsch, Self-duality of serial rings, Comm. Algebra 14 (1986), 581-589.

[49] Weimin Xue, Artinian duo rings and self-duality, Proc. Amer. Math. Soc. 105 (1989), 309-313.

[50] Weimin Xue, Morita duality and artinian left duo rings, Bull.
     Austral. Math. Soc. 39 (1989), 339-342.

[51] Weimin Xue, Exact modules and serial rings, J. Algebra 134
     (1990), 209-221.

[52] Weimin Xue, Rings with Morita Duality, Lect. Notes Math. 1523,
     Springer-Verlag, Berlin, 1992.

[53] Weimin Xue, Quasi-duality of rings and a generalization
     (Chinese. English summary), J. Fujian Normal Univ. (Natural
     Sci.), 9 (1993), no.2, 9-12.

[54] Weimin Xue, Morita duality and some kinds of ring extensions,
     Algebra Collq. 1 (1994), 77-84.

[55] Weimin Xue, On Kasch duality, Algebra Colloq. 1 (1994),
     257-266.

[56] Weimin Xue, On a generalization of Azumaya´s exact rings, J.
     Algebra 172 (1995), 78-89.

[57] Weimin Xue, Characterizations of Morita duality, Algebra
     Colloq. 2 (1995), 339-350.

[58] Weimin Xue, Quasi-duality, linear compactness and Morita
     duality for power series rings, Can. Math. Bull. 39 (1996).

[59] K. Yamagata, On Morita duality for additive group valued
     functors, Comm. Algebra 7 (1979), 367-392.

[60] Shengui Zhang, A characterization of Morita duality on rings
     with local units (Chinese. English summary), J. Fujian Normal
     Univ. (Natural Sci.), 6 (1990), no.2, 34-38.

Some Results on Homological Algebra and Module Theory

ZHOU Boxun
  Department of Mathematics, Nanjing University, Nanjing, China

TONG Wenting
  Department of Mathematics, Nanjing University, Nanjing, China

In China, the study of homological algebra and module theory began in 1979. In that summer, a workshop on homological algebra and the theory of categories was held at Harbin. In that workshop, one of the authors, Zhou, delivered twenty lectures on theory of categories and homological algebra. The book [115] is the revised monograph of these lectures. Since then, the study of these topics has continued to expand in China. Here we shall give a summary of the major works of Chinese mathematicians related to these areas.

In this paper, all rings considered are associative with an identity element, all modules are assumed to be unital.

1. On tensor product of K-algebras

In [111], Zhou considered the tensor product of some well-known K-algebras, where K is a commutative ring. The following results are obtained:

Theorem 1.1 [111] If K is a field, R_1 and R_2 are division rings as K-algebras, and there exist $\alpha \in R_1 \setminus K$, $\beta \in R_2 \setminus K$ such that they have the same minimum polynomial over K, then $R_1 \underset{K}{\otimes} R_2$ can't be a division ring.

Theorem 1.2 [111] If R_1 and R_2 are simple rings, as K-algebras over a field K, with centers $\sum_1$ and $\sum_2$ respectively, and $F = K(\sum_1, \sum_2)$, then $R_1 \otimes R_2$ is simple if and only if for any finitely many elements $t_1, t_2, \cdots, t_n \in \sum_2$, whenever they are linearly independent over K, they are linearly independent over $\sum_1$.

The theorem above improved a theorem of Azumaya and Nakayama obtained in 1944 (see Proc. Imp. Acad. Tokyo. 20(1944), 348-352).

In [113], Zhou defined the tricomplexes, the tricocomplexes and their total complexes, some results on the homology (cohomology) modules of tricomplexes (tricocomplexes) are obtained. In addition [113] also extended the Künneth Theorem of $\otimes$ of bicomplexes to that of tricomplexes.

If K is a field, R, S are K-algebras, L and M are a left R-module and a left S-module respectively, then $L \underset{K}{\otimes} M$ is a left $R \underset{K}{\otimes} S$-module.

Denote by $\mathrm{lpd}_R L$, $\mathrm{lpd}_S M$ and $\mathrm{lpd}_{R \otimes S} L \otimes M$ the left projective dimensions. In [112], Zhou obtained two equivalent conditions for $\mathrm{lpd}_{R \otimes S} M \otimes L = \mathrm{lpd}_R L + \mathrm{lpd}_R M$. Denote by lgD the left global homological dimension of a ring. Zhou also obtained a sufficient condition for $\lg D(R \otimes S) \geqslant \lg D(R) + \lg D(S)$.

For a field K and two K-algebras R, S, Wang F. G. [62] obtained the following results:

(1) If R is right Noetherian, S is right coherent, A is a right R-module, and B is a left S-module, then

$$\text{lfd}_{R \otimes S} \text{Hom}_K (A, B) \geqslant \text{rid}_R A + \text{lfd}_S B,$$

where lfd, rid denotes the left flat dimension and the right injective dimension, respectively;

(2) If one of the following conditions (a), (b) and (c) holds: (a) $[R:K] < \infty$ and A is a finitely generated right R-module, (b) $[R:K] < \infty$ and R is right coherent, (c) $[S:K] < \infty$ and R is right Noetherian, then

$$\text{lfd}_{R \otimes S} \text{Hom}(A, B) = \text{rid}_R A + \text{lfd}_S B,$$

where A is a right R-module and B is a left S-module;

(3) If $[R:K] < \infty$, A is a finitely generated right R-module, and B is a left S-module, then

$$\text{lpd}_{R \otimes S} \text{Hom}_K (A, B) = \text{rid}_R A + \text{lfd}_S B.$$

where lpd denotes the left projective dimension;

If $[R:K] < \infty$, S is left Noetherian and right coherent, A, B are denumerably generated, then

$$\text{lpd}_{R \otimes S} \text{Hom}_K (A, B) \leqslant \text{rid}_R A + \text{lpd}_S B + 2.$$

Let A be a left R-module. If $\text{Ext}_R^1 (N, A) = 0$ for any finitely presented left R-module N, then A is called an FP-injective module. The FP-injective dimension of A, denoted by FP-$\text{id}_R A$, is defined to be the least nonnegative integer n, such that $\text{Ext}_R^{n+1} (N, A) = 0$ for any finitely presented left R-module N. Wang F. G. [58] obtained the following result:

If K is a field, R and S are left coherent K-algebras, A is a left R-module, and B is a left S-module, then

$$\text{FP-id}_{R \otimes S} (A \otimes B) \geqslant \text{FP-id}_R A + \text{FP-id}_S B,$$

and the equality holds when $[R:K] < \infty$.

Let M_j, $j = 1, 2$, be two vector spaces over a field K, A_j an irreducible algebra of linear transformations in M_j, Δ_j the centralizer of M_j, S_j the socle of A_j, (M_j, M_j') a pair of dual vector spaces associated with A_j and $M = M_1 \underset{K}{\otimes} M_2$, $M' = M_1' \underset{K}{\otimes} M_2'$. Xu Y. H. [86] obtained the following theorems.

Theorem 1. 3 [86] (An extension of the Azymaya-Nakayama Theorem).

(1) The centralizer of M' (resp. M) as left $A_1 \underset{K}{\otimes} A_2$-module (resp. right $A_1 \underset{K}{\otimes} A_2$-module) is $\Delta_1 \underset{K}{\otimes} \Delta_2$;

(2) The lattice of left $A_1 \underset{K}{\otimes} A_2$-submodules of M' (resp. right $A_1 \underset{K}{\otimes} A_2$-submodules of M) is isomorphic to the lattice of left (resp. right) ideals of $\Delta_1 \underset{K}{\otimes} \Delta_2$;

(3) The lattice of right (resp. left) $\Delta_1 \bigotimes_K \Delta_2$ -modules of M' (resp. M) is isomorphic to the lattice of all such right (resp. left) ideals of $A_1 \otimes A_2$, which are contained in $S_1 \bigotimes_K S_2$;

(4) $\Delta_1 \bigotimes_K \Delta_2$ and $S_1 \bigotimes_K S_2$ are simple rings. $\Delta_1 \bigotimes_K \Delta_2$ is artinian if and only if $S_1 \bigotimes_K S_2$ has minimal one-sided ideals;

(5) Let C denote a full subcategory of mod- $A_1 \bigotimes_K A_2$, whose ob C consists of all objects N with $N(S_1 \bigotimes_K S_2) = N$; then the category mod -$\Delta_1 \bigotimes_K \Delta_2$ and C are equivalent. The pair of functors $- \bigotimes_{\Delta_1 \otimes \Delta_2} M$ and $- \bigotimes_{A_1 \otimes A_2} M'$ defines an equivalence of $\Delta_1 \bigotimes_K \Delta_2$ —mod and a full subcategory of $A_1 \bigotimes_K A_2$- mod;

(6) Let A_j' and A_j'' be irreducible algebras of linear transformations in M_j, $A_1' \bigotimes_K A_2' \subset A_1'' \bigotimes_K A_2''$. Then the center of $A_1' \bigotimes_K A_2'$ is contained in the center of $A_1'' \bigotimes_K A_2''$, and the center of $A_1'' \bigotimes_K A_2''$ is isomorphic to a subring of the center of $\Delta_1 \bigotimes_K \Delta_2$.

Theorem 1.4 [86] Let M be a vector space over a field K, A an irreducible algebra of linear transformations in M, Δ the centralizer of M, and S the socle of A. Let A^* be a K-algebra and be faithful as an A^*-bimodule. Denote by $(M, M)'$ a pair of dual vector spaces over Δ associated with A; then

(1) The lattice of right $A \otimes A_1^*$-submodules of $M \bigotimes_K A^*$ is isomorphic to the lattice of $\Delta \otimes A_1^*$-unitary right ideals of $\Delta \bigotimes_K A^*$, the lattice of left $A \bigotimes_K A_1$-submodules of $M' \bigotimes_K A^*$ is isomorphic to the lattice of $\Delta \bigotimes_K A_1^*$-unitary left ideals of $\Delta \bigotimes_K A^*$, where $A_1^* = \{(m, a^*) \mid m \in K, a^* \in A^*\}$ with two operations as usual as follows: $(m, a^*) + (n, b^*) = (m+n, a^* + b^*)$, $(m, a^*)(n, b^*) = (mn, mb^* + na^* + a^* b^*)$;

(2) The lattice of left (right) $\Delta \bigotimes_K A_1^*$-submodules of $M \bigotimes_K A^*$ (resp. of $M' \bigotimes_K A^*$) is isomorphic to the lattice of $S \bigotimes_K A_1^*$-unitary left (resp. right) ideals of $A \bigotimes_K A^*$;

(3) The lattice of $\Delta \bigotimes_K A_1^*$-unitary ideals of $\Delta \bigotimes_K A^*$ is isomorphic to the lattice of $S \bigotimes_K A_1^*$-unitary ideals of $A \bigotimes_K A^*$.

As applications of Theorem 1.3 and Theorem 1.4, Xu Y. H. [86] obtained some equivalent categories, and gave some properties of the tensor products of primitive algebras. In addition, Xu Y. H. [86] discussed also the tensor products of primitive algebras with γ-socle.

2.　On tensor products and direct sums of IBN rings

A ring R is called an IBN (invariant basis number) ring if $R^m \simeq R^n$, as left R-modules, implies $m = n$. Let K be a commutative ring, and let rings $R_1, R_2, \cdots, R_n$ be K-algebras. In [49] Tong obtained the following results:

Theorem 2.1 [49]

(1) If $R_1 \bigotimes_K R_2 \bigotimes_K R_3 \cdots \bigotimes_K R_n$ is an IBN ring, then $R_1, R_2, \cdots, R_n$ are IBN rings;

(2) $R_1 \oplus R_2 \oplus \cdots \oplus R_n$ is an IBN ring if and only if there exists j, $1 \leqslant j \leqslant n$, such that R_j is an IBN ring.

Wang F. G. [67] obtained the following results:

Theorem 2.2 [67]

(1) If K is a field, then $R_1 \underset{K}{\otimes} R_2$ is an IBN ring if and only if R_1 and R_2 are IBN rings;

(2) If K is a commutative local ring with the maximal ideal M, and M is nil, then $R_1 \underset{K}{\otimes} R_2$ is an IBN ring if and only if R_1 and R_2 are IBN rings;

If K is not a commutative local ring with nil maximal ideal, then there exist two K-algebras R_1, R_2 such that R_1, R_2 are the IBN rings, but $R_1 \underset{K}{\otimes} R_2$ is not.

In [120] Zhu X. S. gave the following definition of the strongly IBN rings. Let R be a ring, C the center of R. If $R^m \simeq R^n$, as C-modules, implies $m = n$, then R is called a strongly IBN ring. Zhu X. S. obtained the following result:

Theorem 2.3 [120] Let K be a commutative ring, R a K-algebra.

(1) If R is a strongly IBN ring, then $R \underset{K}{\otimes} R^{op}$ is an IBN ring, where R^{op} is the opposite ring of R, hence R is an IBN ring;

(2) If $(R \underset{K}{\otimes} R^{op})^m \simeq (R \underset{K}{\otimes} R^{op})^n$, as K-modules, implies $m = n$, then $R \underset{K}{\otimes} R$ is an IBN ring;

(3) If G is a monoid, $S = R[G]$, and $R^m \simeq R^n$, as K-modules, implies $m = n$, then $R^{op} \underset{K}{\otimes} S$ is an IBN ring.

Let

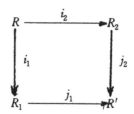

be a Cartesian square for ring homomorphisms, and j_1 or j_2 be surjective. Wang F. G. [67] obtained the following result:

Theorem 2.4 [67] If the Grothendieck group of R, $K_0 R \simeq Z$, then R' is an IBN ring if and only if R_1 and R_2 are IBN rings.

3. Homological dimensions over coherent rings

The class of coherent rings is a very important class of rings. It is well known that the following classes of rings are subclasses of the class of coherent rings: Euclidean domains, PID, Artinian semisimple rings, von Neumann regular rings, QF rings, Dedekind domains, hereditary rings, semihereditary ring, Noetherian rings, and Artinian rings.

Some Chinese mathematicians obtained the following results on homological dimensions over the coherent rings.

Let R be a commutative ring, $\text{fpD}(R)$ be the finitely presented dimension of R, defined by Ho Kuen Ng in 1984 (see Pacific J. of Math. , 113 (1984), 417—431), $\text{gD}(R)$ the global homological dimension of R, and $\text{wD}(R)$ the weak dimension of R. $R \in (a,b,c)$ means that $\text{wD}(R) = a$, $\text{gD}(R) = b$, and $\text{fpD}(R) = c$. For commutative rings, we use also SH, H, N, RE, SS, C to denote the class of semihereditary rings, the class of hereditary rings, the class of Noetherian rings, the class of (von Neumann) regular rings, the class of Artinian semisimple rings, the class of coherent rings, respectively.

It is clear that $\text{SS} = (0,0,0)$, $\text{N} = (0,0,0) \cup (m,m,0)$, $\forall\ m > 0$. In [51] Tong obtained the following result:

Theorem 3. 1 [51]

$\text{RE} = (0,0,0) \cup (0,m,m+1)$, $m > 0$.

$\text{H} = (0,0,0) \cup (1,1,0) \cup (0,1,2) \cup (1,1,2)$.

$\text{SH} = (0,0,0) \cup (1,1,0) \cup (0,m,m+1) \cup ((1,m,m+1) \cap C)$, $m > 0$.

If R is any ring, G is any group, Tong [57] proved the following theorem:

Theorem 3. 2 [57]

(1) If the group ring RG is left coherent, then R is coherent;

(2) If RG is left semihereditary, then so is R;

(3) If R is a commutative ring, G is a finite Abelian group, and $|G|^{-1} \in R$, then RG is semihereditary if and only if R is semihereditary.

For the FP-injective modules over commutative coherent rings, Wang F. G. [60] gave the following local-gobal property.

Theorem 3. 3 [60] If R is a commutative coherent ring, and A is an R-module, then the following statements are equivalent:

(1) A is FP-injective;

(2) For every multiplicative subset S of R, $S^{-1}A$ is an FP-injective $S^{-1}R$-module;

(3) For every prime ideal P of R, A_P is an FP-injective R_P-module;

(4) For every maximal ideal M of R, A_M is an FP-injective R_M-module.

For a left R-module A, recall the definition (Ho Kuen Ng, 1984)

$\text{fpd}_R(A) = \inf\{n\,|\,P_{n+1} \to P_n \to \cdots \to P_0 \to A \to 0$ is exact, P_j is projective,

$j = 0,1,2,\cdots,n+1$, and P_{n+1}, P_n are finitely generated$\}$.

In [12] Ding defined

$\text{fgd}_R(A) = \inf\{n\,|\,P_n \to P_{n-1} \to \cdots \to P_0 \to A \to 0$ is exact, P_j is projective,

$j = 0,1,2,\cdots,n$, and P_n are finitely generated$\}$

Clearly, $\text{fgd}_R(A) \leqslant \text{lpd}_R(A)$ and $\text{fgd}_R(A) \leqslant \text{fpd}_R(A)$. Ding [12] gave the following theorem:

Theorem 3. 4 [12]

(1) If R is left coherent and $\text{fgd}_R(A) > 0$, then $\text{fgd}_R(A) = \text{fpd}_R(A)$.

(2) For any ring R, the following statements are equivalent:

(a) R is left coherent;

(b) For every projective left R-module P and any submodule Q of P, $\text{fgd}_R(Q) = \text{fpd}_R(Q)$;

(c) For every finitely presented left R-module M and any submodule N of M, $\text{fgd}_R(N) = \text{fpd}_R(N)$.

As usual, for a left R-module X, denote $X^* = \text{Hom}_R(X, R)$, the dual module of X. In [13] Ding obtained the following theorem:

Theorem 3.5 [13]

(1) If R is both right and left coherent, then the following statements are equivalent:

(a) If A is a finitely presented torsionless left R-module, then A is reflexive;

(b) For any finitely presented right R-module B, $\text{Ext}_R^2(B, R) = 0$;

(c) For any injective left R-module E, $\text{lfd}(E) \leqslant 1$, where $\text{lfd}(E)$ denotes the flat dimension of E.

(2) If R is both right and left coherent, then the following statements are equivalent:

(a) For any finitely presented right R-module M, M^* is reflexive;

(b) For any finitely presented torsionless left R-module A, $(\text{Ext}_R^1(A, R))^* = 0$;

(c) For any finitely presented left R-module B, $(\text{Ext}_R^2(B, R))^* = 0$.

(3) If R is commutative coherent, then R is a quasi-Frobenius ring if and only if the flatness of M^* implies the injectivity of M for any R-module M.

Let 1fpD denote the left finitely presented dimension of rings. Li Y. L. [35] obtained the following theorem on change of rings.

Theorem 3.6 [35]

If $R \to R_1$ is a surjective ring homomorphism, and by this homomorphism, R_1 is a right flat R-module, then $1\text{fp}D(R) \geqslant 1\text{fp}D(R_1)$.

In addition, Li Y. L. proved the following result:

Theorem 3.7 [35] Let R be a left coherent ring, which is a subring of a ring R', and $1\text{fpD}(R) < \infty$. If R' is both a faithfully flat right R-module and a finitely generated left R-module, then

$$1\text{fpD}(R) \leqslant 1\text{fpD}(R') + \max\{t, s\},$$

where $t = 1\text{fpd}_R R'$, $s = \sup\{\text{lpd}_R A \mid A \text{ is a finitely presented } R'\text{-module}\}$.

Zhao [110] gave the following results on commutative coherent rings.

Theorem 3.8 [110] Let R be a commutative connected coherent ring.

(1) If every principal ideal in R has finite projective dimension, then R is a domain; hence a commuative connected hereditary (semihereditary) ring is a Dedekind (Prüfer) domain.

(2) If every finitely generated ideal in R has finite projective dimension, then R is a GCD domain if and only if every finitely generated projective ideal in R is a principal ideal.

Cheng and Zhao [9] obtained the following result.

Theorem 3. 9 [9] Let R be a commutative coherent ring in which all finitely generated projective R-modules are free, and $wD(R) < \infty$; then R is a GCD domain.

For other classes of rings, Yang obtained the following results.

Theorem 3. 10 [96] Let R be a semilocal ring without left nilpotent minimal ideals; then

$$lgD(R) = lgD(R/SocR).$$

Theorem 3. 11 [95] Let R be a commutative ring and u an algebraic element with monic minimum polynomial over R; then

$$gD(R[u]) = gD(R),$$

where $R[u]$ is an algebraic extension ring over R.

4. Some results on module theory

Let R be a ring, M a left R-module. If for every exact sequence of left R-modules $0 \to M \to B \to C \to 0$ and for every finitely copresented left R-module E,

$$0 \to \operatorname{Hom}_R(C,E) \to \operatorname{Hom}_R(B,E) \to \operatorname{Hom}_R(M,E) \to 0$$

is exact, then M is said to be coflat. Chen J. N. [1] obtained the following results.

Theorem 4. 1 [1]

(1) Let $0 \to A \to B \to C \to 0$ be an exact sequence of left R-modules. If A, C are flat (resp. coflat), then B is flat (resp. coflat);

(2) Let $0 \to A \to B \to C \to 0$ be a pure exact sequence of left R-modules. Then C is flat when B is flat, A is coflat when B is coflat.

Theorem 4. 2 [1] The following statements are equivalent:

(1) Every quotient module of any injective left R-module is coflat;

(2) Every quotient module of any coflat R-mdoule is coflat.

Let R be a ring, L and M be left R-modules. An epimorphism $f:L \to M$ is called n-split $(n \in N)$ if for every n-generated submodule M_0 of M there exists a homomorphism $g:M_0 \to L$ such that fg is the identity map of M_0. A left R-module M is siad to be n-projective if every epimorphism onto M is n-split. Clearly, M is finitely projective (see G. Azumaya, Finite splitness and finite projectivity, J. Alg. 106 (1987), 114−134) if and only if M is n-projective for each positive integer n. Zhu S. L. [119] obtained the following result.

Theorem 4. 3 [119] Let R be a ring, and n a positive integer; then the following conditions are equivalent:

(1) Every flat left R-module is n-projective;

(2) For each descending chain $I_1 \geqslant I_2 \geqslant \cdots$ of finitely generated right ideals of $M_n(R)$, the ascending chain $\operatorname{Ann}(I_1) \leqslant \operatorname{Ann}(I_2) \leqslant \cdots$ in $M_n(R)$ terminates;

(3) For each descending chain sequence $A_1, A_2, \cdots$ of matrices over R, where the number of rows of A_1 equals n, the ascending chain $1(A_1) \leqslant 1(A_1 A_2) \leqslant \cdots$ in $R^{(n)}$ terminates, where $1(A)(A \in R^{n \times m})$ denotes the set

$$\{(r_1, r_2, \cdots, r_n) \in R^{(n)} \mid (r_1, r_2, \cdots, r_n)A = 0\};$$

(4) Every flat left $M_n(R)$-module is singly projective.

Let $_RM_S$ be a bimodule. If M has a composition series with factors being balanced, i. e. , the canonical ring homomorphisms

$$\lambda_: R \to \text{End}(M_S) \text{ and } \rho_: S \to \text{End}(_R M)$$

are surjective, then $_R M_S$ is called exact. We call a one-sided-modoule $_R M$ exact if the bimodule $_R M_{\text{End}(_R M)}$ is an exact bimodule. A module is uniserial in case its submodules are linearly ordered by inclusion, and an Artinian ring is serial in case each of its indecomposable projective modules is uniserial. Xue [92] obtained the following theorem.

Theorem 4. 4 [92] Every module over a serial ring is exact.

A well known result of Small states that if M is a Noetherien left R-module having endomorphism ring S then any nil subring of S is nilpotent. Fisher showed that if M is left Artinian then any nil ideal of S is nilpotent (see J. W. Fisher, Nil subrings of endomorphism ring of modules, Proc. Amer. Math. Soc. , 34 (1972), 75—78). He gave a bound on the indices of nilpotency of nil subrings of the endomorphism rings of Noetherian modules and raised the dual question of whether there are such bounds in the case of Artinian modules. He gave an affirmative answer when the module is also assumed to be finitely generated. In [73] Wu, with J. S. Golan, gave a bound on indices of nilpotency of nil subrings of the endomorphism ring of a left R-module which is τ-torsionfree with respect to some torsion theory τ on R-mod. As a special case, they obtained an affirmative answer to Fisher's question, and gave the following result.

Theorem 4. 5 [73] If M is an Artinian (resp. Noetherian) object in a Grothendieck category having endomorphism ring S, then any nil subring of S is nilpotent and there exists a bound on the indices of nilpotency of all such nil subrings.

A ring is called a right SF ring if all simple right R-modules are flat. A ring R is called a left P. P. ring if every principal left ideal of R is projective. Chen J. L. [4] obtained the following result.

Theorem 4. 5 [4] Let R be a right SF ring. If R belongs to one of the following classes of rings: (1) left P. P. rings, (2) right semi-Artinian rings, (3) right nonsingular rings of finite Goldie dimension; then R is a von Neumann regular ring.

A left R-module M is called τ-flat if and only if for every τ-finitely presented left R-module N, any homomorphism $N \to M$ factors through a finitely generated free left R-module, where $\tau = (T_\tau, F_\tau)$ is a hereditary torsion theory. In [46] Tang H. D. , Yang T. H. and Chen J. L. obtained the following result.

Theorem 4. 7 [46] Let R be a ring.

(1) If $R \in F_\tau$, then any τ—flat left R-module is flat.

(2) A finitely generated τ-flat left R-module M is projective if and only if $Q_\tau(R) \bigotimes_R M$ is a projective left $Q_\tau(R)$-module, where $Q_\tau(R)$ is the quotient ring, on τ, of R.

Let R be an associative ring, not necessarily with identity. A left R-module U is called a quasiprogenerator if it satisfies the following conditions: (1) $_R U$ is quasiprojective; (2) $_R U$

is self-generated, that is, for any submodule N of U, $U\,\mathrm{Hom}(U,N)=N$; (3) $_RU$ has a finite spanning set. Let $\overline{\mathrm{Gen}_R U}$ denote the full subcategory of the category of left R-modules consisting of all left R-modules generated by $_RU$ and their submodules, $\Delta=\mathrm{End}_R U$.

In [100] Yao obtained the following result.

Theorem 4. 8 [100] Let $_RU$ and $_RU$ be quasiprogenerators, $\Delta=\mathrm{End}_R U$, $\Delta'=\mathrm{End}_R U'$ and φ be a ring isomorphism from Δ to Δ'. Then there exists an equivalence functor F from $\overline{\mathrm{Gen}_R U}$ to $\overline{\mathrm{Gen}_R U'}$ such that $_RU'=F(_RU)$, $\varphi(a)=F(a)$ for any $a\in\Delta$.

References

1. Chen Jianai, On coflat modules and the weak global dimension of rings, J. Math. Res. Exposition, 4 (1984), 13—16, MR 88a:16046.

2. Chen Jianlong, some remarks of quasi-flat modules, Ibid, 9 (1989), 9—13, MR:90d: 16029.

3. Chen Jianlong, On regular rings and GP-injective rings, Ibid, 11(1991), 272—274, MR 92c:16002.

4. Chen Jianlong, On von Neumann regular rings and SF-rings, Math Japonica, 36(1991), 1123—1127, MR 92m:16001.

5. Chen Jianlong, On left A-injective rings, Chinese Quart. J. of Math. , 7(1992), 25—31.

6. Chen Jianlong, Some characterizations of FP-injective rings and IF rings, J. Math. Res. Exposition, 12(1992), 395—400.

7. Chen Zhizhong, On the Krull-Remak-Azumaya theorem: a connection between modules and category theory. Ibid. , 10(1990), 601—607, MR 92a:16003.

8. Chen Fuchang, Zhao Yicai, and Tang Gaohua, Homological properties of coherent semilocal rings, Proc. Amer. Math. Soc. , 110(1990), 39—44, MR 90m:13013.

9. Chen Fuchang and Zhao Yicai, On the structure of coherent FP-rings, Chinese J. , Contemporary Math. , 12(1991), 159—163.

10. Deng Peimin, Representaion of an anti-automorphism of ring with quasiprogenerator, J. Math. Res. Exposition, 9(1989), 521—522.

11. Deng Peimin, Generalization of Golan's 19th problem, Ibid. , 119(1991), 562—564.

12. Ding Nanqing, Finitely generated dimension of modules, J. of Nanjing Univ. Math. Biquarterly, no. 1, 1989, 107—111, MR 90k:16031.

13. Ding Nanqing, Duality in coherent rings, Ibid. , 1(1990), 60—66, MR 92a:16006.

14. Ding Nanqing, Dual modules of specific modules, J. Math. Res. Exposition, 10 (1990), 337—340, MR 91f:16012.

15. Ding Nanqing, f. f. p. dimension, Acta Math. Sinica, 34(1991), 340—347, MR 92f: 16006.

16. Ding Nanqing, GQF-rings, Chin. Ann. Math. , 13A(1992), 230—238.

17. Ding Nanqing, On projective equivalence, J. of Nanjing Univ. Math. Biquarterly, no. 1, 1992, 81—84.

18. Ding Nanqing and Chen Jianlong, The flat dimensions of injective modules, Manuscripta Math. , 78(1993), 165—177.

19. Fan Yun, Homomorphism modules of modules and their homological dimensions, J. Math. (Wuhan), 4(1984), 255—266, MR 86c:13014.

20. Guo Jinyun, Conditions for algebras with radical sequared zero to have the property that indecomposable modules are determined by their composition factors, I, Acta Math. Sinica, 31(1988), no. 1, 117—124, MR 90i: 16017a; I , Ibid. , 31(1988), no. 2, 181—191, MR 90i:16017b.

21. Guo Jinyun, Necessary and sufficient conditions for the indecomposable modules of a self-injective algebra to be determined by their composition factors, Ibid. , 32(1989), no. 6, 810—823, MR 92a:16017.

22. Guo Jinyun, The structure of indecomposable modules of an algebra stably equivalent to a hereditary algebra, Chin. Ann. Math. 12A(1991), no 3, 270—281, MR 92k: 16006.

23. Guo Shanliang, A note on Fuller categories, Chinese Sci. Bull. , 36(1991), no. 11, 886—888, MR 92k:16009. .

24. Hu Shuan, On the functor of the tensor product of left modules (I), J. Math. Res. Exposition, no. 1,3(1983), 21—28, MR 85i:16030a; (I), Ibid. no. 2, 3(1983), 1—7, MR 85i:16030b.

25. Hu Shuan, Spectral sequences and injective modules, J. of Nanjing Univ. Math. Biquarterly, no. 2,3(1986), 97—104, MR 88c:16039.

26. Hu Shuan, Tensor products of homogeneous distinguishable completely reducible algebras of linear transformations, J. Math. Res. Exposition, 8(1988), no. 3,333—337, MR 89h:16023.

27. Huang Changling, A note on the Kronecker product of irreducible algebras, Fudan Xuebao, 123(1984), no. 2, 215—216, MR 86d:16009.

28. Li Huishi, Semiprime modules and supernilpotent radical, Acta Math. Sinica, 29 (1986), no. 2, 213—216,MR 87m:16020.

29. Li Huishi and Van Osytaeyen, Zariskian filtrations, Comm. Alg. , 17(1989), no. 12, 2945—2970, MR 90m:16004.

30. Li Huishi, Note on pure module theory over Zarikian filtered ring and the generalized Roos theorem, Ibid. , 19(1991), no. 3,843—862.

31. Li Wei, On the global dimension of primitive rings, J. of Nanjing Univ. Math. Biquarterly, 1984, no. 1, 67—70. MR 85j:16037.

32. Li Wei, Maximal ring of quotients of tensor product of algebras over a field, Acta Math.

Sinica, 29(1986), no. 5,647—650.

33. Li Wei, The Jacobson radical of tensor product of algebras, Chin. Ann. Math. , Ser. A, 8(1987), no. 2, 183—188, MR 89g:16014.

34. Li Yuanlin, Some properties of socles, J. of Nanjing Univ. Math. Biquarterly, 5 (1988), no. 2, 266—273, MR 90a:16016.

35. Li Yuanlin, Some theorems on change of rings of the finitely presented dimension, J. of Nanjing Univ. Math. Biquarterly, 1990, no. 1, 75—83.

36. Liu Shaoxue, Rings and Algebras, Science Press, China, 1983.

37. Liu Shaoxue, Note on a condition for Goldie rings, Beijing Shifan Daxue Xuebao 1984, no. 4,1—2, MR 86j:16017.

38. Liu Shaoxue, Luo Yunlun, Tang Aiping, Xiao Jie, Guo Jinyun, Some results on modules and rings, Bull. Soc. Math. Bely. B. 3(1987). no. 2, 181—193, MR 89c: 16008.

39. Liu Shaoxue, The Baer radical and Levitzki radical of additive categories, Beijing Shifan Daxue Xuebao, 1987, no. 4, 13—20, MR 89e: 16016.

40. Liu Shaoxue, Wedderburn-Artin theorem for additive categories, Chinese Sci. Bull. , 33 (1988), 1501—1503, MR 90d:18003.

41. Liu Yingsheng, The tensor products of left modules and their linear mappings, J. Math. Res. Exposition, 1(1981), no. 2, 21—37.

42. Pan Shizhong, On embedding a finitely generated module in a free module, Acta Math. Sincia, 30(1987),no. 6, 810—819, MR 89f:16019.

43. Pan Shizhong, On injective simple modules over a Noetherian ring, J. Math. Res. Exposition, 11(1991), no. 2, 172, MR 92d:13006.

44. Pan Shizhong, Commutative quasi-Frobenius rings, Comm. Alg. , 19(1991), no. 2, 663—667, MR 92f:13019.

45. Tang Huaiding and Chen Jianlong, On left A-injective rings, J. of Nanjing Univ. Math. Biquarterly, 1990, no. 2, 163—169.

46. Tang Huaiding, Yang Tonghai, Chen Jianlong, Flat module and ML module relative to hereditary torsion theory, J. of Math. (PRC), 12(1992), no. 2, 213—220.

47. Tong Wenting, Clifford algebra over a ring, J. Math. Res. Exposition, 1(1981), no. 1, 1—10. MR 83i: 15028, Zb1 576:15022.

48. Tong Wenting, Generalized duality of ring modules and their tensor products, Ibid. , 4 (1984), no. 2, 5—10, MR 87e:16055, Zb1 556:16010.

49. Tong Wenting, Some results on IBN rings, J. of Nanjing Univ. Math. Biquarterly, 1984, no. 2, 217—223, MR 87d:16031, Zb1 564:16004.

50. Tong Wenting, Grothendieck groups and their applications, Ibid. , 1986, no. 1,1—11, MR 88a:18024, Zb1 636:16012.

51. Tong Wenting, Some homological dimensions and semihereditary rings, Ibid. , 1988,

no. 1, 11—19, MR 89j: 13011, Zbl 695:13013.

52. Tong Wenting, Local characterizations of left Noetherian rings and left ideals, J. Math. Res. Exposition, 8(1988), no. 3, 338—340, MR 89h: 16015, Zbl 688: 16016.

53. Tong Wenting, On commutators in rings and their tensor products, J. of Math. (PRC), 8(1988), no. 3, 281—286, MR 90k: 16025.

54. Tong Wenting, On Euler characteristic of modules, Chin. Ann. of Math. , 10B (1989), 58—63, CMCI. 10, 58(1989), MR 90k: 13005, Zbl. 686: 16018.

55. Tong Wenting, Rings in which finitely generated projective modules are free, J. Math. Res. Exposition, 9(1989), no. 3, 319—323, MR 90h: 16040.

56. Tong Wenting, PF-rings and the Grothendieck groups of group rings, Ibid, 10(1990), no. 2, 157—162, MR 91e: 13013.

57. Tong Wenting, Some homological dimensions on group rings, Chin. Ann. of Math. 13A(1992), 39—46, MR 94a: 16013.

58. Wang Fanggui. The FP-injective dimensions of modules over coherent rings, J. of Nanjing Univ. Math. Biquarterly, 1988, no. 1, 141—146.

59. Wang Fanggui, The Grothendieck groups of categories of finitely presented modules over coherent rings, Ibid, 1988, no. 2, 225—229.

60. Wang Fanggui, The FP-injective modules over domains, Ibid. , 1989, no. 1, 90—93.

61. Wang Fanggui, A remark on FP-injective hulls of modules, Ibid. , 1989, no. 1, 140—143.

62. Wang Fanggui, The homological dimension of the homomorphic module $Hom(A,B)$, J. Math. Res. Exposition, 9(1989), no. 3, 355—360.

63. Wang Fanggui, Some results on discriminant group over a commuative ring, J. of Nanjing Univ. Math. Biquarterly, 1990, no. 1, 56—59.

64. Wang Fanggui, The heredity of IBN rings, Ibid. , 1990, no. 1, 118—121.

65. Wang Fanggui, Inverse limit and FP-injective hulls of modules, J. Math Res. Exposition, 10(1990), no, 4, 541—544.

66. Wang Fanggui, Some characterization on coflat modules, J. of Nanjing Univ. Math. , Biquarterly, 1991, no. 1, 66—71.

67. Wang Fanggui, Tensor products of IBN albegras, Chin. Ann. Math. , 12A(1991), 66—71.

68. Wang Fanggui, Cohn rings and their Grothendieck groups, J. Math. Res. Exposition, 12(1992), no. 2, 287—291.

69. Wang Jinzhou, Automorphism groups of projective modules, J. Nanjing Univ. Math. Biquarterly, 7(1990), no. 2, 250—260, MR 92f: 13013.

70. Wu Pinsan, A class of algebras with zero divisors, Beijing Shifan Daxue Xuebao, 1983, no. 3, 7—10, MR 86b: 13003.

71. Wu Pinsan, The structure of a class of rings with finite number of left zero divisors,

Ibid. , 1983, no. 3, 1—6, MR 86b: 16016.

72. Wu Quanshui, On the structure of homogeneous completely reducible modules, J. Wuhan Univ. Nature Sci. Ed. , 1986, no. 1, 18—22, MR 87k:16032.

73. Wu Quanshui, J. S. Golan, On the endomorphism ring of a module with relative chain condition, Comm. Alg. , 18(1990), no. 8, 2595—2609, MR 91i. 16058.

74. Xi Changchang, Nil MHR-rings and Z-radical rings, Chin. Ann. Math. , 6A(1985), no. 6, 687—688, MR 87g: 16013.

75. Xi Changchang, Some problems on MHR-rings, Ibid. , 8A(1987), no. 4, 530—532, MR 89g:16017.

76. Xi Changchang, The vector space category for an indecomposable projective module of a tubular algebra, Manuscripta Math. , 69(1990), no. 3, 223—235, MR 92c: 16011.

77. Xi Changchang, The category of vector spaces for an indecomposable projective module of an algebra, J. Alg. , 139(1991), no. 2, 355—363. MR. 92i:16013.

78. Xiao Jie, On indecomposable moduules over a representation-finite trivial extension algebra, Sci. China, Ser. A. 34(1992), no. 2, 129—137, MR 92c:16033.

79. Xu Jinzhong, Yi Zhong, Discussion of some statements about homological dimension, J. Math. Res. Exposition, 5(1985), no. 4, 1—8, MR 87m: 13015.

80. Xu Jinzhong, Flatness and injectivity of simple modules over a commutative ring, Comm. algebra, 19(1992), no. 2, 535—537, MR 92c: 13008.

81. Xu Yansong, IBN rings and tensor products, Nanjing Daxue Xuebao Ziran Kexue Ban, 21(1985), no. 4, 571—576, MR 87g: 16029.

82. Xu Yansong, A remark on the extensibility of isomorphisms of miniumum one-sided ideals, Chinese Sci. Bull. , 30(1985), no. 12, 1571—1573, MR 87h:16029.

83. Xu Yansong, Injective dimension of tensor products, Acta Math. Sinica, 30(1987), no. 1, 139—144, MR 88g: 16025.

84. Xu Yansong, Coherent rings and IF rings are characterized by PF-injective modules, J. Math. Res. Exposition, 6(1986), no. 1, 21—26, MR 88i: 16031.

85. Xu Yonghua, On the structure of homogeneous completely reducible modules, Chin. Ann. Math. , 5B(1984), no. 1, 59—66, MR 86b: 16020.

86. Xu Yonghua, On the tensor products of algebras, J. Alg. , 113(1988), no. 1, 40—70, MR 89e: 16014.

87. Xu Yonghua, An equivalence between ring F and infinite matrix subring over F, Chin. , Ann. Math. , 11B(1990), no. 1, 66—69, MR 91d:16011.

88. Xue Weimin, Morita duality and Artinian left duo rings, Bull. Austral. Math. Soc. , 39(1989), no. 3, 339—342, MR 90b: 16033.

89. Xue Weimin, A note on exact hereditary rings, Saitama Math. J. , 1989, 1—4, MR 91d:16014.

90. Xue Weimin, On weakly left duo rings, Riv. Mat. Univ. Parma, (4) 15(1989), 211

−217, MR 91f: 16003.

91. Xue Weimin, On PP rings, Kobe J. Math. , 7(1990), no. 1, 77−80, MR 91i:16007.

92. Xue Weimin, Exact modules and serial rings, J. Alg. , 134(1990), 209−221.

93. Xue Weimin, Linearly compact modules over perfect rings, Adv. in Math. (China), 20 (1991) no. 1, 75−76, MR 92b: 16031.

94. Xue Weimin, A note on two questions of K. Varadarojan, Comm. Alg. , 19(1991), no. 5, 1445−1447, MR 92d: 16015.

95. Yang Jinghua, The global homological dimensions of algebraic extension rings, J. of Nanjing Univ. Math. Biquarterly, 7(1990), no. 1, 122−124, MR 92c:16012.

96. Yang Jinghua, On homological dimensions of semilocal rings, Ibid. , 8(1991), no. 2, 205−208, MR 93b:16016.

97. Yang Jinghua, On the global dimension of residue rings, J. Math. Res. Exposition, 11 (1991), no. 4, 569−573.

98. Yao Dongyuan, Homological dimensions of perfect rings, J. of Nanjing Univ. Math. Biquarterly, 3(1986), no. 1, 23−30, MR 88a: 16047.

99. Yao Musheng, On a correspondence between lattices of ideals of rings and lattices of submodules of modules, Chin. Ann. Math. , 6A(1985), no. 5, 551−557, MR 87e: 16082.

100. Yao Musheng, Equivalence of categories of modules and isomorphisms of rings, Sci. Sinica, Ser A, 30(1987), no. 1, 12−18, MR 88h: 16055.

101. Yi Zhong, On quasiprojective and quasünjective modules, J. Math. (Wuhan), 7 (1987), no. 1, 59−62, MR 88i:16029.

102. Yu Yongxi, Hom functors from a regular category, J. Math. Res. Exposition, 6 (1986), no. 2, 1−4, MR 87k: 18001.

103. Yu Yongxi, On the bijective half-functors, Ibid. , 9(1989), no. 3, 325−335, MR 90j: 18002.

104. Yue Qin, Homological properties of PS-rings, Ibid. , 12(1992), no. 1, 101−106.

105. Zhang Jule, P-injectivity and Artinian semisimple rings, Ibid. , 11(1991), no. 4, 579 −585, MR 92i:16008.

106. Zhang Yinbo, The modules in any component of the AR-quiver of a wild hereditary Artinian algebra are uniquely determined by their composition factors, Acta Math. Sinica (N. S.), 6(1990), no. 2, 97−99, MR 91d: 16023.

107. Zhang Yinbo, The structure of stable components, Canad. J. Math. , 43(1991) no. 3, 652−672, MR 92f: 16017.

108. Zhao Yicai, A test theorem on coherent GCD domains, Proc. Amer. Math. Soc. , 115 (1992), no. 1, 47−49, MR 92h:13018. .

109. Zhao Yicai, A note on coherent rings of dimension two, Ibid. , 115(1992), no. 4, 935 −937, MR 92j:13013.

110. Zhao Yicai, On commutative indecomposable coherent regular rings, Comm. Alg., 20 (1992), no. 5, 1389—1394, MR 93b: 13037.

111. Zhou Boxun, The tensor products and categories of left ring modules, Nanjing Daxue Xuebao Ziran Kexue Ban, 1979, no. 1, 1—20, MR 82j: 16045.

112. Zhou Boxun, On the tensor product of left modules and their homological dimensions, J. Math. Res. Exposition, 1981, 17—24, MR 82c: 16017.

113. Zhou Boxun, On the tensor products of left modules and tricomplexes, Nanjing Daxue Xuebao Ziran Kexue Ban, 1981, no. 2, 239—253, MR 85k: 16044.

114. Zhou Boxun, A brief survey to the founding and development of the algebraic K-theory, J. of Nanjing Univ. Math. Biquarterly, 1987, no. 1, 91—98.

115. Zhou Boxun, Homological Algebra, Sci. Press, China, 1988.

116. Zhu Shenglin, On the equavalences of quotient category Mod-(R,F), with module category mod-S, Comm. Alg., 16 (1988), no. 8, 1639—1661, MR 90b: 16048.

117. Zhu Shenglin, Maximal quotient rings of endomorphisms of quasigenerators, J. Math. Res. Exposition, 9(1989), no. 2, 251—256, MR 90g: 16024.

118. Zhu Shenglin, A note on Fuller's theorem, Chin. Ann. Math., 10B(1989), no. 2, 272—276, MR 90j: 16055.

119. Zhu Shenglin, On rings over which every flat left module is finitely projective, J. Alg., 139(1991), no. 2, 311—321. MR 92j: 16002.

120. Zhu Xiaosheng, The strongly IBN rings and their enveloping algebras, J. of Nanjing Univ. Math. Biquarterly, 5(1988), no. 2, 254—258, MR 89m: 16031.

121. Zhu Xiaosheng, Some properties of Grothendieck groups of the quotient rings of IBN rings, J. Math. Res. Exposition, 9(1989), no. 3, 375—381.

Progress in Algebraic K-Theory in China

ZHOU Boxun
　　Department of Mathematics, Nanjing University, Nanjing, China

TONG Wenting
　　Department of Mathematics, Nanjing University, Nanjing, China

1. Introduction

One of the authors, Zhou, delivered an address to the Second National Symposium on Algebra of China held in Nov. 1986, which is a brief survey of algebraic K-theory, mainly introduced the works of Serre, Swan, Quillen and Milnor (see [29]). He organized a seminar on algebraic K-theory at Nanjing University. As a consequence, his students obtained a series of results in this theory. For instance, some results on PF rings in which all finitely generated projective modules are free; the structure of K_0-groups of some group rings; and some applications of K_2-groups to algebraic number theory. Liu Mulan and Li Fuan of the Academia Sinica obtained some interesting results on K_1-groups, Steinberg groups and the higher ordered K-groups of group rings. Besides at Nanjing and Beijing, the algebraists of Northeast Normal University at Changchun also have done some works in this theory.

In this paper, all rings are associative with unity, and all modules are unitary.

2. Rings in which All Finitely Generated Projective Modules are Free

It is well known that, by Quillen-Suslin theorem, all finitely generated projective S-modules are free for any PID ring R and $S=R[x_1,\cdots,x_n]$ ([30]). In Tong [10] the ring in which all finitely generated projective modules is free is called a PF ring, and the following results were obtained:

Theorem 2.1 ([10])　For any ring R the following statements are equivalent:

(1) $K_0R\overset{f}{\simeq}Z$, the additive group of integers, and $f([R])=1$;

(2) R is an IBN ring and all finitely generated projective R-modules are stably free;

(3) $K_0R^{n\times n}\overset{g}{\simeq}Z$ and $g([R^{n\times1}])=1$, hence $g([R^{n\times n}])=n$;

(4) $K_0R\overset{h}{\simeq}Z$, $h([R])\neq0$ and all finitely generated projective R-modules are stably free.

In Th. 2.1, [10] gives some characterizations of the semihereditary rings. For domains, the commutative rings without zero-divisors other than zero, we have the following

Theorem 2.2 ([10])　If R is a domain, then the following statements are equivalent:

(1) R is a Prüfer ring and $K_0 R \overset{f}{\simeq} Z$, $f([R])=1$;

(2) R is a Prüfer ring and there exists a ring isomorphism $K_0 R \simeq Z$;

(3) The following classes of R-modules are coincident: the stably free R-modules; the finitely generated projective R-modules; the finitely generated flat R-modules; the torsion-free R-modules.

In addition, in [10], it was proved that a ring R is a skew field iff $K_0 R \overset{f}{\simeq} Z$ with $f([R])=1$ and R is artinian semisimple.

For the commutative ring R with $K_0 R \overset{f}{\simeq} Z$ as ring isomorphism, Tong [12] obtained the following results:

Theorem 2.3 ([12])   Let R be a commutative ring with $K_0 R \overset{f}{\simeq} Z$; then:

(1) $f(M)=\chi(M)$ for any finitely generated projective R-module M, where $\chi(M)$ denotes the Euler characteristic of M.

(2) For any finitely generated projective R-module M, $f([M]) \geqslant 0$, and $f([M])=0$ iff $\mathrm{Ann}_R(M) \neq 0$, where $\mathrm{Ann}_R M$ is the annihilator of M in R.

For a ring with weak dimension $\leqslant 1$, Tong [13] obtained a result on the zero-divisors as a part of the following theorem:

Theorem 2.4 ([13])   Let $R \in \mathrm{PF} \cap \mathrm{IBN}$, and $0 \neq f : R \rightarrow R$ be an endomorphism of the left (right) R-module R.

(1) If $WD(R) \leqslant 1$, then f is injective and $\mathrm{Im} f \simeq R$, hence R has no zero-divisors other than zero;

(2) If $R \in \mathrm{PF} \cap \mathrm{UFD}$, then f is injective and $\mathrm{Im} f \simeq R$;

(3) If $f^2 = f$, then f is an isomorphism.

For IBN rings, Tong [11] gave the following result.

Theorem 2.5 ([11])   If R, R_i are rings and $R = \bigoplus_{i=1}^{n} R_i$, then the following conditions are equivalent:

(1) $[R]$ has finite order in $K_0 R$;

(2) $[R_i]$ has finite order in $K_0 R_i$, $i=1, \cdots, n$.

If one of the equivalent conditions above holds, then $R = \langle R, R \rangle$ if and only if $R_i = \langle R_i, R_i \rangle$, $i=1, \cdots, n$, where $\langle S, S \rangle$ denotes the commutator ideal of the ring S.

It is well known that any commutative artinian ring is a finite direct sum of some local rings. A ring R is said to be local-decomposable provided that R is a finite direct sum of local rings. We denote this as $R \in \mathrm{LD}$. Tong [14] obtained the following result.

Theorem 2.6 ([14])   Let R be a commutative ring, then

(1) If $R \in LD$, then $R \in UCP$, the class of rings with the unimodular column property;

(2) If $R \in LD$, then the following statements are equivalent: (a) R is a local ring; (b) $R \in PF$; (c) All finitely generated projective R-modules are stably free; (d) $K_0 R \simeq Z$.

(3) R is local if and only if $R \in LD \cap PF$, if and only if $R \in LD$ and all finitely generated projective R-modules are stably free.

Furthermore, in [9], Tian discussed the existence of the unimodular elements of projective modules over the polynomial rings. His next result generalizes a theorem due to S. M. Bhatwadekar and M. Roy (see Math. Z. 183(1983), 87—94).

Theorem 2. 7 ([9]) Let R be a commutative noetherian ring, P a projective $R[x_1, \cdots, x_n]$-module. If $\mathrm{rank}(P) \geqslant \dim(R/J(R)) + 2$ and $\mathrm{ht}\, O_P(P_0) \geqslant \dim R + 1$, where $P_0 \in P$ and $O_P(P_0) = \{\varphi(P_0) \mid \varphi \in P^*\}$, then there exists a unimodular element in P, and the canonical mapping $U_m(P) \rightarrow U_m(P/(x_1, \cdots, x_n)P)$ is surjective.

3. Some Applications for the Group Rings

Let R be a ring, G a group and RG the group ring of G over R. It is well known that RG reflects properties of group G and the ring of coefficients R. Tong [14] obtained the following results:

Theorem 3. 1 ([14]) Let R be a commutative artinian ring, G a finite abelian group. Then the following properties are equivalent:

(1) RG is a PF ring;

(2) RG is a local ring;

(3) G is a p-group, R is a local ring with $\mathrm{ch}(R/J(R)) = p$, where $J(R)$ is the Jacobson radical of R;

(4) All finitely generated projective RG-modules are stably free;

(5) $K_0(RG) \simeq Z$.

Theoreme 3. 2 ([14]) Let R be a commutative semilocal ring. If G is a finite abelian group, or $\mathrm{ch} R/J(R) = p > 0$ and $G = A \otimes B$, where A is an infinite abelian p-group, B is a finite abelian group with $p \nmid |B|$, then $K_0(RG)$ is a finite free abelian group, i. e. , $K_0(RG) \simeq Z^n$.

Theorem 3. 3 ([14]) If R is a commutative local-decomposable ring and G is an infinite cyclic group, then the following statements are equivalent: (1) $K_0(RG) \simeq Z$; (2) $K_0(R) \simeq Z$; (3) All finitely generated projective R-modules are stably free; (4) $R \in PF$; (5) R is a local ring.

Let R be a right noetherian ring with 1, Π a finite abelian group, $R\Pi$ the group ring of Π over R. Let $\mathcal{M}_R$ be the category of finitely generated right R-modules, $G_0(R)$ and

$G_1(R)$ the Grothendieck group $K_0 \mathscr{M}_R$ and the Whitehead group $K_1(\mathscr{M}_R)$ respectively. H. Lenstra (see J. Pure and Appl. Alg., 20(1981), 173—193) has obtained an excellent calculation formula for $G_0(R\Pi)$. It is natural to ask whether Lenstra's formula can be generalized to higher K-groups of the category $\mathscr{M}_R$ or not. Unfortunately this does not seem to be the case. Nevertheless Lenstra's formula is generalized to the group $G_1(R\Pi)$ (see H. Bass, Algebraic K-theory, Benjamin, New York, 1968 p. 543) which does not coincide with Quillen's $K_1(\mathscr{M}_R)$.

Let $X(\Pi)$ denote the set of cyclic quotient groups of Π. If $\rho \in X(\Pi)$ has order n and a generator t, we put $R(\rho) = R\rho/\Phi_n(t)R\rho$, where Φ_n denotes the nth cyclotomic polynomial; the two-sided ideal $\Phi_n(t)R_\rho$ does not depend on the choice of the generator t (see J. Pure and Appl. Alg., 20(1981), 173—193). Write $\Pi = \Pi\Pi_P$ as the direct product of its primary components Π_P. Liu Mulan [6] obtained the following result, following Lenstra rather closely:

Theorem 3. 4 ([6]) $G_1(R\Pi) = \bigotimes_{\rho \in X(\Pi)} (G_1(R(\rho))/H_\rho)$, where H_ρ is generated by the elements $[M, \alpha, (\rho)]$, the class of (M, α) in $G_1(R(\rho))$, $M \in \mathscr{M}_{R(\rho)}$, $\alpha \in \mathrm{Aut}_{R(\rho)}M$, such that, for some prime number p dividing the order of ρ, $pM = 0$ and $(\rho)_p$ acts trivially on M.

D. L. Webb (see J. Pure and Appl. Alg., 35(1985), 197—223) computed $G_0(Z(\Pi \rtimes \Gamma))$, where $\Pi \rtimes \Gamma$ is the semi-direct product of Π and Γ, Γ is an automorphism group of Π which stabilizes all subgroups of Π (e. g., this is the case for dihedral and quaternary extensions of finite cyclic groups). In [7], Liu Mulan and Li Fuan, following Lenstra's track exactly, proved that Lenstra's formula remains valid for $G_0(R(\Pi \rtimes \Gamma))$ and $G_1(R(\Pi \rtimes \Gamma))$, i. e.,

Theorem 3. 5 ([7]) $G_i(R(\Pi \rtimes \Gamma)) \simeq \bigotimes_{\rho \in X(\Pi)} (G_i(R(\rho)) \# \Gamma)/H_\rho, i = 0, 1$, where $\#$ is the symbol of crossed product.

In addition, Song obtained some applications of K_0-groups for the semigroups. Let S be a semigroup with unit and zero elements, $P(S)$ the category of finitely generated projective S-systems. In [8], Song defined $K_0(S) = K_0(P(S))$ and proved the following

Theorem 3. 6 ([8])   Let S be a semigroup with unit element and zero element. Then

(1) $K_0(S)$ is a free abelian group and $\mathrm{rank}(K_0(S)) = \mathrm{rank}S - 1$;

(2) If S is commutative, then $K_0(S) \simeq Z \oplus rK_0(S)$.

4.   Some Results on K_0-Groups

Let R be a ring, $F(R)$ denote the category of finitely presented left R-modules, $F_0(R)=K_0(F(R))$, the Grothendieck group of $F(R)$. Wang [16] generalized Grothendieck's corresponding results and proved the following theorems:

Theorem 4. 1 ([16])   If R is a left coherent ring and any $M\in F(R)$ has finite projective dimension, then $K_0(R)\simeq F_0(R)$.

Theorem 4. 2 ([16])   If both R and $R[X]$ are left coherent, then

(1) $F_0(R)\simeq F_0(R[X])$;

(2) $K_0(R)\simeq K_0(R[X])$ when any $M\in F(R)$ has finite projective dimension;

(3) $F_0(R[x])\simeq F_0(R[X, X^{-1}])$ when R is commutative.

Let R be a commutative ring, P a finitely generated projective R-module. If there exists a symmetric bilinear mapping $b:P\times P\rightarrow R$, then (P,b) is called a bilinear R-module. It is clear that the mapping b induces an R-module homomorphism $d_b:P\rightarrow P^*=\mathrm{Hom}_R(P,R)$ such that $d_b(x)(y)=b(x,y)$, $x,y\in P$. If d_b is an isomorphism, then (P,b) is also called a non-singular bilinear R-module. For two bilinear R-modules (P_1,b_1), (P_2,b_2), if an R-module isomorphism $f:P_1\rightarrow P_2$, such that $b_1(x,y)=b_2(f(x),f(y))$, $x,y\in P_1$, then f is called isometric, (P,b) is said to be a Disc-module if rank $P=1$. The set of isometric classes of Disc-modules, with respect to $\otimes$, forms a group $\mathrm{Disc}(R)$. Wang [17] obtained the following results:

Theorem 4. 3 ([17])   Let Γ be a directed set, $\{R_1, \Psi_{ij}\}$ a direct system over Γ of commutative rings, and $R=\lim\limits_{\rightarrow} R_i$; then

$$\mathrm{Disc}(R)=\lim\limits_{\rightarrow}\mathrm{Disc}(R_i).$$

Theorem 4. 4 ([17])   Let R' be the integral closure of a commutative ring R, $S=\{x\in R'\,|\,\forall\ P\in\mathrm{Spec}R$, there exists $y\in R_P$ such that $\dfrac{x}{1}-y\in J(R_P')\}=R$. If R is a domain, or a noetherian reduced ring, then

$$\mathrm{Disc}(R)\simeq\mathrm{Disc}(R[x]).$$

Theorem 4. 5 ([17])   If R is a commutative ring and $2\in U(R)$, the group of units of R, then

$$\mathrm{Disc}(R)\simeq\mathrm{Disc}(R[x]).$$

Theorem 4. 6 ([17])   If R is a commutative semihereditary ring, or a commutative coherent ring with $glD(R)\leqslant2$, then

$$\mathrm{Disc}(R)\simeq\mathrm{Disc}(R[x]).$$

In addition, in [22], Xu obtained a result on the Morita equivalence with the

Grothendieck groups, that is

Theorem 4.7 ([22]) Let $(R, M, M', R', \tau, \mu)$ be a Mortia context, $\bar{R}$ its Morita context ring. If R, R' are two local rings, then $K_0\bar{R} \simeq Z$ iff $\bar{R}$-Mod and R-Mod are equivalent. In this case, $R \simeq R'$ and $\bar{R} = R^{2 \times 2}$.

For the primitive rings, an important class of rings, Wang [18] obtained the following result:

Theorem 4.8 ([18]) Let R be a (left) primitive ring, T a simple faithful (left) R-module. If $R \neq \mathrm{Soc}R$, and $\Pi : R \to R/\mathrm{Soc}R$ is the canonical ring homomorphism, then

(1) The homomorphism $K_0\Pi : K_0R \to K_0(R/\mathrm{Soc}R)$ is surjective;

(2) For $K_0\Pi$ of (1), $\mathrm{Ker}(K_0\Pi)$ is a cyclic group with a generator $[T]$;

(3) If $\mathrm{Soc}R = 0$, then there exists a left primitive ring R_1 with a simple faithful (left) R_1-module T_1, such that $\mathrm{Soc}R_1 \neq 0$ and R is a homomorphic image of R_1, and $K_0R_1 \simeq K_0R \oplus N$, where $N = \mathrm{Ker}(K_0R_1 \to K_0R)$ is a cyclic group with a generator $[T_1]$.

5. Some Applications for IBN Rings

P. M. Cohn defined the following three classes of rings (see P. M. Cohn, Some remarks on the invariant basis property, Topology, 5(1966), 215−228). If $R^m \simeq R^n$, as left R-modules, implies $m = n$, then R is called an IBN ring, denoted by $R \in \mathrm{IBN}$. If $R^m \simeq R^n \oplus K$ implies $m \geqslant n$, then R is called an $\mathrm{IBN_1}$ ring, denoted by $R \in \mathrm{IBN_1}$. If $R^m \simeq R^m \oplus K$ implies $K = 0$, then R is of class $\mathrm{IBN_2}$. It is clear that $\mathrm{IBN_2} \subsetneqq \mathrm{IBN_1} \subsetneqq \mathrm{IBN}$.

K. R. Goodearl obtained the following results (see K. R. Goodearl, Partially ordered Grothendieck groups, Lecture Notes in Pure and Appl. Math, 91, Ed. by H. B. Srivastava, Marcel Dekker, 1984, 71−90):

(1) For any ring R, K_0R is a pre-ordered abelian group where $x \leqslant y$ for $x, y \in K_0R$ means that there exists a finitely generated projective left R-module A such that $y - x = [A]$;

(2) If all the matrix rings $R^{n \times n}$ over R are directly finite, i. e. , $yx = 1$ for x, $y \in R^{n \times n}$ implies $xy = 1$, then K_0R is a partially ordered abelian group for the pre-order defined in (1).

In [15] Tong obtained the following results:

Theorem 5.1 ([15]) If $R \in \mathrm{IBN}$, and $f : K_0R \simeq Z$ such that $f([R]) = 1$, then K_0R is a totally ordered abelian group with the ordering defined in (1).

Theorem 5. 2 ([15])   For any ring R, the following statements are equivalent; (1) $R \in IBN_2$; (2) $[P] > 0$ in $K_0 R$ for any non-zero finitely generated projective left R-module P; (3) $[P] = 0$ if and only if $P = 0$, for any finitely generated projective left R-module P.

Theorem 5. 3 ([15])   For any ring R, the following statements are equivalent; (1) $R \in IBN_1$; (2) $Z[R]$ is a totally ordered abelian group with the natural ordering and $R \in IBN$; (3) $Z[R]$ is a partially ordered abelian group with the natural ordering and $R \in IBN$; (4) $n[R] \geqslant m[R]$ for the ordering defined in (1) above iff $n \geqslant m$ and $R \in IBN$.

Using the following symbols, [15] gave some inclusion relations of some classes of rings.

$DC = \{R | M \simeq N$ when $M \oplus A \simeq N \oplus A$ for any f. g. projective modules $M, N, A\}$

$DF = \{R | R$ is a directly finite ring$\}$

$SFF = \{R |$ all stably free left R-modules are free$\}$

$PSF = \{R |$ all f. g. projective left R-modules are stably free$\}$

$TK_0 = \{R | K_0 R$ is totally ordered with the ordering of (1)$\}$

$PK_0 = \{R | K_0 R$ is partially ordered with the ordering of (1)$\}$

$S = \{R | K_0 R = 0\}$

Theorem 5. 4 ([15]) $DC \subseteq IBN_2 \cap SFF = IBN \cap SFF \subseteq IBN_2 \subseteq PK_0 \cap DF \subsetneqq PK_0 \subseteq IBN_1 \cap S \subseteq IBN \cap S$, $IBN_1 \cap PSF \subseteq TK_0 \subseteq PK_0$.

Theorem 5. 5 ([15])   (1) If $R \in PSF \backslash S$, then $R \in IBN_1$ iff $R \in TK_0$ iff $R \in PK_0$;

(2) If $R \in SFF \backslash S$, then the following statements are equivalent; (a) $R \in IBN_1$; (b) $R \in IBN$; (c) $R \in IBN_2$; (d) $R \in PK_0$.

(3) If $R \in PF \backslash S$, then the following statements are equivalent:

(a) $R \in IBN$; (b) $R \in IBN_1$; (c) $R \in IBN_2$; (d) $R \in PK_0$; (e) $R \in TK_0$.

Consider the following Cartesian square of ring homomorphisms

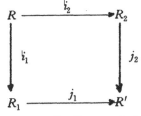

where j_2 or j_1 is surjective. Wang [19] proved the following result.

Theorem 5. 6 ([19])   If $K_0R \simeq Z$, then $R' \in$ IBN if and only if $R_i \in$ IBN for $i = 1, 2$.

Using some properties of K_0-groups, Wang [19] obtained the following theorem.

Theorem 5. 7 ([19])   (1) Let R be a commutative local ring with nil maximal ideal, then $R \underset{F}{\otimes} S \in$ IBN if and only if $R \in$ IBN and $S \in$ IBN;

(2) If F is a commutative ring, but is not local with nil maximal ideal, then there exist F-algebras $R, S \in$ IBN such that $R \underset{F}{\otimes} S \notin$ IBN.

Theorem 5. 8 ([15])   (1) If $K_0R \simeq Z$ and for any f. g. projective left R-module P, there exists $n \geqslant 1$ such that P^n is free then $R \in$ IBN$_2$;

(2) If $R \in$ PF, then $K_0R \simeq Z$ if and only if $R \in$ IBN$_2$;

(3) If $K_0R \simeq Z$, then $R \in$ IBN iff for any f. g. projective left R-module P, there exists $n \geqslant 1$ such that P^n is stably free;

(4) If $R \in$ PSF and $K_0R \simeq Z$, then $R \in$ IBN$_2$ iff for any f. g. left R-module P, there exists $n \geqslant 1$ such that P^n is free.

Xu Y. S. [23] proved the following result.

Theorem 5. 9 ([23])   For any ring R,

(1) K_0R is a torsion group if and only if for any ring S Mortia equivalent to R, $[S]$ is a torsion element in K_0S;

(2) $R \in$ IBN$_1$ if and only if for any ring S Mortia equivalent to R, $[S]$ is not torsion in K_0S.

Zhu [31] gave a property of K_0R for $R \in$ IBN.

Theorem 5. 10 ([31])   If $R \in$ IBN, then K_0R has a system of generators $\{x_i \mid \text{ord } x_i = \infty \text{ in } K_0R\}$.

6.   Some Results on K_1-Groups and K_2-Groups

Let R and R' be commutative rings with 1. A system from R to R' is a triple $(\varphi, \alpha, \sigma)$, where

$\varphi: \text{Max}R \rightarrow \text{Max}R'$ is a bijection such that $R/M = R'/\varphi(M)$ for all $M \in \text{Max}R$;

α is an element in R' satisfying $\alpha^2 = 1$ and $2(\alpha - 1) = 0$;

$\sigma: R \rightarrow R'$ is a bijection such that,

$$x \in M \text{ iff } \sigma(x) \in \varphi(M),$$
$$\sigma(x)(\sigma(x) - 1)(\alpha - 1) = 0,$$
$$\sigma(x + y) = \sigma(x) + \sigma(y) + \sigma(x)\sigma(y)(\alpha - 1),$$
$$\sigma(xy) = \sigma(x)\sigma(y),$$

for all $x, y \in R$. If $a \neq 1$, $(\varphi, \alpha, \sigma)$ is said to be non-trivial.

Denote $U = U(R)$, $U' = U(R')$, where $U(R)$ is the group of units in R. From the definition of $(\varphi, \alpha, \sigma)$, we see that σ yields a group isomorphism $U \simeq U'$. Li Fuan proved the following

Theorem 6.1 ([3]) σ yields an isomorphism $K_1 R \simeq K_1 R'$ when R and R' are two commutative local rings and there exists a system $(\varphi, \alpha, \sigma)$ from R to R'.

For two commutative local rings R and R', with residue field F_2, [3] obtained an equivalent condition for the existence of $(\varphi, \alpha, \sigma)$.

Let R be a commutative ring with 1, $n \geqslant 3$ an integer, and H a subgroup of $GL_n(R)$. It is known that H is normalized by the elementary linear group $E_n(R)$ if and only if $E_n(R, A) \subset H \subset GL'_n(R, A)$ for a unique ideal A of R. This is called the sandwich theorem. A. Bak obtained in 1982 the following generalized form of the sandwich theorem: H is normalized by $E_n(R, B)$ for an ideal B of R iff $E_n(R, AB^{24}) \subset H \subset GL'_n(R, A)$ for an ideal A of R under a stable range condition on R. In 1985 at Beijing, he conjectured that the stability condition is unnecessary if 24 is replaced by some sufficiently large number. In 1987, Li Fuan and Liu Mulan [5] proved the conjecture for the exponent 40:

Theorem 6.2 ([5]) H is normalized by $E_n(R, B)$ for an ideal B of R if and only if $E_n(R, AB^m) \subset H \subset GL_n'(R, A)$ for an ideal of A of R, where m is a sufficiently large number.

Let R be a commutative ring, M a f. g. R-module. Wang Jingzhou [21] introduced the following definitions:

$$^M M^* = \{^P \Phi \mid P \in M, \ \Phi \in \mathrm{Hom}_R(M, R), \quad ^P \Phi \in \mathrm{End}_R M \text{ such that } \quad ^P \Phi(m) = P\Phi(m), \ m \in M\}$$

$$T_R^1(M) = \{^P \Phi \mid P \in {}^M M^*, \text{ such that } \Phi(P) = 0\}$$

$T^1 G_R(M) = \langle I_M + T_R^1(M) \rangle$, where $\langle X \rangle$ denotes the subgroup generated by X in $\mathrm{Aut}_R M$. It is clear that $T^1 G_R(M) \lhd \mathrm{Aut}_R M$, i. e. , $T^1 G_R(M)$ is a normal subgroup of $\mathrm{Aut}_R M$. Wang [21] gave a simpler proof of the following

Theorem 6.3 ([21])   Let R be a commutative noetherian ring with j-dim$R$ $= d$, P a finitely generated projective R-module and rank$P \geqslant d+1$. Then

$$\mathrm{Aut}_R(P \oplus R^n) = \mathrm{Aut}_R(P) T^1 G_R(P \oplus R^n).$$

Let R be an arbitary ring with 1, I an ideal of R, and $*: R \to R$ an involutive endomorphism such that $I^* \subseteq I$. If ε is a central element of R and $\varepsilon\varepsilon^* = 1$, then denote $I_\varepsilon = \{x - \varepsilon x^* \mid X \in I\}$ and $I^\varepsilon = \{x = -\varepsilon x^* \mid x \in I\}$. Assume $\wedge$ is a subgroup of

I such that $I_e \subseteq \wedge \subseteq I^e$. If $a_i - 1$, $b_i \in I$ and $Ra_i + Rb_i = R$ implies that there exists $x \in \wedge$ such that $a_i + xb_i \in GL(1, I)$, then we say that I has the $\wedge$-n-fold condition.

Let $F_n = \begin{pmatrix} 0 & I_n \\ \varepsilon^* I_n & 0 \end{pmatrix}$, $U^\varepsilon(2n, I) = \{A \in GL(2n, I) \mid AF_n A^* = F_n\}$. In You Hong [25] the following notations are used:

$E^\varepsilon(2n, R)$: the subgroup generated by all elementary unitary matrices of $U^\varepsilon(2n, I)$;

$E^\varepsilon(2n, R, I)$: the normal subgroup generated by all elementary matrices over I;

$$U^\varepsilon(I) = \bigcup_n U^\varepsilon(2n, I); \quad E^\varepsilon(R, I) = \bigcup_n E^\varepsilon(2n, R, I);$$

$$K_1 U^\varepsilon(R, I) = U^\varepsilon(I) / E^\varepsilon(R, I);$$

$\tilde{E}^\varepsilon(2n, R, I)$: the normal subgroup generated by $[U^\varepsilon(2n, I), E^\varepsilon(2n, R)]$, $[F_n, A]$ $(A \in U^\varepsilon(2n, I))$, and the invertible matrices of the form $(I_{2n} + YX\tilde{Y})(I_{2n} + \tilde{Y}XY)^{-1}$, where $Y = y \oplus I_{n-1} \oplus I_n$, $X \in M_{2n}(I)$, $\tilde{Y} = I_n \oplus y^* \oplus I_{n-1}$, $\forall y \in R$ and $I_{2n} + YX\tilde{Y}$, $I_{2n} + \tilde{Y}XY \in U^\varepsilon(2n, I)$;

$$D_{(1+xy)(1+yx)^{-1}} = \left\{ \begin{pmatrix} (1 + xy)(1 + yx)^{-1} & 0 \\ 0 & [(1 + xy)(1 + yx)^{-1}]^{* -1} \end{pmatrix} \right.$$
$\left. | 1 + xy \in GL(1, I), \ x \in I, \ y \in R \right\}$;

S_ε: the subgroup generated by $\{1 + xy \mid x = -\varepsilon x^*, \ y = -\varepsilon^* y^*\} \subset GL(1, I)$;

T: the subgroup generated by $\{yy^* \mid y \in GL(1, I)\} \subset GL(1, I)$;

$$V(R, I) = \{(1 + xy)(1 + yx)^{-1} \mid x \in I, \ y \in R\}.$$

You Hong [25] obtained the following results:

Theorem 6.4 ([25]) If an ideal I of R has the $\wedge$-2-fold condition, then

(1) $U^\varepsilon(2, I) / \tilde{E}^\varepsilon(2, R, I) = U^\varepsilon(2n, I) / E^\varepsilon(2n, R, I) = K_1 U^\varepsilon(R, I)$,

   $\tilde{E}^\varepsilon(2, R, I) = D_{(1+xy)(1+yx)^{-1}} [U^\varepsilon(2, R), U^\varepsilon(2, I)] E^\varepsilon(2, R, I)$.

(2) $K_1 U^\varepsilon(R, I) = GL(1, I) / S_\varepsilon TV(R, I)$.

(3) $U^\varepsilon(2, I) / E^\varepsilon(2, R, I) = GL(1, I) / S_\varepsilon$.

For any commutative ring R, denote $U(R)$ by $R^\cdot$. Let $\{x_{ij}\}$ be the set of generators of the Steinberg group $St(R)$ of R and let

$w_{ij}(u) = x_{ij}(u) x_{ji}(-u^{-1}) x_{ij}(u), \quad u \in R^\cdot$

$h_{ij}(u) = w_{ij}(u) w_{ij}(-1), \quad u \in R^\cdot$

$\langle a, b, c \rangle_{ij} = x_{ij}(-a) x_{ji}(b) x_{ij}(-c) x_{ji}(1 - ab) x_{ij}(-(1 - bc)) w_{ij}(1)$,

where $a, b, c \in R$, $1 - a - c + abc = 0$.

F. J. Keune proved that $\langle a, b, c \rangle_{ij}$ is independent of the choice of i, j, and he defined $\langle a,b,c \rangle_* = \langle a,b,c \rangle_{ij}$ (see J. of Pure and Appl. Alg. , 22(1981), 131 — 141).

Let I be an ideal of R. If $aR+(1+b)R=R$, $a,b \in I$ implies that there exists $t \in (R,I)^{\cdot} = \{u \in R^{\cdot} \mid u \equiv 1 \pmod{I}\}$ such that $a+(1+b)_t \in (R,I)^{\cdot}$, then R is said to be unit stable for the ideal I. If R is unit stable for all ideals, then R is called a unit stable ring. In [24], You Hong obtained the following.

Theorem 6. 5 ([24]) If R is unit stable for an ideal I, then

(1) $St(R,I)$, the $\ker(St(R) \to St(R/I))$, is generated by all $\langle a, b,c \rangle$. $(b \in I)$, $h_{ij}(u)$ $(u \in (R,I)^{\cdot})$, $x_{ij}(p)$ $(p \in I)$, and $x_{ij}(1)x_{ji}(q)x_{ij}(-1)$ $(q \in I)$;

(2) $K_2(R,I)$ is generated by all $\langle a,b,c \rangle$. $(b \in I)$.

It is well known that the K_2 groups are closely related to Steinberg groups. Li Fuan [1] discussed in 1989 isomorphisms between two unstable Steinberg groups over commutative rings. For commutative rings R, S, the following results are obtained:

Theorem 6. 6 ([1])

(1) If $\wedge : St_m(R) \to St_n(S)$ is an isomorphism, then $m=n$;

(2) If $\wedge : St_3(R) \to St_3(S)$ is an isomorphism and both $St_3(R)$ and $St_3(S)$ are central extensions of $E_3(R)$ and $E_3(S)$, respectively, then there exists a $1-1$ correspondence between Max R and Max S: $J \leftrightarrow M$, which makes the following diagram commute:

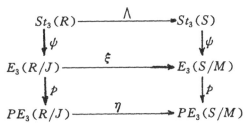

where ψ is the composite of the canonical homomorphisms, ξ and η are the isomorphisms induced by $\wedge$;

(3) If $n \geqslant 4$, then

(a) $St_n(R) \simeq St_n(S)$ iff $R \simeq S$;

(b) $\{\wedge \mid \wedge : St_n(R) \simeq St_n(S)\} \xrightarrow{\quad 1-1 \quad} \{\lambda \mid E_n(R) \overset{\lambda}{\simeq} E_n(S)\}$

(c) $St_n(R) \simeq St_n(S)$ implies $K_{2,n}(R) \simeq K_{2,n}(S)$, where $K_{2,n}(R)=\ker(St_n(R)$ $\xrightarrow{\varphi} E_n(R))$, φ is the canonical homomorphism sending $x_{ij}(a)$ to the elementary matrix $e_{ij}(a)$;

(4) If $n=3$, then (c) of (3) holds also, and the induced mappings in (b) of (3) are injective.

Let R be a finitely generated commutative Z-algebra with Krull dimension d, Π a finite group. Li [2] proved the following

Theorem 6.7 ([2]) $St_n(R\Pi)$ is finitely presented when $n\geq 4$. If, in addition, $n\geq d+3$ and $K_1(R\Pi)$ and $K_2(R\Pi)$ are finitely generated, then $E_n(R\Pi)$ and $GL_n(R\Pi)$ are also finitely presented.

Let A be a ring with 1, $\varphi:St_n(A)\to E_n(A)$ the canonical homomorphism sending the generators $x_{ij}(a)$ of $St_n(A)$ to $e_{ij}(a)$. L (resp. U) the subgroup of $St_n(A)$ generated by $\{x_{ij}(a)\,|\,a\in A,\ i>j\}$ (resp., $\{x_{ij}(a)\,|\,a\in A,\ i<j\}$). For any $u\in GL_n(A)$ and $i\neq j$, write $w_{ij}(u)=x_{ij}(u)x_{ji}(-u^{-1})x_{ij}(u)$, where $u\in A^\cdot$, the group of units of A. Let W be the subgroup of $St_n(A)$ generated by all the $w_{ji}(u)$. The letters L,U and W are also used to denote the subgroups $\varphi(L)$, $\varphi(U)$ and $\varphi(W)$ of $E_n(A)$ respectively, e. g. , $w_{ij}(u)=e_{ij}(u)e_{ji}(-u^{-1})e_{ij}(u)$ in $E_n(A)$. Several authors studied various normal forms for elements of $St(A)$ and $St_n(A)$. Among those, R. W. Sharpe has given a beautiful decomposition $St(A)=LPLU$, where P is a certain subgroup of $St(A)$ (see J. Alg. , 68(1981), p. 453). Unfortunately, Sharpe's decomposition fails for the unstable Steinberg group in any dimension even if A is the ring of integers or a field with more than three elements, where the stable Steinberg group is the direct limit of the $St_n(A)$ in a natural way.

In Li [4] the author discussed the decomposition of elements in $St_n(A)$ replacing the group P in Sharpe's decomposition above by the subgroup W.

Denote the set of unimodular elements in A^n by V_n. Let V'_n (resp. V_n'') be the set of the columns of elements in $GL_n(A)$ (resp. in $E_n(A)$); then $GL_n(A)$ (resp. $E_n(A)$) acts transitively on V_n' (resp. V_n'') and $V_n\supseteq V_n'\supseteq V_n''$. In general, however, $V_n\neq V_n'\neq V_n''$. An element $(a_1,\cdots,a_n)'$ in V_n is said to be stable if there exist $b_1,\cdots,b_n\in A$ such that $(a_1+b_1a_n,\cdots,a_{n-1}+b_{n-1}a_n)'\in V_{n-1}$.

Let d be a non-negative integer. The ring A is said to satisfy $(S)_d$ if for all $n\geq d+2$, every element in V_n is stable. Li [4] defined two weaker stability conditions. The ring A is said to satisfy $(S')_d$ (resp. , (S'')) if for all $n\geq d+2$, every element in V_n' (resp. V_n'') is stable. It is clear that

$$(S), \Rightarrow (S')_d \Rightarrow (S'')_d$$
$$\Downarrow \qquad \Downarrow \qquad \Downarrow$$
$$(S)_{d+1} \Rightarrow (S')_{d+1} \Rightarrow (S'')_{d+1}$$

Li [4] obtained the following results:

Theorem 6.8 ([4]) Let A be a commutaive ring; then the following statements are equivalent:

(1) A satisfies $(S)_0$;

(2) $E_n(A) = LWLU$ for all $n \geqslant 2$;

(3) $E_n(A) = LWLU$ for some $n \geqslant 2$;

(4) $St_n(A) = K_{2,n}(A)LWLU$ for all $n \geqslant 2$;

(5) $St_n(A) = K_{2,n}(A)LWLU$ some $n \geqslant 2$;

Here $K_{2,n}(A) = \mathrm{Ker}(St_n(A) \to E_n(A))$. In particular, $St_n(A) = CLWLU$ when $(S)_0$ holds and $n \geqslant 3$, where C is the center of $St_n(A)$. For skew fields, Li [4] obtained better decompositions:

Theorem 6.9 ([4]) The following statements are equivalent:

(1) A is a skew field;

(2) $St(A) = LWL$;

(3) $St_n(A) = LWL$ for all $n \geqslant 2$;

(4) $St_n(A) = LWL$ for some $n \geqslant 2$;

(5) $E(A) = LWL$;

(6) $E_n(A) = LWL$ for all $n \geqslant 2$;

(7) $E_n(A) = LWL$ for some $n \geqslant 2$.

Here W denotes the subgroup generated by $\{w_{ij}(p(a,b,c)) \mid a+c+abc \in A^{\cdot}\}$ in $St_n(A)$, $p(a,b,c) = a+c+abc$.

Now let $p(a,b) = 1+ab$, W' the subgroup generated by $\{w_{ij}(p(a,b)) \mid 1+ab \in A^{\cdot}\}$ in $St_n(A)$. In [26], You Hong generalized Theorem 6.8 to the noncommutative rings. He also obtained the following result:

Theorem 6.10 ([26]) If R is an arbitrary ring, then the following statements are equivalent:

(1) R is a local ring;

(2) $E_n(R) = LUW'$ for all $n \geqslant 2$;

(3) $E_n(R) = LUW'$ for some $n \geqslant 2$;

(4) $St_n(R) = K_2(n,R)LUW'$ for all $n \geqslant 2$;

(5) $St_n(R) = K_2(n,R)LUW'$ for some $n \geqslant 2$.

As for applications of algebraic K-theory to algebraic number theory, Qin Hourong obtained some results. Let $F=Q(\sqrt{d})$ be a quadratic field, O_F the ring of integers in F. Qin [34], [35] and [36] got a method which can be used to determine the 4-rank of K_2O_F. He gave 4-rank K_2O_F tables for quadratic fields $F=Q(\sqrt{d})$ whose discriminats have at most three odd prime divisors and listed many real quadratic fields with the 2-sylow subgroups of K_2O_F being isomorphic to $Z/2Z \oplus Z/2Z \oplus Z/4Z$.

By improving Tate's method, Qin Showed:

Theorem 6.11 ([32])   Let $O_F=Z(\sqrt{-6})$. Then $\#(K_2O_F)=1$.

Theorem 6.12 ([37])   Let $O_F=Z[\dfrac{1+\sqrt{-35}}{2}]$. Then $K_2O_F \simeq Z/2Z$.

Qin also got the following results.

Theorem 6.13 ([33])   Let F be a field of characteristic $\neq 2$. Then any element of order 4 in K_2F has the form $\{a, a^2+1\}\{-1,b\}$, where $a, a^2+1, b \in F^{\cdot}$.

Theorem 6.14 ([33])   Let $\Phi_n(x)$ be the nth cyclotomic polynomial and $G_{2^{\cdot}}(Q)=\{\{a, \Phi_{2^{\cdot}}(a)\} | a \in Q^{\cdot}\}$. Then $G_{2^{\cdot}}(Q)$ is a subgroup of K_2Q if and only if $n=1$ or 2.

REFERENCES

1   Li Fuan, Isomorphisms of Steinberg groups over commutative rings, Acta Math. Sinica (New Ser.), 5(1989), No. 2, 146—158.

2   Li Fuan, Finite presentability of Steinberg groups over group rings, Acta Math. Sinica (New Ser.), 5(1989), No. 4, 287—301.

3   Li Fuan, Homological meaning of system (φ, α,σ), Acta Math. Sinica (New Ser.), 7(1991), No. 4, 348—353.

4   Li Fuan, Decomposition of Steinberg Groups, Chinese Science Bulletin, 37 (1992), No. 15, 1244—1248.

5   Li Fuan and Liu Mulan, A generalized sandwich theorem, K-Theory, 1 (1987), No. 2, 171—183, MR 88h: 20062.

6   Liu Mulan, The group $G_1(R\Pi)$ for Π a finite Abelian group, J. of Pure and Appl. Algebra, 24(182), 287—291.

7   Liu Mulan and Li Fuan, G_0 and G_1 of a class of group rings $R(\Pi \rtimes \Gamma)$, Chinese Science Bulletin, 31(1986), No. 11, 721—724.

8   Song Guangtian, Algebraic *K*-theory method on semigroups (*I*)-Grothendieck groups of semigroups, Acta Math. Sinica, 33(1990), No. 3, 309—322.

9   Tian Qingchun, Projective modules over polynomial rings, Acta Sci. Nat. Univ. Jilinensis, 96(1991), No. 2, 10—12.

10   Tong Wenting, Grothendieck groups and their applications, J. Nanjing Univ. Math. Biq. , 3(1986), No. 1, 217—223 MR 88a:18024, Zbl 636: 16012.

11   Tong Wenting, On commutators in rings and their tensor products, J. Math, 8(1988), No. 3, 281—286, MR 90$_k$:16025.

12   Tong Wenting, On Euler characteristic of modules, Chin, Ann. Math. , 10B (1) (1989), 58—63, MR 90k:13005, CMCI, 10, 58(1989).

13   Tong Wenting, Rings in which finitely generated projective modules are free, J. Math. Res. Exposition, 9(1989), No. 3, 319—323.

14   Tong Wenting, *PF*-rings and the Grothendieck groups of group rings, J. Math. Res. Exposition, 10(1990), No. 2, 157—162, MR 91e:13013

15   Tong Wenting, Invariant basis number rings and K_0-groups, Proceed. of the First China-Japan International Symposium of Ring Theory, 148 — 150, Okayama, Japan, 1992.

16   Wang Fanggui, The Grothendieck groups of categories of finitely presented modules over coherent rings, J. Nanjing Univ. Math, Biq. , 5(1988), No. 2, 225—229.

17   Wang Fanggui, Some results on discriminant group over a commutative ring, J. Nanjing Univ. Math. Biq. , 7(1990), No. 1, 56—59.

18   Wang Fanggui, Grothendieck groups of primitive rings, Acta Math. Sinica, 34(1991), No. 5, 645—652

19   Wang Fanggui, Tensor products on IBN algebras, Chin. Ann. Math. , 10A (1990), 66—71.

20   Wang Fanggui, Cohn rings and their Grothendieck groups, J. Math. Res. Exposition 12(1992), No. 2, 287—291

21   Wang Jingzhou, On automorphism groups of projective modules, J. Nanjing Univ. Math. Biq. , 7(1990), No. 2, 250—291.

22   Xu Kejian, Morita equivalence and Grothendieck group, Northeastern Math. J. , 4(1988), No. 1, 81—89.

23   Xu Yansong, A remark on the torsion subgroup of K_0, *J*. Nanjing Univ.

Math. Biq. , 2(1985), No. 2, 216—217.

24 You Hong, $K_2(R,I)$ over stable rings, Chinese Science Bulletin, 34(1989), No. 20, 1526—1529.

25 You Hong, Prestabilization of K_1U^t over a $\wedge$-2-fold ring, Chinese Science Bulletin, 26(1991), No. 11, 811—814.

26 You Hong, Some remarks on decompositions of Steinberg groups, Chinese Science Bulletin, 37(1992), No. 18, 1645—1649

27 You Hong, On the defining relations of GL_2 over 1-fold rings, J. Math. Res. Exposition, 12(1992), No. 3, 385—390.

28 You Hong, Stable rings and their two-dimensional linear groups, Chin. Ann. Math. , 7 A(1986), 255—266.

29 Zhou Boxun, A brief survey to the founding and development of the algebraic K-theory, J. Nanjing Univ. Math. Biq. , 4(1987), No. 1, 91—88.

30 Zhou Boxun, Homological Algebra, Science Press, Beijing, 1988.

31 Zhu Xiaosheng, Some properties of Grothendieck Groups of the quotient rings of IBN rings, J. Math. Res. Exposition, 9(1989), No. 3 375—381.

32 Qin Hourong, Computation of $K_2Z[\sqrt{-6}]$, J. pure and applied albegra, 96 (1994), 133—146.

33 Qin Hourong, Elements of finite order in K_2 of fields, Chinese Science Bulletin, 39(1994), No. 6, 449—451.

34 Qin Hourong, 2-Sylow subgroups of K_2O_F for real quadratic fields F, Science in China (A), 37(1994), 1302—1313.

35 Qin Hourong, The 2-Sylow subgroups of the tame kernel of imaginary quadratic fields, Acta Arith. , LXIX, 2(1995), 153—169.

36 Qin Hourong, 4-rank of K_2O_F for real quadratic fields F, Acta Arith. LXXII, 4(1995), 323—333.

37 Qin Hourong, Computation of $K_2Z[\dfrac{1+\sqrt{-35}}{2}]$. Chin. Ann. Math. (B). 1 (1996), 63—72.

Printed and bound by CPI Group (UK) Ltd, Croydon, CR0 4YY

21/10/2024

01777093-0013